Modelling Phase Equilibria

Wiley Series in Chemical Engineering

Bird, Stewart and Lightfoot: TRANSPORT PHENOMENA
Brownell and Young: PROCESS EQUIPMENT DESIGN: VESSEL DESIGN
Felder and Rousseau: ELEMENTARY PRINCIPLES OF CHEMICAL PROCESSES, 2nd Edition
Franks: MODELING AND SIMULATION IN CHEMICAL ENGINEERING
Froment and Bischoff: CHEMICAL REACTOR ANALYSIS AND DESIGN, 2nd Edition
Gates: CATALYTIC CHEMISTRY
Henley and Seader: EQUILIBRIUM-STAGE SEPARATION OPERATIONS IN CHEMICAL ENGINEERING
Hill: AN INTRODUCTION TO CHEMICAL ENGINEERING KINETICS AND REACTOR DESIGN
Jawad and Farr: STRUCTURAL ANALYSIS AND DESIGN OF PROCESS EQUIPMENT, 2nd Edition
Levenspiel: CHEMICAL REACTION ENGINEERING, 2nd Edition
Malanowski and Anderko: MODELLING PHASE EQUILIBRIA: THERMODYNAMIC BACKGROUND AND PRACTICAL TOOLS
Reklaitis: INTRODUCTION TO MATERIAL AND ENERGY BALANCES
Sandler: CHEMICAL AND ENGINEERING THERMODYNAMICS, 2nd Edition
Seborg, Edgar and Mellichamp: PROCESS DYNAMICS AND CONTROL
Smith and Corripio: PRINCIPLES AND PRACTICE OF AUTOMATIC PROCESS CONTROL
Ulrich: A GUIDE TO CHEMICAL ENGINEERING PROCESS DESIGN AND ECONOMICS
Welty, Wicks and Wilson: FUNDAMENTALS OF MOMENTUM, HEAT AND MASS TRANSFER, 3rd Edition

Modelling Phase Equilibria

Thermodynamic Background and Practical Tools

Stanisław Malanowski
Andrzej Anderko
Institute of Physical Chemistry
Polish Academy of Sciences
Warszawa, Poland

John Wiley & Sons, Inc.
New York • Chichester • Brisbane • Toronto • Singapore

Library of Congress Cataloging in Publication Data:

Malanowski, Stanisław.

Modelling phase equilibria: thermodynamics background and practical tools / Stanisław Malanowski and Andrzej Anderko.

p. cm. -- (Wiley series in chemical engineering)

Includes bibliographical references and index.

ISBN 0-471-57103-2

1. Phase rule and equilibrium. 2. Thermodynamics. I. Anderko, Andrzej. II. Title. III. Series.

QD503.M35 1992

660'.296--dc20 91-36125

CIP

Printed in the United States of America

10 9 8 7 6 5 4 3 2 1

Printed and bound by Quinn - Woodbine, Inc.

About the Authors

Stanisław Malanowski is Professor of chemical and engineering thermodynamics at the Institute of Physical Chemistry of the Polish Academy of Sciences where he is also Head of the Department of Basic Organic Raw Materials. He graduated from the University of Warsaw in 1953 and received his Ph.D. from the Institute of Physical Chemistry in 1960. His research interests include measurements of vapor–liquid equilibria and other thermodynamic properties, methods for the representation of phase equilibria and fluid properties with emphasis on excess function models and computation of phase equilibria and separation processes. Stanisław Malanowski has published about 100 papers, mostly in *Fluid Phase Equilibria*, *Bulletin of the Polish Academy of Sciences*, *CODATA Bulletin*, *Chemical Engineering Science*, *Transactions of the Faraday Society*, *Journal of Chemical Thermodynamics*, *Thermochimica Acta*, *Journal of Thermal Analysis*, *Chemical Engineering Communications* and *Chemical and Biochemical Engineering Quarterly*. He also authored two books: *Vapor–Liquid Equilibrium* (Polish Scientific Publishers, Warsaw, 1974), *Handbook of the Thermodynamics of Organic Compounds* (with R. M. Stephenson, Elsevier, New York, 1987) and edited *Thermodynamics of Fluids: Measurement and Correlation* (with A. Anderko, World Scientific Publishing Co., Singapore, 1990). He is a member of Editorial Boards of *Fluid Phase Equilibria* and *Chemical and Biochemical Engineering Quarterly*. Stanisław Malanowski has taught graduate and undergraduate courses in chemical thermodynamics at the University of Connecticut (Storrs, Connecticut), Imperial College (London, Great Britain), Technical University of Denmark (Lyngby, Denmark), Institute of Physical Chemistry of the Polish Academy of Sciences (Warsaw, Poland) and Nicolaus Copernicus University (Toruń, Poland).

Andrzej Anderko is Assistant Professor at the Institute of Physical Chemistry of the Polish Academy of Sciences in Warsaw. He currently is a postdoctoral fellow at the University of California, Berkeley, in the research group of Professor Kenneth S. Pitzer. He graduated from the Warsaw Technical University in 1987 and obtained his Ph.D. from the Institute of Physical Chemistry in 1989. His research subjects include mostly the development of models for fluid phase equilibria and fluid properties with emphasis on equations of state for complex fluids as well as study of association phenomena and experimental measurements of vapor–liquid equilibria. Andrzej Anderko has published 20 papers in *Fluid Phase Equilibria*, *AIChE Journal*, *Chemical Engineering Science*, *Berichte der Bunsengesellschaft*, *Bulletin of the Polish Academy of Sciences*, *Journal of the Chemical Society* (*Faraday Transactions*), *Journal of Chemical Thermodynamics* and edited a book *Thermodynamics of Fluids: Measurement and Correlation* (with S. Malanowski, World Scientific, Singapore, 1990).

Contents

2. Pure-Component Properties Relevant to Phase Equilibrium Calculations 67

3. The Gamma–Phi Method 93

4. Equation-of-State Methods 153

Preface

The knowledge of phase equilibria and their dependence on pressure and temperature is very important for the understanding of various phenomena occurring in nature and industrial processes. Phase equilibria appear in our environment in the macro- as well as the microscale. They are essential for the design and optimization of various technologies. Investigation of phase equilibria, especially in multicomponent systems, is of utmost importance in numerous branches of science and engineering.

Almost all contemporary technical processes involve multicomponent mixtures. Such mixtures are of particular importance in the chemical and related industries. Thus, the parameters of phase equilibria are crucial for designing, optimizing and performing separation and purification processes. There is a growing tendency to obtain high-purity compounds or mixtures of constant composition. The cost of the separation of mixture components and purification of products represents a significant fraction of the total cost of the process. Another area where phase equilibria in multicomponent mixtures play an important role is protection of the environment. Industrial wastes—solid, liquid and gaseous—produced by any kind of industry have one thing in common: They are multicomponent mixtures.

In the present state of knowledge a similar degree of purity of a substance can be achieved with the use of various separation methods. Selection of the method finally applied is based on a cost analysis. In addition, environmental protection factors and the reliability of the process and materials used are involved in the decision making. Until the energy crisis in the early seventies the separation processes were responsible for 80 percent of the total investment spending by the chemical industry. Distillation was almost exclusively used as a separation process at that time. New methods introduced since then have changed the situation. In the years that followed the cost of separation processes as a fraction of total running and investment costs was showing a decreasing tendency. Simultaneously, the costs of fulfilling regulations reinforcing the obligatory standards of environmental protection were growing. The regulations are now more restrictive and a better knowledge of multicomponent phase equilibria is necessary to meet them.

The functional relations between temperature, pressure, phase composition and volume are essential for the representation of phase equilibria in systems formed by one or more substances. Such relations are provided by thermodynamics, a science considering those physical phenomena that necessitate at least two notions for their description: heat and temperature. In any case thermodynamics is not a model for the representation of the properties of matter. The properties of a substance depend only on its nature and are usually various for different substances. Investigation of the properties of various substances is not a subject of thermodynamics, but it is necessary for all practical applications of thermodynamics including phase equilibria.

The first chapter of this book is devoted to the thermodynamic principles of phase equilibria. The purpose of this chapter is to provide a rigorous and self-containing, though concise, outline of the thermodynamic background of practical methods. Thermodynamics makes it possible to determine the smallest set of primary data that is sufficient for the representation of the investigated system with desired accuracy. This is the advantage of using thermodynamics for the analysis of phase equilibria. The primary data are, in principle, experimental. They can be found in the published literature or measured directly by the investigator. In some cases it is possible to predict the data on the basis of the knowledge of the properties of matter. The relations provided by thermodynamics and the, small as possible, set of primary experimental and, sometimes, predicted data are necessary to determine a complete set of thermodynamic functions. These functions make it possible to compute the work delivered by the system, accompanying heat effects and conditions of stability and equilibrium. The theory of thermodynamics was established during the last 150 years and can be rigorously and correctly deduced by an almost unlimited number of various methods. For example, an excellent deduction of thermodynamics has been given by Münster (1970). The multiplicity of feasible approaches leads to some disorder and is sometimes misleading, especially when students try to use various texts coming from different scientific schools. To avoid any confusion this book will present a rigorous development of the laws of thermodynamics starting from fundamental principles, accompanied by a short description of the mathematics necessary for the clear understanding of basic concepts. Numerous erroneous statements found in various texts on phase equilibrium thermodynamics have their roots in misunderstandings of the mathematical background of thermodynamics.

Chapters 2–4 present practical methods used to calculate phase equilibria. The methods can be roughly divided into two groups: the so-called gamma–phi and equation-of-state methods. The first group employs separate models for the calculation of liquid phase activity coefficients (γ) and gas phase fugacity coefficients (ϕ). Moreover, it requires separate techniques to calculate the pure-component properties that are relevant to phase equilibrium computations. On the other hand, the equation-of-state method uses a homogeneous model for all fluid phases and is equally applicable, at least in principle, to pure-component and mixture properties. In Chapter 5 we focus on methods for analyzing thermodynamic consistency of experimental data.

There are two appendices to this book. Appendix A provides the basic definitions and the vocabulary of thermodynamics. Appendix B introduces the basic concepts of mathematics. The appendices are given to unify the vocabulary used.

Modelling of all types of phase equilibria is now an active area of research. A growing number of articles is published each year. Several textbooks and reference books are available (e.g., Pitzer and Brewer, 1961; Van Ness and Abbott, 1982; Prausnitz et al., 1986; Sandler, 1989; Rowlinson and Swinton, 1982; Chao and Greenkorn, 1975; Null, 1970; Modell and Reid, 1983; Walas, 1985; Bittrich, 1981; Reid et al., 1987 and others). Our monograph is intended to present a critical review of up-to-date methods with emphasis on practical applications. We hope it will bridge a gap between textbooks and articles in scientific and technical journals.

To arrive at practical solutions to real phase equilibrium problems, it is almost always necessary to develop relatively large computer programs. To elaborate such programs, the knowledge of both thermodynamic background and models of equilibrium behavior is essential. This explains the selection of material in this book. Emphasis was put on a good understanding of the well-established thermodynamic tools and on the novel, state-of-the-art methods to represent the real fluid behavior. It was the authors' intention to provide a comprehensive state-of-the-art review of phase equilibrium models. We have not provided, however, many numerical examples of calculations keeping in mind that practical calculations always entail the use of large (while not always complicated) computer programs.

The models covered in this book are relevant to the modelling of nonelectrolyte mixtures over wide ranges of pressure, temperature, composition and molecular variety. Electrolytes and polymers are not included. The huge amount of information about phase equilibria accumulated in the literature precludes, however, an exhaustive coverage of all relevant methods. On the other side, it is human nature to give preference to the subject of one's own work. Although we tried to be objective, we are aware that it was unavoidable to leave out some valuable methods. Nevertheless, we attempted to merge an objective description of various methods with our personal opinions about their usefulness. We hope that our book will be of value for graduate students, chemical engineers and all those who are interested in phase equilibria.

We are pleased to express our gratitude to Professor Stanley I. Sandler and Dr. Jacek Gregorowicz who read the manuscript carefully and helped to introduce many valuable corrections. Also, our thanks are due to Ms. Jadwiga Szemioth, Ms. Barbara Wiśniewska and Mr. Zbigniew Fraś for their assistance in the preparation of the typescript and figures.

Stanisław Malanowski
Andrzej Anderko

References

Bittrich, H. J. (Ed.), 1981. *Modellierung von Phasengleichgewichten als Grundlage von Stofftrennprozessen*, Akademie Verlag, Berlin.

Chao, K. C. and R. A. Greenkorn, 1975. *Thermodynamics of Fluids*, Dekker, New York.

Modell, M. and R. C. Reid, 1983. *Thermodynamics and Its Applications*, 2nd ed., Prentice-Hall, Englewood Cliffs, N.J.

Münster, A., 1970. *Classical Thermodynamics*, Wiley-Interscience, London.

Null, H. R., 1970. *Phase Equilibrium in Process Design*, Wiley-Interscience, New York.

Pitzer, K. S. and Brewer, L., 1961. Revised edition of *Thermodynamics* by G. N. Lewis and M. Randall, McGraw-Hill, New York.

Prausnitz, J. M., R. N. Lichtenthaler and E. G. Azevedo, 1986. *Molecular Thermodynamics of Fluid Phase Equilibria*, 2nd ed., Prentice-Hall, Englewood Cliffs, NJ.

Reid, R. C., J. M. Prausnitz and B. E. Poling, 1987. *The Properties of Gases and Liquids*, 4th ed., Wiley, New York.

Rowlinson, J. S. and F. L. Swinton, 1982. *Liquids and Liquid Mixtures*, 3rd ed., Butterworth, London.

Sandler, S. I., 1989. *Chemical and Engineering Thermodynamics*, 2nd ed., Wiley, New York.

Van Ness, H. C. and M. M. Abbott, 1982. *Classical Thermodynamics of Nonelectrolyte Solutions with Applications to Phase Equilibria*, McGraw-Hill, New York.

Walas, S. M., 1985. *Phase Equilibria in Chemical Engineering*, Butterworth, Boston.

Modelling Phase Equilibria

1

Thermodynamic Principles

1.1 INTRODUCTION

First, let us introduce some useful definitions. Our discussions will be limited to the equilibrium thermodynamics. *Equilibrium* will be defined as the state in which the thermodynamic variables of the system are independent of time. A system spontaneously tends toward this state. The number of variables that are necessary to describe the equilibrium state is smaller than that for a corresponding nonequilibrium state. For example, for a mole of a pure gas any two of the three variables—pressure, volume and temperature—are sufficient while for a nonequilibrium state two derivatives of these properties are necessary (e.g., pressure and temperature gradients). As *variables of state* we shall denote a set of variables that are sufficient to determine the state of equilibrium. A quantity whose differential is a complete differential of the variables of state is called a *function of state*. The abstract space that corresponds to variables of state is called the *coordinate* or *phase space*.

1.2 FUNDAMENTALS OF PHENOMENOLOGICAL THERMODYNAMICS

1.2.1. Introduction

There are two independent ways to deduce the laws of thermodynamics: *phenomenological* and *statistical*.

The phenomenological approach has the following properties:

1. The axioms, traditionally called laws of thermodynamics, are introduced as a sum of accumulated experience of macroscopic phenomena in the form of easy to use mathematical expressions.

2. Only measurable properties defined for macroscopic systems are considered. The molecular and atomic structure of matter is ignored.
3. The characteristic properties of substances under consideration are introduced in the form of parameters.

Besides thermodynamics, the phenomenological deduction is used in hydrodynamics and electrodynamics.

The statistical deduction of thermodynamics is based on the principle that the same laws as in the phenomenological method are introduced by means of microscopic properties. The methods of mathematical statistics and classical and quantum mechanics are used for this purpose.

In this book the principles of thermodynamics will be derived using the phenomenological method due to its simplicity. The methods based on, or inspired by, statistical mechanics will be used only for the computation of numerical values of thermodynamic properties for actual systems. Such a method, called by Prausnitz (1969) "molecular thermodynamics" is very useful for many applications.

The mathematics necessary for applied thermodynamics is relatively simple. It is prerequisite to know elements of the vector and differential calculuses, ordinary differential equations and properties of integrals in the potential field. From the point of view of mathematics, thermodynamics is considered in the so-called phase space. This is a vector space constructed specially for a system under consideration so that the basis of the vector space is sufficient for the representation of thermodynamic properties. For example, for the representation of the properties of a mole of a perfect gas any two variables are sufficient that are chosen arbitrarily from the three available pressure, temperature and volume variables.

Numerical methods, especially statistical methods for verifying the experiments and for fitting experimental data to nonlinear mathematical models, are also necessary to solve practical problems.

Traditionally, the phenomenological thermodynamics is divided into two branches: the classical thermodynamics and the thermodynamics of irreversible processes (nonequilibrium thermodynamics).

Classical thermodynamics does not consider the time and position variables. All processes considered can be approximated by an infinite number of sequential equilibrium processes. The investigated systems consist of any number of phases and the location of particular phases is not important. Natanson (cf. Guminski, 1982) proposed a new name for the classical thermodynamics that has a better logical background—thermostatics. Unfortunately, this name has not gained any popularity.

In principle, thermodynamics of irreversible processes is much more general due to the variables it uses (i.e., time and position coordinates). A statement can be made that classical thermodynamics is a subset of the thermodynamics of irreversible processes. For the sake of simplicity it is much more convenient to use classical thermodynamics whenever it is possible and to introduce the methods of irreversible thermodynamics in cases when it is unavoidable.

1.2.2. Theorems of Thermodynamics

The traditional course of thermodynamics presents the three laws of thermodynamics in the form of equations that are conclusions drawn from the experimental knowledge accumulated over centuries. This method is not internally consistent. Some additional statements are necessary. In this course the theorems of thermodynamics will be derived in a different way enabling all the classical thermodynamics relations to be properly deduced.

A rigorous derivation of classical thermodynamics was proposed by Caratheodory in 1909, at the time when the development of uniform and systematic theories was very popular in all branches of science. The developed theorems were analogous to the already existing theorems of theoretical mechanics.

The method of Caratheodory will be used in this course. The formulation of axioms and definitions is influenced by those given by Münster (1970). The description is much simplified and only very few examples are given, mostly those necessary for the understanding of the relations of phase equilibrium thermodynamics.

Thermodynamic considerations are always made for a part of the physical world, called a *system*. The system is separated from its surroundings by some conceptual boundaries. These boundaries are called walls, jackets or barriers.

According to fluid mechanics (hydrostatics), the volume of a one-phase macroscopic system in the state of equilibrium, not influenced by external forces, is determined exclusively by its mass and pressure. Introduction of the heat and temperature variables leads to the following theorems:

Theorem I (based on experience)***:*** Processes exist that lead to a change in volume (V) even though pressure (P) is constant. These are called *thermal processes*.

Theorem I leads to an immediate conclusion that a system considered in thermodynamics needs two independent variables to define its state. Such independent variables are called *variables of state*. Pressure and volume, which are defined in mechanics, can be arbitrarily chosen for this purpose. When the system can exchange energy with its surroundings, the walls surrounding it are called *heat conducting* or *diathermic*. When the state of a system can be changed only by nonthermal (i.e., mechanical) processes (e.g., movement of a piston), the surrounding wall is called *adiabatic*.

Theorem II (based on experience)***:*** If two bodies are separated by a diathermic wall and are isolated adiabatically from the surroundings, they are in equilibrium with each other only if one equation of the form

$$\mathscr{F}(P^{\mathrm{I}}, V^{\mathrm{I}}; P^{\mathrm{II}}, V^{\mathrm{II}}) = 0 \qquad (1.2\text{-}1)$$

is obeyed.

Definition #1: The equilibrium defined by Theorem II is called thermal equilibrium because according to Theorem I, it can be attained only by a thermal process.

Theorem III (based on experience, sometimes called the zeroth law of thermodynamics)***:*** Two systems each being independently in the state of thermal equilibrium with a third system are also in thermal equilibrium with each other.

Definition #2: Each of two systems, I and II, being in the state of thermal equilibrium is characterized by a measurable property

$$t = t(P, V) \tag{1.2-2}$$

such that the algebraic identities of $t = t^{\mathrm{I}}$ and $t = t^{\mathrm{II}}$ entail the identity $t^{\mathrm{I}} = t^{\mathrm{II}}$. This property is called empirical temperature.

Theorem IV (based on experience)***:*** The same amount of work is always necessary to change a system's state at adiabatic conditions from an initial state (1) to a final state (2).

Theorem IV is equivalent to a statement that the amount of work is independent of the equilibrium or nonequilibrium states passed during the process. It depends only on the initial and final states of the system.

Definition #3: Work (W) done on the system in an adiabatic process is called the increase in internal energy (U) of this system.

The change of internal energy during an adiabatic process of changing a system's state from 1 to 2 is given by

$$U_2 - U_1 = \int_1^2 dW \tag{1.2-3}$$

According to the third definition the *internal energy* of a system is a function of state. Its value depends only on the initial and final states and is independent of the path of the process, i.e., the intermediate states passed by the system. Equation (1.2-3) defines the value of the internal energy of a system except for an integration constant.

Definition 3 directly leads to a conclusion that not all changes can be made in an adiabatic way. For a nonadiabatic process the work done on a system depends on the intermediate states of the process (i.e., the path of the process). At the same time the change of internal energy in this process is independent of the path and is the same as in an equivalent adiabatic process. The internal energy is a function of state while work is not.

Definition #4: For any process of a change from state 1 into state 2, the difference between the increase of internal energy (U) and the work (W) done on the system is called the absorbed heat (Q).

The definition can be written in the form of a difference equation

$$dU = \delta Q + \delta W \tag{1.2-4}$$

Theorem IV together with Definitions 3 and 4 is equivalent to the *first law of thermodynamics* in its traditional formulation. The subsequent two theorems formulate the conditions that should be fulfilled to transform the differences from the right side of Eq. (1.2-4) into complete differentials in the phase space.

Theorem V (based on experience)***:*** For each state P_0 of a system there exist certain states of the system that cannot be reached by an adiabatic process starting in the state P_0.

Theorem VI (based on experience)***:*** When an adiabatic process occurs without any change in the system's volume, the internal energy of the system always increases.

The last two theorems are equivalent to the *second law of thermodynamics*. There is one more theorem that is very important for the thermodynamics of phase equilibria.

Theorem VII (based on experience, called the Nernst theorem)***:*** It is impossible to cool a system to the absolute zero temperature by any process repeated a finite number of times.

For simplicity, the names of the first, second and third laws of thermodynamics will be used in further parts of this work. The first law will be composed of Theorem IV and Definitions 3 and 4. The second law means the simultaneous use of Theorems V and VI, and the third law means Theorem VII.

1.2.3. Traditional Representation of the Laws of Thermodynamics

It is questionable who first formulated the laws of thermodynamics in a proper way. Boyle (1627–1691), one of the precursors of thermodynamics, believed in the caloric theory of heat. About a century later Thompson (1753–1814) proved experimentally that the caloric theory of heat was erroneous, but this paper remained unnoticed. Later on, Carnot (1796–1832) stated in his paper "Rèflexions sur la puissance motorice du feu" that the efficiency of a heat engine depended only on the difference in temperature and did not depend on the applied media. That was equivalent to the second law of thermodynamics. In 1878 Carnot's brother communicated to the French Academy that he had found the formulation of the first law of thermodynamics in the papers of his deceased brother. In 1842 Mayer formulated the principle of conservation of heat, and in 1845 Joule published his experiments of 1840–1845 proving the equivalence of work and energy. In 1848 Thomson defined, on the basis of Carnot's work, a scale of temperature independent of the media used. In 1850

Clausius published a paper combining the previously published work of Carnot with the principle of conservation of energy and formulated the principles of the second law of thermodynamics.

Both laws were explicitly formulated for the first time in 1851 by Thomson on the basis of the work of Carnot and Joule:

First law: When equal amounts of mechanical work are produced from thermal sources or disappear in thermal effects, equivalent amounts of heat disappear or appear.

Second law: It is impossible to construct a cyclic engine that performs mechanical work by merely cooling a heat reservoir.

This formulation of the second law is called the principle of the impossibility of perpetual motion of the second kind. The first law can be written in the form of the principle of the impossibility of perpetual motion of the first kind:

It is impossible to construct a machine that performs mechanical work without consuming an equivalent amount of heat.

In 1854 Clausius introduced the concept of entropy and presented the second law in the following form:

It is impossible for heat to flow spontaneously inside an isolated system from a colder to a warmer part.

It can be proven (Münster, 1970) that all these formulations of the first and second law are equivalent. This is a characteristic example for the whole thermodynamics. The logic of this subject can be constructed in an almost unlimited number of ways. The reasons for choosing a particular way are either coming from local tradition or are motivated by the author's desire to express thermodynamics by his original method.

The Thomson formulation of the second law leads to the following conclusion. Transformation of heat into work in a continuously working heat engine is possible only in a process carried out between two heat containers. An amount Q_1 of heat is continuously withdrawn from container 1 and an amount Q_2 is continuously transmitted into container 2. Only the difference, $Q_1 - Q_2$, is transferred into work. The heat efficiency (η) of the engine is determined by the formula

$$\eta = \frac{Q_1 - Q_2}{Q_1} \tag{1.2-5}$$

Similarly, a direct conclusion can be drawn from the Clausius formulation of the second law that the heat efficiency of a reversible cyclic process cannot be lower than the efficiency of any other process based on the same temperature difference.

Carnot's formulation of the second law stating that the efficiency of a reversible thermal process proceeding between two heat containers (1 and 2) depends only on their temperatures T_1 and T_2 and Thomson's definition of efficiency (1.2-5) give rise to the following relation:

$$\frac{Q_2}{Q_2} = \frac{T_2}{T_1} \tag{1.2-6}$$

This equation was traditionally used to define the temperature scale. Equation (1.2-6) can be written for the boiling t_b and freezing t_0 temperatures of water:

$$\frac{Q_2}{Q_1} = \frac{t_0}{t_b + t_0} \tag{1.2-7}$$

where the present values of t_b and t_0 (ITS-90) are $t_0 = 273.15$ (melting temperature of water) and $t_b = 99.975$ (difference between the boiling and melting temperature of water.

The name *entropy* was introduced by Clausius in 1865. In the same work he published his famous statement: "The energy of the universe is constant. The entropy of the universe tends towards maximum."

In 1869 Massieu introduced the concept of characteristic functions, from which all thermodynamic properties can be derived by differentiation.

In 1875 Gibbs published a paper, with the Clausius statement of 1865 as a motto, containing the generalization of thermodynamics to multicomponent, multiphase systems. He introduced chemical reactions and derived the equilibrium conditions for many special cases from general assumptions. Gibbs was the first who recognized and was able to describe the generality and complexity of thermodynamics.

It is very interesting that the American thermodynamics developed by Gibbs remained unknown in Europe for many years. In 1882 Helmholtz, not aware of Gibbs's results, introduced the free energy concept and the equation now known as the Gibbs–Helmholtz equation. In 1886 Duhem introduced once more the equation known now as the Gibbs–Duhem equation.

In 1887 Planck introduced the concept of reversible and irreversible processes. In 1906 Nernst published his theorem, later known as the third law. The basic development of thermodynamics was closed in 1909 by Carathèodory when he published his new axiomatic basis of thermodynamics. The majority of later papers are devoted to the applications of thermodynamics to specific purposes. One such purpose was the representation of the properties of phases in equilibrium.

1.2.4. Introduction of the Gibbs Fundamental Equation

The differential of heat (Q) can be written, according to its definition, for any system in an m-dimensional phase space in the following way:

$$dQ = \sum_{i=1}^{m} X_i \, dx_i \tag{1.2-8}$$

where variables X_i are functions of all or some of the variables of state and x_i are generalized coordinates. Equation (1.2-8) is called the *Pfaff differential equation*. For simplicity it can be written as a scalar product.

$$dQ = \mathbf{X} \cdot \mathbf{x} \tag{1.2-9}$$

Comparison with Eq. (1.2-4) shows that the mathematical formulation of the first law leads to the Pfaff equations. Expression (1.2.9) can be integrated along any path in the m-dimensional phase space. The result will depend on the path of integration because the differentials of heat (Q) and work (W) are not always complete in the phase space [cf. Definition 4 and Eq. (1.2-4)].

According to the vector calculus, the necessary and sufficient condition for a differential to be complete or, equivalently, that the line integral of $\mathbf{X}$ vanishes along a closed curve is

$$\text{rot}\, \mathbf{X} = 0 \tag{1.2-10}$$

Now it is necessary to find an integrating factor $1/\tau$ for which the expression dQ/τ will be complete integral.

It can be proven (Caratheodory, 1909; Münster, 1970) that for the Pfaff differential form of heat dQ (1.2-8) the integrating factor $1/\tau$ exists when and only when there is a solution to the Pfaff differential equation for dQ in the form

$$S(x_1, \ldots, x_m) = S \tag{1.2-11}$$

This solution defines a family of one-parameter surfaces [$(m-1)$-dimensional forms] in the m-dimensional phase space. By definition, $d\tau$ is a complete differential and can be written in the form

$$d\tau = \frac{dQ}{\tau(x_1, \ldots, x_m)} \tag{1.2-12}$$

Let us introduce a one-parameter (τ) function in the form

$$S = S[\tau(x_1, \ldots, x_m)] \tag{1.2-13}$$

Upon substitution of (1.2-12) the differential of S can be written as

$$dS = \frac{dS}{d\tau}\, d\tau = \frac{dS}{d\tau}\, \frac{dQ}{\tau} \tag{1.2-14}$$

For the function $T(x_1, \ldots, x_m)$ defined as

$$T(x_1, \ldots, x_m) = \tau(x_1, \ldots, x_m) \frac{d\tau}{dS} \tag{1.2-15}$$

Eq. (1.2-13) can be transformed to

$$S[\tau(x_1, \ldots, x_m)] = \frac{dQ}{T(x_1, \ldots, x_m)} \tag{1.2-16}$$

Now it is evident that the reciprocal of the function $T(x_1, \ldots, x_m)$ is an integrating factor in the phase space for the differential of heat. In the same way other integrating factors can be constructed for the Pfaff differential form.

For a better understanding of the equation introduced two examples will be demonstrated.

Example 1.2-1 What conditions should be fulfilled to transform dQ into a complete differential in a three-dimensional phase space?

SOLUTION According to Eq. (1.2-9) the complete differential can be written in the following form:

$$dQ = \mathbf{X} \cdot d\mathbf{x}$$

In agreement with the vector calculus (Eq. B.2-21 in Appendix B) a sufficient condition for dQ to be a complete differential is

$$\text{rot}\,\mathbf{X} = 0 \tag{1.2-10}$$

Consequently, dQ will be a complete differential when the vector $\mathbf{X}$ is a gradient of a scalar function. ■

Example 1.2-2 What are the conditions for the existence of an integrating factor with two independent variables $1/\tau$ transforming dQ/τ into a complete integral?

SOLUTION For solving this problem it is necessary to transform Theorem V into a form given by Caratheòdory: "If a Pfaff differential dQ has the property that in the vicinity of $P_0(\mathbf{x}^0)$ there are other points $P_1(\mathbf{x})$ which are not accessible from $P_0(\mathbf{x}^0)$ along such a path that $dQ = 0$ then an integrating factor exists for dQ."

This condition can be written in a vector form:

$$\mathbf{X} \cdot \text{rot}\,\mathbf{X} = 0 \tag{P1.2-1}$$

For two independent variables the Pfaff differential equation is

$$dQ = X\,dx + Y\,dy = 0 \tag{P1.2-2}$$

The integrating factor for this equation is

$$\tau = \tau(x, y) \tag{P1.2-3}$$

Apart from fulfilling the requirement that (P1.2-3) should be a solution to (P1.2-2), the complete differential of the integrating factor should be equal to zero

$$d\tau = \frac{\partial \tau}{\partial x} dx + \frac{\partial \tau}{\partial y} dy = 0 \tag{P1.2-4}$$

Equations (P1.2-2) and (P1.2-4) lead to

$$\frac{dy}{dx} = -\frac{X}{Y} = -\frac{d\tau/dx}{dy/dx} \tag{P1.2-5}$$

Now a function can be obtained

$$t(x, y) = \frac{X}{d\tau/dx} = \frac{Y}{d\tau/dy} \tag{P1.2-6}$$

as, according to the first assumption, the following relation can be written

$$d\tau = \frac{dQ}{t(x, y)} = \frac{X}{t} dx + \frac{Y}{t} dy \tag{P1.2-7}$$

A function $dQ/t(x, y)$ of two independent variables x and y is always an integrating factor. For a general, multivariable case the deduction is more complicated and the shape of the integrating factor corresponds to Eq. (P1.2-6). ■

The following equation represents the differential of work (W):

$$dW = -P\,dV + \sum_{j=2}^{l} Y_j\,dy_j \tag{1.2-17}$$

The first term on the right represents the work due to a change of volume ($P\,dV$), called the *volumetric work*, and all the other $(l-1)$ terms ($Y_j\,dy_j$) represent other possible types of work performed within a closed phase during a quasi-static (i.e., proceeding through a sequence of equilibrium states) change of state caused by external fields Y_j (e.g., gravitational, magnetic, electric etc.). The number of work variables does not influence the general thermodynamic description. Only the dimension of the appropriate phase space in which Eq. (1.2-13) is a function of state changes. For simplicity, only the work due to a change of volume will be used for a general discussion. Other work coordinates can be introduced when necessary.

The complete differential of the internal energy (dU) in the case when only volumetric work is considered is as follows:

$$dU = dQ - P\,dV \tag{1-2-18}$$

Substitution of (1.2-17) gives

$$T\,dS = dU + P\,dV \tag{1.2-19}$$

Any contact equilibrium can be reached and maintained within an adiabatically isolated system. Let us describe the state of equilibrium for a system consisting of k phases and n components. The functions of state in Eq. (1.2-19), i.e., the internal energy (U) and volume (V), can be treated, in general, as parameters $\mathcal{P}_{E,i}$ satisfying the principle of conservation of the total $\mathcal{P}_E$ value for all k phases of the system.

$$\sum_{\alpha=1}^{k} \mathcal{P}_{E,i}^{(\alpha)} = \text{const}\,. \tag{1.2-20}$$

The parameters $P_{E,i}$ depend on the size of the system considered, i.e., on the amount of material involved. They are known as *extensive properties* of the system. Properties such as temperature (T) and pressure (P) do not involve the amount of matter in any way, and they are known as *intensive properties*. In general, the variables that are not related to the amount of the system and can be measured locally are called *intensive variables*. According to Münster's (1970) formulation, they express the "intensity" of the system. They are homogeneous functions of the zeroth order in relation to the amount of substance. To specify completely the system, its mass must be specified in addition to volume and entropy. This is called the "extent" of the system. Variables that relate to the extent of the system are of the first order in relation to mass and are called *extensive variables*.

The processes leading to equilibrium are adiabatically irreversible. Therefore, the entropy of a system is at maximum in the state of equilibrium for given values of intensive parameters $\mathcal{P}_{I,i}$. The intensive and extensive parameters are conjugated through the function (1.2-13)

$$\mathcal{P}_{I,i} = \frac{\partial S}{\partial \mathcal{P}_{E,i}} \tag{1.2-21}$$

According to Definition 2 each contact equilibrium between the phases is determined by the intensive parameters $\mathcal{P}_{I,i}$ having constant values for all k coexisting phases

$$\mathcal{P}_{I,i}^{(1)} = \mathcal{P}_{I,i}^{(2)} = \cdots = \mathcal{P}_{I,i}^{(\alpha)} \tag{1.2-22}$$

There exists a function of state (S) defined by Eq. (1.2-16) for each phase α containing n components. Besides the "volumetric work" it is necessary to take into account the work of external fields, surface tension etc. by introduction of additional work coordinates. They can be defined so that the differential dy_j of the work coordinate y_j multiplied by a "work coefficient" (generalized force) Y_j gives the work performed in a closed phase (α) during a quasi-static change of state. Function S is a function of the variables of state: *internal energy of phase* α ($U^{(\alpha)}$) *volume of phase* α ($V^{(\alpha)}$), additional work coordinates (y_j) for l generalized forces Y_j and the number of moles ($N_i^{(\alpha)}$) of each of n various components.

$$S^{(\alpha)} = S^{(\alpha)}(U^{(\alpha)}, V^{(\alpha)}, y_2^{(\alpha)}, \ldots, y_i^{(\alpha)}, N_1, \ldots, N_n) \tag{1.2-23}$$

Function S in an extensive property and is called the entropy of phase α. The following statements can be proven on the basis of the preceding assumptions.

The differential of entropy is defined by

$$T\,dS^{(\alpha)} = dU^{(\alpha)} + P\,dV^{(\alpha)} - \sum_{j=2}^{l} Y_j^{(\alpha)}\,dy_j^{(\alpha)} + \sum_{i=1}^{n} \mu_i\,dN_i^{(\alpha)} \tag{1.2-24}$$

where T is the empirical temperature.

The intensive parameters defined by (1.2-21) are

$$\frac{1}{T} = \mathcal{P}_{I,1}^{(\alpha)} = \left(\frac{\partial S^{(\alpha)}}{\partial U^{(\alpha)}}\right)_{V^{(\alpha)}, y_{2,\ldots,1}, N_{1,\ldots,n}} \tag{1.2-25}$$

$$\frac{P}{T} = \mathcal{P}_{I,2}^{(\alpha)} = \left(\frac{\partial S^{(\alpha)}}{\partial V^{(\alpha)}}\right)_{U^{(\alpha)}, y_{2,\ldots,1}, N_{1,\ldots,n}} \tag{1.2-26}$$

and for all above $(l+1)$ terms

$$\frac{\mu_i^{(\alpha)}}{T} = \mathcal{P}_{I,l+1+i}^{(\alpha)} = \left(\frac{\partial S^{(\alpha)}}{\partial N_i^{(\alpha)}}\right)_{V^{(\alpha)}, y_{2,\ldots,1}, N_{j\neq i}} \tag{1.2-27}$$

where $\mu_i^{(\alpha)}$ was introduced by Gibbs and is called the *chemical potential of component* i in the phase α. In contemporary thermodynamics the chemical potential is usually related to the number of moles of the substance. It can be also related to mass or the number of molecules.

The entropy of the whole system of k different phases is a sum of entropies of these phases:

$$S = \sum_{\alpha=1}^{k} S^{(\alpha)} \tag{1.2-28}$$

For an adiabatically isolated system in equilibrium the differential of entropy satisfies the relation

$$(dS \geq 0)_{\text{adiabatic isolated}} \tag{1.2-29}$$

Equation (1.2-24) together with conditions (1.2-25)–(1.2-29) is called the *Gibbs fundamental equation*. It contains all thermodynamic information on the system under consideration. The fundamental equation is applicable not only to reversible processes. The criterion for applicability is that the internal state of the phase at each stage of a process is completely described by the indicated variables. This means that no variable of state can be omitted in the formulation of the equation for a particular system.

Entropy of the phase (1.2-23) is, by definition, a continuous and monotonous function of all variables of state. Therefore, it can be rearranged to yield an explicit equation for the internal energy

$$U^{(\alpha)} = U^{(\alpha)}(S^{(\alpha)}, V^{(\alpha)}, y_2^{(\alpha)}, \ldots, y_i^{(\alpha)}, N_1, \ldots, N_n) \tag{1.2-30}$$

A differential form of Eq. (1.2-30) is

$$dU^{(\alpha)} = T\,dS^{(\alpha)} - P\,dV^{(\alpha)} + \sum_{j=2}^{l} Y_j^{(\alpha)}\,dy_j^{(\alpha)} + \sum_{i=1}^{n} \mu_L^{(\alpha)}\,dN_i^{(\alpha)} \tag{1.2-31}$$

Equations (1.2-23) and (1.2-30) are equivalent. Either is called the *fundamental equation*. Equation (1.2-33) is termed the *entropy representation* and Eq. (1.2-30) is the *energy representation*. The fundamental equation represents a surface in an $(n + l + 1)$-dimensional space. A point on this surface represents stable equilibrium states of the system. A quasi-static process can be represented by a curve on this surface. Processes that are not quasi-static are not identified with points on this surface. At present, the energetic representation of the Gibbs fundamental equation is used almost exclusively in the phase equilibrium thermodynamics. This equation will be used in this work for further developments.

The fundamental equation can be also used for open (nonisolated) systems. The following procedure can be applied. An adiabatically isolated system consists of two subsystems, I and II. The pressure (P^{I} and P^{II}) and temperature (T^{I} and T^{II}) in both subsystems are different. Let us designate the internal energy of the whole system by U and that of the subsystems I and II by U^{I} and U^{II}, respectively. After removing the wall separating systems I and II, the pressure and temperature will change while U remains constant. This process can serve as a model of a change of state in an open phase I (the subsystem I). The change of internal energy of phase I will be

$$\Delta U^{\mathrm{I}} = U - U^{\mathrm{I}} \tag{1.2-32}$$

Internal energy U and U^{I} can be measured and, subsequently, the left side of the equation can be computed by means of the integrated from of the fundamental equation (1.2-30). This means that the value of the internal energy can be computed from the first law (1.2-4).

The fundamental equation was given for the first time by Gibbs (1875). It is very useful for introducing and demonstrating the methods used in phase equilibrium thermodynamics. All necessary formulas can be deduced from this equation only with the use of mathematics and without additional assumptions.

1.2.5. Properties of the Gibbs Fundamental Equation

Let us consider the Gibbs fundamental equation [(1.2-23) or (1.2-30)] omitting all work variables except for the volumetric work. Such simplification does not influence the discussion and is used merely to simplify notation. For the energy representation the equation has the form

$$U = U(S, V, N_1, \ldots, N_n) \tag{1.2-33}$$

and for the entropy representation it becomes

$$S = S(U, V, N_1, \ldots, N_n) \tag{1.2-34}$$

For simplicity, let us introduce vector notation. A vector of the numbers of moles of n substances $N_1, \ldots, N_n$ will be denoted by $\mathbf{N} = (N_1, \ldots, N_n)$. The energy representation is mostly used in the thermodynamics of phase equilibria. The entropy representation is used only in very special cases.

Let us summarize all properties of the fundamental equation:

1. The fundamental equation is a *characteristic function*. This means that the equation contains all the thermodynamic information about the system. It should be emphasized that the properties of a characteristic function are not only the properties of internal energy or entropy, but those of a complete set of independent variables used for this equation. Let us consider an example.

Example 1.2-3 What is the effect of independent variables on a thermodynamic function?

SOLUTION Let us consider internal energy U of a single component system as a function of independent variables: temperature (T), volume (V) and number of moles (N)

$$U = U(T, V, \mathbf{N}) \tag{P1.2-8}$$

The differential of this function is

$$dU = \left(\frac{\partial U}{\partial T}\right)_{V,\mathbf{N}} dT + \left(\frac{\partial U}{\partial V}\right)_{T,\mathbf{N}} dV + \left(\frac{\partial U}{\partial N}\right)_{V,T} dN \tag{P1.2-9}$$

According to Theorem I, dU is a complete differential and a potential in the phase space. However, (P1.2-9) is not a characteristic function due to the fact that entropy does not appear in (P1.2-9) as an independent variable. In addition, the intensive variables pressure (P) and chemical potential (μ) cannot be obtained by differentiation of (P1.2-9). Equations (1.2-33) and (P1.2-9) are not equivalent. Equation (P1.2-9) contains much less information than (1.2-33). Equation (P1.2-9) can be derived from the fundamental equation (1.2-33) while an opposite transformation is impossible. Equation (P1.2-9) is very useful, however, for the definition of the molar heat capacity at constant volume (C_v). For one mole of substance the definition is

$$C_v = \left(\frac{\partial U}{\partial T}\right)_{V,\mathbf{N}=1} \tag{P1.2-10}$$

The heat capacity is, by definition, an intensive property. ■

This example shows that the set of variables given for the fundamental equation in the energetic (1.2-33) and entropic (1.2-34) representation is the only one sufficient to represent U or S as characteristic functions. If other sets of variables are used, the properties of characteristic functions vanish.

2. The Gibbs fundamental equation is a homogeneous function of the first order for all its independent variables. All independent variables are extensive properties. In agreement with (1.2-20) each extensive variable $\mathcal{P}_{E,i}$ representing the property i for a system of k phases (subsystems) is equal to the sum of its values for all phases (subsystems) forming the system.

$$\sum_{\alpha=1}^{k} \mathcal{P}_{E,i}^{(\alpha)} = \mathcal{P}_{E,i} \tag{1.2-35}$$

The extensive parameters are the internal energy ($U = \mathcal{P}_{E,1}$), entropy ($S = \mathcal{P}_{E,2}$), volume ($V = \mathcal{P}_{E,3}$) and the number of moles of all components of an n-component system ($N_1 = \mathcal{P}_{E,4}, \ldots, N_n = \mathcal{P}_{E,3+n}$).

3. The intensive parameters $\mathcal{P}_I$ are determined for the energetic representation as partial derivatives of internal energy (U) against corresponding extensive parameters $\mathcal{P}_E$

$$\mathcal{P}_{I,i} = \left(\frac{\partial U}{\partial \mathcal{P}_{E,i}} \right) \tag{1.2-36}$$

A pair of an intensive and an extensive parameter coupled together by Eq. (2.36) is called *conjugate parameters.*

The differential form of the fundamental equation (1.2-31) defines the intensive parameters of the energetic representation. They are:

Temperature (T)

$$\left(\frac{\partial U}{\partial S} \right)_{V,\mathbf{N}} = T \tag{1.2-37a}$$

Pressure (P)

$$\left(\frac{\partial U}{\partial V} \right)_{S,\mathbf{N}} = -P \tag{1.2-37b}$$

Chemical potential μ_i *of each component* i

$$\left(\frac{\partial U}{\partial N_i} \right)_{V,S,N_{j \neq i}} = \mu_i \tag{1.2-38}$$

An advantage of the energetic representation is the use of intensive parameters that are directly measurable. According to eqs. (1.2-25)–(1.2-27) the intensive parameters of entropic representation are: $1/T$, P/T and $-\mu_i/T$. Such parameters can only be computed from measurable properties rather than measured directly.

The second partial derivatives of the fundamental equation represent also measurable properties. The derivatives computed according to the general scheme

$$\left(\frac{\partial^2 \mathscr{F}}{\partial x\, \partial y}\right) = \left(\frac{\partial^2 \mathscr{F}}{\partial y\, \partial x}\right) \tag{1.2-39}$$

lead to some relations between parameters that are very useful for thermodynamic computations.

Example 1.2-4 What are the relations between the first derivatives of P, T, S and U?

SOLUTION From Eqs. (1.2-31) and (1.2-36) we obtain

$$\left(\frac{\partial^2 U}{\partial V\, \partial S}\right) = \left[\frac{\partial}{\partial V}\left(\frac{\partial U}{\partial S}\right)\right] = \left(\frac{\partial T}{\partial V}\right)_{S,\mathbf{N}} \tag{P1.2-11}$$

and from Eqs. (1.2-31) and (1.2-37) we get

$$\left(\frac{\partial^2 U}{\partial S\, \partial V}\right) = \left[\frac{\partial}{\partial S}\left(\frac{\partial U}{\partial V}\right)\right] = -\left(\frac{\partial P}{\partial S}\right)_{V,\mathbf{N}} \tag{P1.2-12}$$

The following relation is obtained according to (1.2-39) by comparing (P1.2-12) and (P1.2-13):

$$\left(\frac{\partial T}{\partial V}\right)_{S,\mathbf{N}} = -\left(\frac{\partial P}{\partial S}\right)_{V,\mathbf{N}} \tag{P1.2-13}$$

Relations of this type are called Maxwell relations and will be further discussed. ■

4. The fundamental equation is a homogeneous function of the first order in relation to all its extensive variables. In agreement with the definition of a derivative, each intensive variable of this equation is a homogeneous function of the zeroth order of all extensive variables of the fundamental equation. The corresponding functions

$$\left(\frac{\partial U}{\partial \mathscr{P}_{E,i}}\right) = \mathscr{P}_{I,i} = \mathscr{P}_{I,i}(\mathscr{P}_{E,1}, \ldots, \mathscr{P}_{E,r}) \tag{1.2-40}$$

are called *equations of state* and are very important in the thermodynamics of phase equilibria.

1.2.6. Thermodynamic Potentials

It has been shown that the fundamental equation contains all possible information about the system considered. However, the use of the fundamental equation for practical purposes is almost impossible. All independent (extensive) variables of this equation, except for the number of moles (N), are either difficult [like volume (V)] or impossible to measure directly [like entropy (S) or internal energy (U)]. To avoid the difficulties it is necessary to establish functions that contain the same information as the fundamental equation and depend on easy-to-measure variables like pressure (P) and temperature (T). In addition, the mathematical form of the functions used should be simple.

Functions that are potentials in the phase space are most useful for this purpose. The value of a linear integral in the potential field in the phase space depends only on the initial and final state and does not depend on the integration path. This is in agreement with the first law (Definition 3). This property simplifies considerably the thermodynamic calculations. The integral represents the work performed in the potential field. All transformations of the thermodynamic functions should be performed using the potentials in phase space (*coordinate space*). These potentials should be *functions of state*, i.e., their differentials should be complete.

The *Legendre transformation* is the only method enabling the Gibbs fundamental equation in the energy, as well as entropy representation, to be changed into other thermodynamic potentials preserving its mathematical content and the corresponding physical information. This method transforms one or more extensive independent variables into conjugate intensive parameters. The Legendre transformation is described in Appendix B.

Let us write the fundamental equation in an r-dimensional coordinate space for a system consisting of n chemical individuals using the generalized extensive parameters $\mathcal{P}_{E,i}$ as independent variables in Eqs. (1.2-23) and (1.2-30) (the integral form)

$$U = U(\mathcal{P}_{E,1}, \ldots, \mathcal{P}_{E,r}) \tag{1.2-41}$$

and Eqs. (1.2-24) and (1.2-31) (the differential form)

$$dU = \sum_{i=1}^{r} \mathcal{P}_{I,1}\, d\mathcal{P}_{E,i} \tag{1.2-42}$$

where E denotes extensive and I intensive parameters. The dimension of the phase space is $r \geq n + 2$. The intensive and extensive parameters are coupled in pairs by the relation (1.2-36)

$$\mathcal{P}_{I,i} = \left(\frac{\partial U}{\mathcal{P}_{E,i}}\right)_{\mathcal{P}_{E,j\neq i}} \tag{1.2-43}$$

According to relations given in Appendix B the k-fold transformation of the internal energy (U) into a thermodynamic potential E_k can be carried out according to the following scheme:

$$E_k(P_{I,1}, \ldots, \mathcal{P}_{I,k}, \mathcal{P}_{E,k+1}, \ldots, \mathcal{P}_{E,r}) = U(\mathcal{P}_{E,1}, \ldots, \mathcal{P}_{E,r}) - \sum_{i=1}^{k} \mathcal{P}_{I,i}\mathcal{P}_{E,i} \tag{1.2-44}$$

The necessary and sufficient condition for such a transformation is that the Jacobi determinant should not vanish

$$\frac{\partial(\mathcal{P}_{I,1},\ldots,\mathcal{P}_{I,k})}{\partial(\mathcal{P}_{E,1},\ldots,\mathcal{P}_{E,k})} \neq 0 \tag{1.2-45}$$

Equation (1.2-44) can be also written in a differential form by taking into account Eq. (1.2-42)

$$dE_k = \sum_{i=k+1}^{r} \mathcal{P}_{I,i}\, d\mathcal{P}_{E,i} - \sum_{j=1}^{k} \mathcal{P}_{E,j}\, d\mathcal{P}_{I,j} \tag{1.2-46}$$

All thermodynamic functions (potentials) obtained by the Legendre transformation (1.2-46) when $k < r$ are homogeneous functions of the first order in relation to extensive parameters and have the same properties of the characteristic function as the fundamental equation. This property vanishes in the case of a full transformation (when $k = r$). For a complete description of the thermodynamic system at least one extensive variable representing the extent of system is necessary.

The partial derivatives of the obtained potentials E_k against the extensive parameters are analogous to the derivatives of the fundamental equation and are equal to the conjugate intensive parameters.

$$\frac{\partial E_k}{\partial \mathcal{P}_{E,i}} = \mathcal{P}_{I,i} \tag{1.2-47}$$

The derivatives against intensive parameters are equal to the negative values of conjugate extensive parameters

$$\frac{\partial E_k}{\partial \mathcal{P}_{I,i}} = -\mathcal{P}_{E,i} \tag{1.2-48}$$

Now the special cases of Eq. (1.2-44) that are used in thermodynamics will be considered. The fundamental equation can be written in the energy (1.2-30) and entropy (1.2-23) representation. First, the potentials of the energy representation will be considered.

Potentials of the Energy Representation

1. $k = 0$. *Internal energy* (U). The internal energy, defined directly by the fundamental equation is also a thermodynamic potential. Its application as a potential is considerably limited due to the fact that it depends only on extensive variables. The properties of the internal energy have been described earlier [Eqs. (1.2-30) and (1.2-31), Example 1.2-3].

2. $k = 1$. Two cases should be distinguished.

A. Enthalpy (H) is obtained by transforming the extensive property volume (V) into the conjugate intensive property pressure (P). Equation (2.41) has the following form for enthalpy:

$$H = H(P, S, \mathbf{N}) \tag{1.2-49}$$

Equation (1.2-44) becomes

$$H = U + PV \tag{1.2-50}$$

and Eq. (1.2-46) takes the form

$$dH = V\,dP + T\,dS + \sum_{i=1}^{n} \mu_i\,dN_i \tag{1.2-51}$$

The partial derivatives computed against the independent variables are

$$\left(\frac{\partial H}{\partial S}\right)_{P,\mathbf{N}} = T\,; \qquad \left(\frac{\partial H}{\partial P}\right)_{S,\mathbf{N}} = V\,; \qquad \left(\frac{\partial H}{\partial N_i}\right)_{S,P,\mathbf{N}_{i\neq j}} = \mu_i \tag{1.2-52}$$

The application of enthalpy as a thermodynamic potential is inconvenient due to the use of entropy as an independent variable. The enthalpy as a function of state can be represented with the use of any complete set of the variables of state. It is useful to employ temperature (T) in place of the conjugate variable entropy.

$$H = H(P, T, \mathbf{N}) \tag{1.2-53}$$

Enthalpy represented by (1.2-53) as a function of P and T is no more a characteristic function. Some information contained in (1.2-49) gets lost. The differential form corresponding to (1.2-53) is

$$dH = \left(\frac{\partial H}{\partial P}\right)_{T,\mathbf{N}} dP + \left(\frac{\partial H}{\partial T}\right)_{P,\mathbf{N}} dT + \sum_{i=1}^{n} \left(\frac{\partial H}{\partial N_i}\right)_{T,P,N_{j\neq i}} dN_i \tag{1.2-54}$$

For a better understanding of the use of enthalpy in thermodynamic computations let us consider an example.

Example 1.2-5 Let us determine the relation between enthalpy and heat (Q) defined by the first law (1.2-4).

SOLUTION For this purpose let us write equation (1.2-51) for one mole of a substance

$$dH = V\,dP + T\,dS + \mu_i\,dN_i \tag{P1.2-14}$$

The last term can be omitted as a constant number of moles is considered. By substituting (1.2-16) we obtain

$$dH = V\,dP + dQ \tag{P1.2-15}$$

The obtained equation leads to a conclusion that the amount of heat delivered to the system at constant pressure is equal to the increase of enthalpy:

$$(dH)_{P,\mathbf{N}} = dQ \quad \blacksquare \tag{P1.2-16}$$

The equations deduced in the example lead to the definition of the *heat capacity at constant pressure* (C_P) in an analogous way to the heat capacity at constant volume, defined previously (P1.2-11). For a one-component system formed by one mole of a substance C_P is given by

$$C_P = \left(\frac{\partial H}{\partial T}\right)_{P,N=1} \tag{1.2-55}$$

Taking into account that most processes in our environment and in technology are carried out at constant pressure, enthalpy is very useful for their interpretation and mathematical modelling.

B. Helmholtz energy (A) is obtained by transforming the extensive property entropy (S) into the conjugate intensive property temperature (T). Equation (1.2-41) for the Helmholtz energy has the form

$$A = A(V, T, \mathbf{N}) \tag{1.2-56}$$

Equation (1.2-44) becomes

$$A = U - TS \tag{1.2-57}$$

and Eq. (1.2-46) takes the form

$$dA = -P\,dV - S\,dT + \sum_{i=1}^{n} \mu_i\,dN_i \tag{1.2-58}$$

The partial derivatives computed against the independent variables are

$$\left(\frac{\partial A}{\partial V}\right)_{T,\mathbf{N}} = -P \qquad \left(\frac{\partial A}{\partial T}\right)_{V,\mathbf{N}} = -S \qquad \left(\frac{\partial A}{\partial N_i}\right)_{T,V,N_{i\neq j}} = \mu_i \tag{1.2-59}$$

The applications of the Helmholtz energy are directly related to its complete differential. The isothermal work due to a volume change in a closed system is equal to a change in the Helmholtz energy. The derivative $(\partial A/\partial V)_{T,\mathbf{N}} = -P$ is equivalent to the so-called *thermal equation of state*. In addition, the Helmholtz energy is very useful for computations involving statistical mechanics due to the set of its independent variables (V and T).

3. $k = 2$. The Gibbs energy (G) is a thermodynamic potential obtained by a twofold transformation of the internal energy. The extensive parameter volume (V) is transformed into its conjugate parameter pressure (P) and the extensive parameter entropy into its conjugate parameter temperature (T).

Equation (1.2-41) for the Gibbs energy has the form

$$G = G(P, T, \mathbf{N}) \tag{1.2-60}$$

Equation (1.2-44) becomes

$$G = U - TS + PV = H - TS = A + PV \tag{1.2-61}$$

and Eq. (1.2-46) is

$$dG = V\,dP - S\,dT + \sum_{i=1}^{n} \mu_i\,dN_i \tag{1.2-62}$$

The partial derivatives computed against the independent variables are

$$\left(\frac{\partial G}{\partial P}\right)_{T,\mathbf{N}} = V \qquad \left(\frac{\partial G}{\partial T}\right)_{P,\mathbf{N}} = -S \qquad \left(\frac{\partial G}{\partial N_i}\right)_{T,P,N_{i\neq j}} = \mu_i \tag{1.2-63}$$

The Gibbs energy is the most important thermodynamic potential in the thermodynamic of phase equilibria. This is a consequence of the fact that all the independent variables of this potential are easily measurable. The only extensive properties involved are the numbers of moles of individual chemical species. A value of the Gibbs energy can be easily computed from the phase equilibrium, enthalpy and molar volume data. Such computations will be described in detail in this work.

4. $k = r - 1$. Grand potential (Ω). This potential is not used in phase equilibrium thermodynamics. It should be mentioned here due to its importance in statistical thermodynamics, where it is used for computations involving the grand canonical ensemble. Equation (1.2-41) for the grand potential has the form

$$\Omega = \Omega(V, T, \mu_1, \ldots, \mu_n) \tag{1.2-64}$$

Equation (1.2-44) is

$$\Omega = U - TS - \sum_{i=1}^{n} \mu_i\,dN_i \tag{1.2-65}$$

and Eq. (2.46) is

$$d\Omega = -P\,dV - S\,dT - \sum_{i=1}^{n} N_i\,d\mu_i \tag{1.2-66}$$

The partial derivatives computed against the independent variables are

$$\left(\frac{\partial \Omega}{\partial V}\right)_{T,\mu} = -P \qquad \left(\frac{\partial \Omega}{\partial T}\right)_{V,\mu} = -S \qquad \left(\frac{\partial \Omega}{\partial \mu_i}\right)_{T,V,\mu_{i\neq j}} = N_i \tag{1.2-67}$$

POTENTIALS OF THE ENTROPY REPRESENTATION

These potentials are also called the Massieu–Planck functions and are obtained by a Legendre transformation of entropy (S) [Eqs. (1.2-44) and (1.2-46)] analogously to the potentials of the energy representation. The potentials can be obtained by k-fold transformations of entropy. The value of k should be smaller than the dimension r of the respective coordinate space. Practically, only two potentials are used, the first one being analogous to the Helmholtz energy and the second one being analogous to the Gibbs energy.

1. $k = 1$. Massieu function. This was the first characteristic function used in thermodynamics. It was introduced in 1865 by Massieu and, due to this fact, Guggenheim proposed its name. This name is not generally accepted. The extensive property *internal energy* (U) is transformed into the conjugate intensive property *reciprocal of temperature* ($1/T$). Equation (1.2-41) for the Massieu function has the form

$$\Phi_1 = \Phi_1\left(V, \frac{1}{T}, \mathbf{N}\right) \tag{1.2-68}$$

Equation (1.2-44) becomes

$$\Phi_1 = S - \frac{U}{T} \tag{1.2-69}$$

and Eq. (1.2-46) takes the form

$$d\Phi_1 = \frac{P}{T}\,dV - U d\left(\frac{1}{T}\right) - \sum_{i=1}^{n} \frac{\mu_i}{T}\,dN_i \tag{1.2-70}$$

The partial derivatives computed against the independent variables are

$$\left(\frac{\partial \Phi_1}{\partial V}\right)_{T,\mathbf{N}} = \frac{P}{T} \qquad \left[\frac{\partial \Phi_1}{\partial(1/T)}\right]_{V,\mathbf{N}} = -U \qquad \left(\frac{\partial \Phi_1}{\partial N_i}\right)_{T,V,N_{i \neq j}} = \frac{\mu_i}{T} \tag{1.2-71}$$

2. $k = 2$. Planck function. This function was frequently used by Planck and due to this Guggenheim proposed this name, which once more is not widely accepted. The extensive property internal energy (U) is transformed into the conjugate intensive property reciprocal of temperature ($1/T$) and the extensive property volume (V) is transformed into the conjugate intensive property *quotient of pressure and temperature*. The properties of the Planck function are given in analogy to (1.2-41) by

$$\Phi_2 = \Phi_2\left(\frac{P}{T}, \frac{1}{T}, \mathbf{N}\right) \tag{1.2-72}$$

Equation (1.2-44) takes the form

$$\Phi_2 = S - \frac{U}{T} - \frac{PV}{T} \tag{1.2-73}$$

and Eq. (1.2-46) becomes

$$d\Phi_2 = -V d\left(\frac{P}{T}\right) - U d\left(\frac{1}{T}\right) - \sum_{i=1}^{n} \frac{\mu_i}{T}\,dN_i \tag{1.2-74}$$

The partial derivatives computed against the independent variables are

$$\left[\frac{\partial \Phi_2}{\partial (P/T)}\right]_{T,\mathbf{N}} = -V \qquad \left[\frac{\partial \Phi_2}{\partial (1/T)}\right]_{P/T,\mathbf{N}} = -U \qquad \left(\frac{\partial \Phi_2}{\partial N_i}\right)_{T,V,N_{i\neq j}} = \frac{\mu_i}{T} \tag{1.2-75}$$

The advantage of the potentials of the entropy representation is the use of the so-called caloric values enthalpy (U) and entropy (S) as variables. The most important applications of the Massieu–Planck functions are in the thermodynamics of irreversible processes. Their detailed description goes beyond the scope of this work. The Massieu–Planck functions are very simply related to the potentials of the energy representation. Simple relations are obtained by comparing (1.2-69) and (1.2-61) or (1.2-73) and (1.2-61)

$$\Phi_1 = -\frac{F}{T} \qquad \Phi_2 = -\frac{G}{T} \tag{1.2-76}$$

These and analogous relations make it possible to use the potentials of the entropy representation without the necessity to introduce them explicitly.

1.2.7. The Gibbs–Helmholtz Equation

The phase equilibrium thermodynamics is, first of all, an applied science. Its main task is to introduce functional relations that make it possible to represent the properties of multicomponent and multiphase systems. As mentioned before, there exist thermodynamic properties that can be easily measured with a high degree of accuracy. Some other properties can be measured with some difficulty and some, like entropy, cannot be measured at all. Therefore, it is very important to introduce relations between various thermodynamic functions. Such relations make it possible to measure or predict the values that are easily measurable or predictable. Subsequently, the values of those properties, which are necessary for further work, can be computed.

The set of available relations should be large. The progress in experimental and theoretical thermodynamics is continuously supplying new, more accurate experimental data, models and correlating equations. Simultaneously with the progress in experiment the thermodynamic properties that lead to the most important information are changing. The choice of the property to be measured depends on the range of pressure, temperature, concentration and on the nature of the substance investigated. The local situation is also very important. This means the skill of experimentalists and theoreticians and the available equipment.

Equations (1.2-76) can serve as examples of relations between the thermodynamic potentials in the entropy and energy representations. Relations between the potentials of the same representation can be obtained by transforming a potential E_l (a function of l intensive independent variables) into a potential E_k (a function of k intensive independent variables) by means of the Legendre transformation. A transformation is possible for any number of independent variables provided that the number of transformation k is smaller than the dimension of the coordinate space r. From the practical point of view only onefold transformation fulfilling the condition $l = k - 1$ is important. The scheme of such transformations is as follows:

$$\begin{aligned} &E_k(\mathcal{P}_{I,1}, \ldots, \mathcal{P}_{I,l}, \mathcal{P}_{I,k}, \mathcal{P}_{E,k+1}, \ldots, \mathcal{P}_{E,r}) \\ &= E_l(\mathcal{P}_{I,1}, \ldots, \mathcal{P}_{I,l}, \mathcal{P}_{E,k}, \mathcal{P}_{E,k+1}, \ldots, \mathcal{P}_{E,r}) \\ &\quad + \mathcal{P}_{I,k}\left[\frac{\partial E_k(\mathcal{P}_{I,1}, \ldots, \mathcal{P}_{I,l}, \mathcal{P}_{I,k}, \mathcal{P}_{E,k+1}, \ldots, \mathcal{P}_{E,r})}{\partial \mathcal{P}_{I,k}}\right] \quad (k = l + 1) \end{aligned} \tag{1.2-77}$$

As a product of the transformation a differential equation of the first order is obtained. The equation interrelates functions defined in various coordinate spaces. The potential E_k is a function of an intensive independent variable $\mathcal{P}_{I,k}$ while the E_l potential is a function of a conjugate extensive independent variable $\mathcal{P}_{E,k}$. Now it is necessary to transform Eq. (1.2-77) into a form in which both potentials belong to the same coordinate space. For this purpose Eq. (1.2-36), determining the relation between conjugated parameters, can be used. This equation shows that the derivatives of potentials versus extensive parameters are equal to conjugate intensive parameters. As derivatives, they are functions of the same set of independent variables as the considered potential. The obtained relations are called equations of state:

$$\frac{\partial E_l}{\partial \mathcal{P}_{E,k}} = \mathcal{P}_{I,k}(\mathcal{P}_{I,1}, \ldots, \mathcal{P}_{I,l}, \mathcal{P}_{E,k}, \ldots, \mathcal{P}_{E,r}) \tag{1.2-78}$$

Rearranging (1.2-78) and substituting $\mathcal{P}_{E,k}$ into (1.2-77), the following relation is obtained:

$$\begin{aligned} &E_k(\mathcal{P}_{I,1}, \ldots, \mathcal{P}_{I,l}, \mathcal{P}_{I,k}, \mathcal{P}_{E,k+1}, \ldots, \mathcal{P}_{E,r}) \\ &= E_l(\mathcal{P}_{I,1}, \ldots, \mathcal{P}_{I,l}, \mathcal{P}_{I,k}, \mathcal{P}_{E,k+1}, \ldots, \mathcal{P}_{E,r}) \\ &\quad + \mathcal{P}_{I,k}\left(\frac{\partial E_k(\mathcal{P}_{I,1}, \ldots, \mathcal{P}_{I,l}, \mathcal{P}_{I,k}, \mathcal{P}_{E,k+1}, \ldots, \mathcal{P}_{E,r})}{\partial \mathcal{P}_{I,k}}\right) \end{aligned} \tag{1.2-79}$$

As a result we obtain Eq. (1.2-79) with the intensive parameter $\mathcal{P}_{I,k}$ instead of the extensive parameter $\mathcal{P}_{E,k}$. In Eq. (1.2-79) both functions E_k and E_l are determined in the same coordinate space. Only E_k has the properties of a characteristic function. Function E_l does not always have the properties of a potential in the phase space. This equation is, once more, an example of the application of a thermodynamic function in such a coordinate space in which this function does not have the properties of a potential.

Among the relations corresponding to Eq. (1.2-79) the most important one is that relating the Gibbs energy (G) to enthalpy (H). In this case Eq. (1.2-79) reduces to

$$G(P, T, \mathbf{N}) = H(P, T, \mathbf{N}) + T\left(\frac{\partial G(P\,T, \mathbf{N})}{\partial T}\right) \tag{1.2-80}$$

This equation is frequently used for the computation of the Gibbs energy as a function of temperature at constant pressure and composition. Equation (1.2-80) is frequently written in the form

$$G(T)_{P,\mathbf{N}} = H(T)_{P,\mathbf{N}} + T\left[\frac{\partial G(T)}{\partial T}\right]_{P,\mathbf{N}} \tag{1.2-81}$$

This equation was proposed by Gibbs in the United States and later by Helmholtz in Germany. Now it is called the *Gibbs–Helmholtz equation*. The values of the enthalpy $H(T)$ can be computed by integrating (1.2-55) with the use of heat capacity (C_P) measured as a function of temperature. The Gibbs energy is computed from Eq. (1.2-81) transformed to

$$\left[\frac{\partial(G/T)}{\partial(1/T)}\right]_{P,\mathbf{N}} = H \tag{1.2-82}$$

and integrated in the following form:

$$G(T) = T\int Hd\left(\frac{1}{T}\right) \tag{1.2-83}$$

The integration constant of (1.2-83) remains to be found. This problem will be discussed in detail. The obtained relation is basically a switching equation between potentials of the energy and entropy representation (Massieu–Planck functions). Further equations analogous to the Gibbs–Helmholtz equation can be obtained with the use of (1.2-79). Transformation of internal energy (U) into enthalpy (H) makes it possible to compute the values of enthalpy with the use of internal energy as a function of pressure:

$$H(P)_{S,\mathbf{N}} = U(P)_{S,\mathbf{N}} + P\left[\frac{\partial H(P)}{\partial P}\right]_{S,\mathbf{N}} \tag{1.2-84}$$

A further equation is

$$A(T)_{V,\mathbf{N}} = U(T)_{V,\mathbf{N}} + T\left[\frac{\partial A(T)}{\partial T}\right]_{V,\mathbf{N}} \tag{1.2-85}$$

which can be used to compute the Helmholtz energy using data representing the temperature dependence of the internal energy (U). The last one (1.2-86) allows the computation of the Gibbs energy (G) using data representing the pressure dependence of the Helmholtz energy:

$$G(P)_{T,\mathbf{N}} = A(P)_{T,\mathbf{N}} + P\left[\frac{\partial G(P)}{\partial P}\right]_{T,\mathbf{N}} \tag{1.2-86}$$

1.2.8. The Gibbs–Duhem Equation

Let us consider the Legendre transformation of the Gibbs fundamental equation in the case when the order of the transformation (k) is equal to the dimension of the phase space (r). The fundamental equation, as mentioned previously, is a homogeneous function of the first order. Therefore, the internal energy (U) can be written according to the Euler theorem (cf. Appendix B) in the form

$$U = \sum_{i=1}^{r} \left(\frac{\partial U}{\partial \mathcal{P}_{E,i}} \right) \mathcal{P}_{E,i} \tag{1.2-87}$$

By substituting (1.2-44) into (1.2-87), the following relation is obtained:

$$U = \sum_{i=1}^{r} \mathcal{P}_{I,i} \mathcal{P}_{E,i} \tag{1.2-88}$$

For $r = k$, substitution of (1.2-46) into (1.2-80) gives

$$E_r = \sum_{i=1}^{r} \mathcal{P}_{I,i} \mathcal{P}_{E,i} - \sum_{i=1}^{r} \mathcal{P}_{E,i} \mathcal{P}_{I,i} \tag{1.2-89}$$

Both sums on the right side of (1.2-89) are identical. The obtained potential E_r obtained by the r-fold Legendre transformation, where r is equal to the dimension of the coordinate space, is always equal to zero. Subsequently, the complete derivative of this potential is also equal to zero. The derivative written in terms of generalized variables (1.2-46) is

$$\sum_{i=1}^{r} \mathcal{P}_{E,i} \, d\mathcal{P}_{I,i} = 0 \tag{1.2-90}$$

The obtained equation is called the *Gibbs–Duhem equation*. If the energetic representation is used and only volumetric work is considered, Eq. (1.2-90) reduces for an n-component mixture to

$$S \, dT - V \, dP + \sum_{i=1}^{n} N_i \, d\mu_i = 0 \tag{1.2-91}$$

In an analogous way the version of the Gibbs–Duhem equation in the entropy representation can be deduced:

$$Ud\left(\frac{1}{T}\right) + Vd\left(\frac{P}{T}\right) - \sum_{i=1}^{n} N_i \, d\mu_i = 0 \tag{1.2-92}$$

Equation (1.2-91) or (1.2-92) can be written, analogously to the fundamental equation, separately for each phase. The Gibbs–Duhem equation for a two- or more phase system is a sum of equations written for all phases of the system.

The Gibbs–Duhem equation is very useful in the thermodynamics of phase equilibria. It gives relations between all intensive parameters for each phase. It proves that for the complete thermodynamic description of the properties of a phase of an n-component system it is not necessary to measure all $n + 2$ intensive parameters (independent variables in the phase space) of this phase. It is sufficient to measure $n + 1$ parameters, and the value of the last, not measured, parameter can be computed with the use of the Gibbs–Duhem equation.

1.2.9. The Phase Rule. Conditions of Phase Equilibria

The main task of the phase equilibrium thermodynamics is to establish mathematical relations, which make it possible to calculate the variables: pressure (P), temperature (T), volume (V) and the number of moles (N_i) of each component i for each phase forming the investigated system as well as for the whole system. Therefore, it is necessary to define the state of equilibrium. The following definition is very useful:

An isolated system is in the state of thermodynamic equilibrium if measurable changes of thermodynamic parameters do not occur in this system.

This definition points to the empirical character of the thermodynamics of phase equilibria. The state of equilibrium is determined with the accuracy of empirical measurements of thermodynamic parameters in each particular case. It is known from experience that thermodynamic processes tend to the state of equilibrium in an asymptotic way. The process of arriving at equilibrium is a nonreversible one, and the driving forces of such process head for zero gradients of the values of intensive parameters (Theorems II and III and Definition 2). Only external forces can stop or reverse the process of approaching the state of equilibrium.

A mathematical formulation of the conditions of thermodynamic equilibrium was given by Gibbs. Gibbs attributed a geometric interpretation to the equilibrium conditions. He showed that, according to the second law, the condition of thermodynamic equilibrium is represented by an extremum on the entropy (S) surface or on the internal energy (U) surface determined for constant values of work variables [V and y_j in Eqs. (1.2-23) and (1.2-30)] in the phase space constructed for the investigated system. *The equilibrium state corresponds to the maximum of entropy and to the minimum of internal energy.* According to tradition let us discuss first the equilibrium condition for entropy.

Entropy (S) is a function of the vector of variables of state $\mathbf{z}$ [$S = S(\mathbf{z})$]. These variables should be always those for which entropy is a potential in the phase space. They were listed in Eq. (1.2-23). The state of equilibrium corresponds to the maximum of $S(\mathbf{z})$ in the point $\mathbf{z}^0$ (Fig. 1.2-1). In this point the first derivative of S vanishes for all variables z_i:

$$\left(\frac{\partial S}{\partial(z_i^0)}\right) = 0 \qquad (1.2\text{-}93)$$

and the second derivative has a negative value

$$\left(\frac{\partial^2 S}{\partial(z_i^0)^2}\right) < 0 \qquad (1.2\text{-}94)$$

To analyze the state of thermodynamic equilibrium, it is convenient to introduce the concept of entropy fluctuations. This means small changes of entropy in the vicinity of the state of equilibrium. Let us denote the entropy in

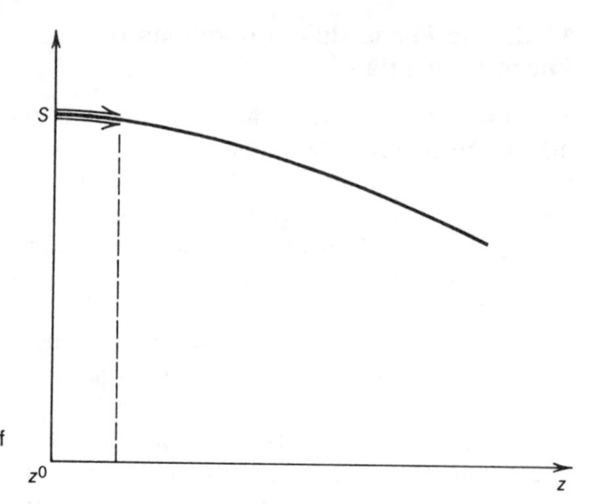

Figure 1.2-1
Entropy as a function of the vector of variables of state **z**.

the state of equilibrium by S^0 and its value after a perturbation not far from the equilibrium state by S. The change of entropy due to the perturbation is

$$\Delta S = S - S^0 \tag{1.2-95}$$

The change of entropy has to be accompanied by a change of the vector of independent variables **z**. The change of entropy can be computed from

$$\Delta S = S(\mathbf{z}) - S(\mathbf{z}^0) \tag{1.2-96}$$

Let us determine the change of the ith independent variable (z_i) accompanying the perturbation of entropy

$$\delta z_i = z_i - z_i^0 \tag{1.2-97}$$

Now we can determine the variation of entropy δS in the vicinity of the equilibrium state as a function of r independent variables represented by the vector **z**. The dimension of the coordinate space is $r + 1$. A first-order variation is

$$\delta S \equiv \sum_{i=1}^{r} \left(\frac{\partial S}{\partial z_i} \right) \delta z_i \tag{1.2-98}$$

A variation of the second order is

$$\delta^2 S \equiv \sum_{i=1}^{r} \sum_{j=1}^{r} \left(\frac{\partial^2 S}{\partial z_i \, \partial z_j} \right) \delta z_i \, \delta z_j \tag{1.2-99}$$

A variation of the third order is

$$\delta^3 S \equiv \sum_{i=1}^{r} \sum_{j=1}^{r} \sum_{k=1}^{r} \left(\frac{\partial^3 S}{\partial z_i \, \partial z_j \, \partial z_k} \right) \delta z_i \, \delta z_j \, \delta z_k \tag{1.2-100}$$

Further variations can be written until the mth order.

To find the complete change of entropy due to the perturbation a Taylor series can be constructed in close vicinity of the point $(\mathbf{z}^0)$ corresponding to the state of equilibrium:

$$\Delta S = \delta S + \left(\frac{1}{2!} \right) \delta^2 S + \cdots + \left(\frac{1}{m!} \right) \delta^m S \tag{1.2-101}$$

where ΔS represents the change of entropy due to the perturbation corresponding to a change from the equilibrium state (S^0) to the perturbed state (S). To satisfy the Gibbs maximum entropy principle, the following relation should be fulfilled:

$$\Delta S < 0 \tag{1.2-102}$$

Entropy is a continuous and smoothly varying function of variables of state $\mathbf{z}$, and the necessary and sufficient condition for a maximum of S is

$$(\delta S)_U = 0 \tag{1.2-103}$$

and

$$(\delta^m S)_U < 0 \tag{1.2-104}$$

where $(\delta S)_U$ is the first-order variation of S at constant internal energy U and $(\delta^m S)_U$ is the lowest-order, nonvanishing variation of S. That means that $\delta S = 0$ and $\delta^2 S \le 0$, but if $\delta^2 S = 0$ then $\delta^3 S \le 0$, but if $\delta^3 S = 0$ then $\delta^4 S \le 0$ and so on until a nonvanishing variation will be found. Condition (1.2-103) is a *condition of thermodynamic equilibrium*, while condition (1.2-104) is the *condition of thermodynamic stability*. According to Eq. (1.2-29), the equilibrium and stability conditions are defined at constant internal energy. This means that they can be satisfied only by an isolated system. A change of state cannot lead to a decrease of entropy. Entropy should be defined for the initial and for the final state of the system, and the change of state cannot be a function of time.

In an analogous way the equilibrium

$$(\delta U)_s = 0 \tag{1.2-105}$$

and stability

$$(\delta^m U)_s > 0 \tag{1.2-106}$$

conditions can be written for the energy representation. A system fulfilling the stability condition in the energy representation does not need to be isolated. A constant value of entropy as an independent variable does not imply that the system is at adiabatic conditions. This condition cannot be applied to an isolated system, when it is impossible to effect a change of internal energy.

The condition (1.2-105) can be cast in an extremum formulation, which is more convenient for further discussions. For a system whose intensive parameters are fixed, the necessary and sufficient condition of equilibrium is

$$\delta U - \sum_{i=1}^{k} P_{I,i} \delta P_{E,i} \geq 0 \quad (P_{E,j} = \text{const for } j > k) \tag{1.2-105a}$$

A detailed proof of this theorem can be found in Münster (1970).

The equilibrium conditions (1.2-103) and (1.2-105) lead to a conclusion that entropy (S) and internal energy (U) have stationary values in the state of thermodynamic equilibrium. Analogous to theoretical mechanics, several types of equilibria can be distinguished. To distinguish the equilibrium types, the series (1.2-101) is important as the types of equilibrium can be classified according to its sign.

TYPES OF EQUILIBRIUM

Stable Equilibrium

The thermodynamic equilibrium is called stable when the following relation for entropy is satisfied for any change of independent variables $\mathbf{z}$ in the immediate vicinity of the equilibrium point

$$(\Delta S)_U < 0 \tag{1.2-107}$$

An analogous relation can be written for internal energy

$$(\Delta U)_S > 0 \tag{1.2-108}$$

These relations follow immediately from the second law. Other types of equilibrium can be defined analogously with theoretical mechanics:

Unstable Equilibrium

There are processes for which

$$(\Delta S)_U > 0 \quad \text{or} \quad (\Delta U)_S < 0 \tag{1.2-109}$$

Such types of equilibrium cannot be experimentally maintained. This state corresponds to nonequilibrium conditions and, therefore, the system always changes automatically toward the equilibrium state.

Neutral Equilibrium

There is a possibility of processes for which

$$(\Delta S)_U = 0 \quad \text{or} \quad (\Delta U)_S = 0 \tag{1.2-110}$$

Metastable Equilibrium

For this type of equilibrium it is impossible to write equations determining the state of the system on the basis of phenomenological thermodynamics. In general, this is a state when an equilibrium is stable with respect to infinitesimally neighboring states but is unstable with respect to other neighboring states accessible by a finite change of parameters. This type of equilibrium is important for the phase equilibrium thermodynamics as it is describing the superheated and subcooled states.

A question arises as to what conditions must be satisfied by a system to prevent it from splitting into two or more macroscopic phases that differ from the original phase. This is the basic question of the thermodynamic stability theory as long as chemical reactions are excluded. Equation (1.2-88) can be written for the internal energy (U) of the original phase denoted by a prime in an n-component system excluding all types of work except volumetric:

$$U' - TS' + PV' - \sum_{i=1}^{n} \mu_i N_i' = 0 \tag{1.2-111}$$

According to the stability condition, the following inequalities can be written for phases 1 and 2

$$U^1 - TS^1 + PV^1 - \sum_{i=1}^{n} \mu_i N_i^1 \geq 0 \tag{1.2-112}$$

$$U^2 - TS^2 + PV^2 - \sum_{i=1}^{n} \mu_i N_i^2 \geq 0 \tag{1.2-113}$$

and, in addition, the conditions of equilibrium (1.2-20) should be fulfilled:

$$S^1 + S^2 = S' \tag{1.2-114}$$

$$V^1 + V^2 = V' \tag{1.2-115}$$

$$N_i^1 + N_i^2 = N_i' \quad \text{for } (i = 1, \ldots, n) \tag{1.2-116}$$

In the case of stable equilibrium the equality sign applies to Eqs. (1.2-112) and (1.2-113) and we get

$$U^1 + U^2 > U' \tag{1.2-117}$$

which means that the internal energy increases when the "new" phases are formed. According to the second law such process cannot be spontaneous. The original phase is thus stable with respect to the new phases.

In the case of unstable equilibrium we get

$$U^1 + U^2 < U' \tag{1.2-117a}$$

which means that the internal energy decreases when the "new" phases are formed. According to the second law such process is spontaneous. The original phase is unstable with respect to the new phases. The original homogeneous system cannot exist as such and will immediately become heterogeneous.

In the case of neutral equilibrium we get

$$U^1 + U^2 = U' \tag{1.2-117b}$$

which means that the formation of new phases does not change the internal energy as long as the system is in equilibrium [conditions (1.2-114)–(1.2-116) are obeyed. This means that the system is in neutral equilibrium and the new phases can exist together with the "old" phase.

This discussion leads to the *Gibbs stability criterion*: If the intensive parameters P, T and μ_i in the expression

$$U - TS + PV - \sum_{i=1}^{n} \mu_i N_i \tag{1.2-118}$$

can be given values such that the expression is zero for the phase under consideration and positive for all other phases, then the considered phase is absolutely stable with respect to separation into other phases. If the expression (1.2-118) is equal to zero for some other phases as well, the phase under consideration is in neutral equilibrium with respect to these other phases; they can coexist with the original phase. Temperature (T), pressure (P) and chemical potential (μ_i) of all i components have common values for all phases coexisting in the state of equilibrium.

The significance of the Gibbs stability criterion follows from the fact that it is obtained directly from the second law by a simple deduction and all other, general and special, stability conditions can be derived from it by purely mathematical methods.

In all considerations of the phase equilibrium thermodynamics it is prerequisite to establish a necessary and sufficient set of independent variables for each investigated case and a similar set of equations describing the system in the equilibrium state. A general relation between the equations describing the equilibrium state and the number of independent variables was established by Gibbs and is known as the *phase rule*.

In Section 1.2.5 it was shown that entropy and enthalpy appearing in the Gibbs fundamental equation are homogeneous functions of the first order of all independent variables. If an assumption is made that there are no effects due to phase boundaries, i.e., the boundaries are thermally conducting, deformable and permeable to all components, no chemical reactions occur, the only existing type of work is due to a volume change and heat as well as mass can be exchanged between phases, then the change of entropy as well as volume and amount of substance of each phase is determined only by equilibrium conditions. There are no chemical reactions and the sum of mole numbers N_k^i of each component k over all ϕ ($i = 1, \ldots, \phi$) phases remains constant in the whole system and is equal to C_k. This condition for an n-component system corresponds to n equations representing the balance of moles of all components in the whole system:

$$\psi_k(\mathbf{N}) = C_k - \sum_{i=1}^{\phi} N_k^i = 0 \tag{1.2-119}$$

Equations (1.2-119) represent constraints and determine, together with the equilibrium condition (1.2-103), the extremum on a surface constructed in the phase space. According to the Lagrange method of indeterminate multipliers, the general equation determining the extremum point taking into account secondary conditions is

$$\text{grad}[S(U, V, \mathbf{N}) + \psi_1(\mathbf{N})\lambda_1 + \cdots + \psi_n(\mathbf{N})\lambda_n] = 0 \tag{1.2-120}$$

where λ_k are indeterminate, constant Lagrange factors. Equation (1.2-120) can be written in the form of scalar equations, and the gradient can be computed by differentiation. After differentiation, $n\phi$ scalar equations of the following form are obtained:

$$\frac{\partial S^i(U, V, \mathbf{N})}{\partial N_k^i} + \lambda_k = 0 \tag{1.2-121}$$

where $i = 1, \ldots, \phi$ and $k = 1, \ldots, n$.

Now the number of independent variables can be computed. There are n Lagrange multipliers λ_k. Entropy is a homogeneous function of the first order, and, consequently, the derivative of entropy is a homogeneous function of the zeroth order. Therefore, Eq. (1.2-121) has only $n - 1$ independent variables in each of the ϕ phases of the system. There are n variable mole numbers and, additionally, internal energy (U) and volume (V). A summation leads to the total number of independent variables:

$$n + \phi(n - 1) + 2$$

For the problem to be solvable the number of equations ($n\phi$) should be equal or smaller than the number of unknowns:

$$n\phi \leq n + \phi(n - 1) + 2 \tag{1.2-122}$$

After rearrangement, the *Gibbs phase rule* is obtained:

$$n + 2 - \phi \geq 0 \tag{1.2-123}$$

When the left side of (1.2-123) is equal to zero, a change of any one independent variable causes a departure from the state of equilibrium. When the left side is greater than zero, it represents the number of freely adjustable variables f:

$$f = n + 2 - \phi \tag{1.2-124}$$

The number f is called the *thermodynamic degree of freedom* and, according to its value, invariant, univariant, bivariant and multivariant types of equilibrium can be distinguished. The number of thermodynamic degrees of freedom decreases by 1 when the number of coexisting phases increases by 1. The last statement is valid as far as the assumptions made in the derivation of the phase rule are valid. Generally speaking, the phase rule is a mnemonic tool allowing fast determination of the number of degrees of freedom of the system. The number obtained is accurate as far as the primary assumptions of the phase rule are adequate to the problem under consideration. A solution to problems of phase equilibrium thermodynamics always involves a precise determination of the independent variables and their number as well as the independent equations determining the state of equilibrium. This is frequently a difficult task.

According to the laws of thermodynamics and the properties of the Gibbs fundamental equation, each extensive parameter $\mathcal{P}_{E,i}$ of a system consisting of ϕ phases is equal to the sum of its values in all phases forming this system [Eq. (1.2-35)]:

$$\mathcal{P}_{E,i} = \sum_{\alpha=1}^{\phi} \mathcal{P}_{E,i}^{(\alpha)} \tag{1.2-35}$$

The Gibbs equilibrium [(1.2-103) and (1.2-105)] and stability [(1.2-104) and (1.2-106)] conditions can be transformed into analogous conditions for any thermodynamic potential. For this purpose it is necessary to establish additional theorems.

Theorem 1.2.9.0: A sufficient and necessary condition of equilibrium in a multiphase system characterized by stable and constant values of intensive parameters: $\mathcal{P}_{I,1}, \ldots, \mathcal{P}_{I,k}$ is expressed by the relation

$$\delta U - \sum_{i=1}^{k} \mathcal{P}_{I,i}\delta\mathcal{P}_{E,i} \geq 0 \tag{1.2-125}$$

for all remaining extensive parameters $\mathcal{P}_{E,j}$ (where $j > k$) having constant values.

For example, $k = 1$ and $\mathcal{P}_{E,1} = S$ for constant values of V and $\mathbf{N}$. Relation (1.2-125) is a sufficient and necessary condition of equilibrium and can be always reduced to condition (1.2-106).

Theorem 1.2.9.1: For a system defined in the phase space of dimension r and characterized by constant values of intensive parameters $\mathcal{P}_{I,1}, \ldots, \mathcal{P}_{I,k}$ for the whole system, the necessary and sufficient condition of equilibrium is

$$(\delta E_k)_{\mathcal{P}_{I,1},\ldots,\mathcal{P}_{I,k},\mathcal{P}_{E,k+1},\ldots,\mathcal{P}_{I,r}} \geq 0 \tag{1.2-126}$$

The proof of Theorem 1.2.9.0. can be obtained by a detailed analysis of all possible cases and is given by Münster (1970). The proof of Theorem 1.2.9.1. can be obtained by a substitution to (1.2-125).

The equilibrium condition for constant values of parameters can be obtained by writing Theorem 1.2.9.1. for all potentials of the energetic representation. The most important conditions are:

1. The equilibrium condition for constant entropy (S), pressure (P) and amount of substance is

$$(\delta H)_{S,P,\mathbf{N}} \geq 0 \tag{1.2-127}$$

2. The equilibrium condition for constant temperature (T), volume (V) and amount of substance is

$$(\delta A)_{T,V,\mathbf{N}} \geq 0 \tag{1.2-128a}$$

3. The equilibrium condition for constant temperature (T), pressure (P) and amount of substance is

$$(\delta G)_{T,P,\mathbf{N}} \geq 0 \tag{1.2-128b}$$

Similar conditions can be written for the entropy representation. Details can be found in Münster (1970). A general conclusion can be drawn that the stationary values of thermodynamic potentials of the energy representation reach a minimum for the stable equilibrium while the potentials of the entropy representation reach a maximum.

1.2.10. Computation of Relations between Thermodynamic Functions

Relations between thermodynamic functions are very important from the practical point of view. The main task is to compute the necessary thermodynamic properties using those properties that are most easily measured. It should be pointed out that the thermodynamic properties that are most easily measured vary with the experimental conditions. For example, in the high-pressure region it is easier to measure the saturation pressure as a function of temperature and compute the enthalpy of vaporization from Eq. (1.2-82) while for the low-pressure region it is much easier to determine accurate values of enthalpy by direct measurements.

Relations between the derivatives of thermodynamic functions can be obtained in a general way using the Gibbs fundamental equation and the functions of state (thermodynamic potentials) obtained by its transformation. Equation (1.2-46),

$$dE_k = \sum_{i=k+1}^{r} \mathcal{P}_{I,i}\, d\mathcal{P}_{E,i} - \sum_{j=1}^{k} \mathcal{P}_{E,j}\, d\mathcal{P}_{I,j} \tag{1.2-46}$$

representing a k-fold transformation of the fundamental equation, can be used for this purpose. Three types of relations are obtained by computing the second derivatives of Eq. (1.2-46) against intensive ($\mathcal{P}_I$) and extensive ($\mathcal{P}_E$) parameters with indices satisfying the conditions: i, $m \leq k$ and j, $n > k$.

Type I

$$\frac{\partial^2 E_k}{\partial \mathcal{P}_{I,i}\, \partial \mathcal{P}_{E,j}} = -\frac{\partial \mathcal{P}_{E,i}}{\partial \mathcal{P}_{E,j}} = \frac{\partial \mathcal{P}_{I,j}}{\partial \mathcal{P}_{I,i}} \tag{1.2-129}$$

Type II

$$\frac{\partial^2 E_k}{\partial \mathcal{P}_{I,i}\, \partial \mathcal{P}_{E,m}} = -\frac{\partial \mathcal{P}_{E,i}}{\partial \mathcal{P}_{I,m}} = -\frac{\partial \mathcal{P}_{E,m}}{\partial \mathcal{P}_{I,i}} \tag{1.2-130}$$

Type III

$$\frac{\partial^2 E_k}{\partial \mathcal{P}_{E,j}\, \partial \mathcal{P}_{E,n}} = -\frac{\partial \mathcal{P}_{I,j}}{\partial \mathcal{P}_{E,n}} = \frac{\partial \mathcal{P}_{I,n}}{\partial \mathcal{P}_{E,j}} \tag{1.2-131}$$

The derivatives of various thermodynamic potentials can be computed using Eqs. (1.2-129)–(1.2-131). It has been found that the following equations are useful.

$k = 0$, Internal energy (only relations of type III are used):

Type III
$$\left(\frac{\partial T}{\partial V}\right)_{S,\mathbf{N}} = -\left(\frac{\partial P}{\partial S}\right)_{V,\mathbf{N}} \tag{1.2-132}$$

Type III
$$\left(\frac{\partial T}{\partial N_i}\right)_{S,V} = \left(\frac{\partial \mu_i}{\partial S}\right)_{V,\mathbf{N}} \tag{1.2-133a}$$

Type III
$$-\left(\frac{\partial P}{\partial N_i}\right)_{S,V} = \left(\frac{\partial \mu_i}{\partial V}\right)_{S,\mathbf{N}} \tag{1.2-133b}$$

$k = 1$, Enthalpy (relations of types I and III are used):

Type I
$$\left(\frac{\partial V}{\partial S}\right)_{P,\mathbf{N}} = \left(\frac{\partial T}{\partial P}\right)_{S,\mathbf{N}} \tag{1.2-134}$$

Type I
$$\left(\frac{\partial V}{\partial N_i}\right)_{S,P} = \left(\frac{\partial \mu_i}{\partial P}\right)_{S,\mathbf{N}} \tag{1.2-135}$$

Type III
$$\left(\frac{\partial T}{\partial N_i}\right)_{S,P} = \left(\frac{\partial \mu_i}{\partial S}\right)_{P,\mathbf{N}} \tag{1.2-136}$$

Helmholtz energy (relations of types I and III are used):

Type I
$$\left(\frac{\partial S}{\partial V}\right)_{T,\mathbf{N}} = \left(\frac{\partial P}{\partial T}\right)_{V,\mathbf{N}} \tag{1.2-137}$$

Type I
$$-\left(\frac{\partial S}{\partial N_i}\right)_{T,V} = \left(\frac{\partial \mu_i}{\partial S}\right)_{V,\mathbf{N}} \tag{1.2-138}$$

Type III
$$-\left(\frac{\partial P}{\partial N_i}\right)_{T,V} = \left(\frac{\partial \mu_i}{\partial V}\right)_{T,\mathbf{N}} \tag{1.2-139}$$

$k = 2$, Gibbs energy (relations of types I and II are used):

Type I
$$\left(\frac{\partial V}{\partial N_i}\right)_{T,P} = \left(\frac{\partial \mu_i}{\partial P}\right)_{T,\mathbf{N}} \tag{1.2-140}$$

Type II
$$-\left(\frac{\partial S}{\partial P}\right)_{T,\mathbf{N}} = \left(\frac{\partial V}{\partial T}\right)_{P,\mathbf{N}} \tag{1.2-141}$$

Type I

$$-\left(\frac{\partial S}{\partial N_i}\right)_{T,P} = \left(\frac{\partial \mu_i}{\partial T}\right)_{P,\mathrm{N}} \tag{1.2-142}$$

Equations (1.2-132)–(1.2-142) are called the Maxwell relations.

Summarizing, the following thermodynamic variables have been introduced so far by considering only the potentials of the energetic representation at a constant amount of substance:

1. Heat (Q)
2. Work (W)
3. Internal energy (U)
4. Volume (V)
5. Pressure (P)
6. Entropy (S)
7. Temperature (T)
8. Enthalpy (H)
9. Helmholtz energy (A)
10. Gibbs energy (G)

It is important to find a method for the computation of derivatives of the type $(\partial x/\partial y)_z$, where x, y and z denote any of the above-mentioned functions. There are 720 of such combinations possible. Thermodynamic considerations are frequently carried out for systems with a constant amount of substance, i.e., either constant composition mixtures or pure substances. In such cases thermodynamic variables can be treated as functions of two independent variables. For such a function $[x = \mathscr{F}(y, w)]$ the following differential can be written:

$$dx = \left(\frac{\partial x}{\partial y}\right)_w dy + \left(\frac{\partial x}{\partial w}\right)_y dw \tag{1.2-143}$$

together with its derivative

$$\left(\frac{\partial x}{\partial y}\right)_z = \left(\frac{\partial x}{\partial y}\right)_w + \left(\frac{\partial x}{\partial w}\right)_y \left(\frac{\partial w}{\partial y}\right)_z \tag{1.2-144}$$

The derivative for the function $[x = \mathscr{F}(y, w)]$ is equivalent to the relation

$$\left(\frac{\partial x}{\partial w}\right)_y \left(\frac{\partial w}{\partial y}\right)_x \left(\frac{\partial y}{\partial x}\right)_w = -1 \tag{1.2-145}$$

In agreement with (1.2-140)–(1.2-145) any of 720 possible derivatives can be expressed by means of three independent, arbitrarily chosen, derivatives. The total number of equations of the type (1.2-141) is about 1.1×10^9. The practical use of such a number of relations is impossible and, therefore, Bridgeman (1926) proposed a systematic treatment. There is no necessity to analyze all possible combinations. It is sufficient to choose as a standard always the same three derivatives and use them to express any derivative under consideration. As standards, Bridgeman chose the derivatives of volume versus pressure and temperature or entropy versus temperature due to simple measurement methods. For simplicity, he defined the following three functions of these derivatives:

Molar heat capacity at constant pressure:

$$C_P = T\left(\frac{\partial S}{\partial T}\right)_P \tag{1.2-146}$$

Coefficient of thermal expansion:

$$\alpha = \frac{1}{V}\left(\frac{\partial V}{\partial T}\right)_P \tag{1.2-147}$$

Isothermal compressibility:

$$\kappa = -\frac{1}{V}\left(\frac{\partial V}{\partial P}\right)_T \tag{1.2-148}$$

Now, a relation known from the differential calculus can be introduced:

$$\left(\frac{\partial x}{\partial y}\right)_z = \frac{(\partial x/\partial w)_z}{(\partial y/\partial w)_z} \tag{1.2-149}$$

Relation (1.2-149) is general and can be used for any functions, not only potentials in the phase space. Bridgeman used it for all the 10 functions mentioned above including heat and work. As the function w in (1.2-149) plays only a complementary role, he introduced the following abbreviations to facilitate memorizing:

$$\left(\frac{\partial x}{\partial w}\right)_z = (\partial x)_z \tag{1.2-150}$$

$$\left(\frac{\partial y}{\partial w}\right)_z = (\partial y)_z \tag{1.2-151}$$

Any partial derivative at constant z can be computed using the abbreviations. For this purpose Bridgeman prepared a table of all possible derivatives $(\partial x)_z$, $(\partial y)_z$ etc. for 10 important thermodynamic variables. The table consists of only 45 derivatives due to the use of an additional relation, which can be easily proven by differentiating Eq. (1.2-143):

$$(\partial x)_z = -(\partial z)_y \tag{1.2-152}$$

A complete table can be found in Bridgeman's (1926) paper, including a table of second derivatives and derivatives for other than volume types of work coordinates. Simplified versions can be found in the textbooks of thermodynamics. The Bridgeman table can be used only for constant composition systems. Goranson (1930) published an extension of Bridgeman's method to binary and ternary variable composition systems and to constant composition quaternary systems.

Shaw (1935) applied the Jacobi determinant method for the computation of derivatives and proposed also a table for this purpose. The method was described in Margenau and Murphy's handbook (1950). Shaw's method is only more elegant from the mathematical point of view.

For practical calculations involving the use of digital computers a modification of Bridgeman's table by Model and Reid (1974) is very useful. To introduce this version it is necessary to define the compressibility factor

$$Z \equiv \frac{PV}{RT} \tag{1.2-153}$$

where R is the universal gas constant. Values of this constant are given in Table 2.2-1. The compressibility factor is an easy-to-measure value and, in addition, it is very widely used in thermodynamic methods for correlation and prediction. It is useful for defining two additional functions connecting the compressibility factor with its derivatives versus pressure and temperature:

$$Z_P \equiv Z - P\left(\frac{\partial Z}{\partial P}\right)_T \tag{1.2-154}$$

$$Z_T \equiv Z + T\left(\frac{\partial Z}{\partial V}\right)_P \tag{1.2-155}$$

The derivatives computed in this way are collected in Table 1.2-1. The derivatives of heat (Q) and work (W) are omitted in the table as they are not always potentials in the coordinate space, and their use can be sometimes misleading.

Example 1.2-6 Let us compute the speed of sound in air at normal conditions.

SOLUTION The speed of sound (w) can be computed from

$$w^2 = \frac{V(\partial P/\partial V)_S}{\rho} \tag{P1.2-17}$$

where $\rho = N/V$ denotes density. For simplicity, the following assumptions are made: (1) the system under consideration contains a constant amount of substance (N); (2) only one type of work is performed—the change of volume. In such a case the Gibbs fundamental equation (1.2-23) reduces to a function of two independent variables. The derivative of pressure versus volume at constant entropy appearing in Eq. (P1.2-17) is, according to (1.2-141),

$$\left(\frac{\partial P}{\partial V}\right)_S = \frac{(\partial S/\partial V)_P}{(\partial S/\partial P)_V} \tag{P1.2-18}$$

The heat capacity is defined as a derivative of heat versus temperature.

Accordingly, the heat capacity at constant volume is

$$C_V = \left(\frac{\partial Q}{\partial T}\right)_V = T\left(\frac{\partial S}{\partial T}\right)_V \tag{P1.2-18a}$$

Table 1.2-1
BRIDGEMAN'S TABLE USING THE COMPRESSIBILITY FACTOR

1. $(\partial T)_P = (\partial P)_T = 1$

2. $(\partial V)_P = -(\partial P)_V = \dfrac{RZ_T}{P}$

3. $(\partial S)_P = -(\partial P)_S = \dfrac{C_P}{T}$

4. $(\partial U)_P = -(\partial P)_U = C_P - RT_T$

5. $(\partial H)_P = -(\partial P) = C_P$

6. $(\partial G)_P = -(\partial P)_G = -S$

7. $(\partial A)_P = -(\partial P)_A = -(S + RZ_T)$

8. $(\partial V)_T = -(\partial T)_V = \dfrac{Z_P RT}{P^Z}$

9. $(\partial S)_T = -(\partial T)_S = \dfrac{RZ_T}{P}$

10. $(\partial U)_T = -(\partial T)_S = \dfrac{RT}{P}(Z_T - Z_P)$

11. $(\partial H)_T = -(\partial T)_U = \dfrac{RT}{P}(Z_T - Z)$

12. $(\partial G)_T = -(\partial T)_G = -V$

13. $(\partial A)_T = -(\partial T)_A = -\dfrac{Z_P RT}{P}$

14. $(\partial S)_V = -(\partial V)_S = \dfrac{R}{P^2}(-C_p Z_p + RZ_T^2)$

15. $(\partial U)_V = -(\partial V)_U = \dfrac{RT}{P^2}(-C_p Z_p + RZ_T^2)$

16. $(\partial H)_V = -(\partial V)_H = \dfrac{RT}{P^2}(-C_p Z_p + RZ_T^2 - RZZ_T)$

17. $(\partial G)_V = -(\partial V)_G = \dfrac{RT}{P^2}(-SZ_p + RZZ_T)$

18. $(\partial A)_V = -(\partial V)_A = \dfrac{SZ_p RT}{P^2}$

19. $(\partial U)_S = -(\partial S)_U = \dfrac{R}{P}(-C_p Z_p + RZ_T^2)$

20. $(\partial H)_S = -(\partial S)_H = \dfrac{ZRC_P}{P}$

21. $(\partial G)_S = -(\partial S)_G = \dfrac{R}{P}(SZ_T - C_P Z)$

22. $(\partial A)_S = -(\partial S)_A = \dfrac{R}{P}(-C_P Z_P + RZ_T^2 + SZ_T)$

23. $(\partial H)_U = -(\partial U)_H = \dfrac{RT}{P}(-ZC_P + ZRZ_T + C_P Z_P + RZ_T^2)$

Table 1.2-1 *(continued)*

24. $(\partial G)_U = -(\partial U)_G = \frac{RT}{P}(-RZZ_T - ZC_P + SZ_T - SZ_P)$

25. $(\partial A)_U = -(\partial U)_A = \frac{RT}{P}(-C_PZ_P + RZ_T^2 + SZ_T - SZ_P)$

26. $(\partial G)_H = -(\partial H)_A = \frac{RT}{P}(-SZ - C_PZ + SZ_T)$

27. $(\partial A)_H = -(\partial H)_A = \frac{RT}{P}[(S + RZ_T)(Z_T - Z) - C_PZ_P]$

28. $(\partial A)_G = -(\partial G)_A = \frac{RT}{P}(SZ - Z_PS + RZZ_T)$

and that at constant pressure is

$$C_P = \left(\frac{\partial Q}{\partial T}\right)_P = T\left(\frac{\partial S}{\partial T}\right)_P \tag{P1.2-18b}$$

For a function of two variables we can write

$$\left(\frac{\partial P}{\partial V}\right)_T = \frac{(\partial T/\partial V)_P}{(\partial T/\partial P)_V} \tag{P1.2-19}$$

Substituting all the above equations we obtain

$$\left(\frac{\partial P}{\partial V}\right)_S = \frac{(\partial S/\partial T)_P}{(\partial S/\partial T)_V}\frac{(\partial T/\partial V)_P}{(\partial T/\partial P)_V} = \frac{C_P}{C_V}\frac{(\partial T/\partial V)}{(\partial T/\partial P)_V} = \frac{C_P}{C_V}\left(\frac{\partial P}{\partial V}\right)_T \tag{P1.2-20}$$

Now, the derivative $(\partial P/\partial V)_T$ can be computed by means of expressions for $(\partial P)_T$ and $(\partial V)_T$ taken from Table 1.2-1:

$$\left(\frac{\partial P}{\partial V}\right)_T = \frac{(\partial P)_T}{(\partial V)_T} = \frac{-P^2}{Z_PRT} = \frac{-PZ}{V(Z - P(\partial Z/\partial P)_T]} \tag{P1.2-21}$$

Substitution into (P1.2-17) gives

$$w^2 = \frac{V^2(\partial P/\partial V)_S}{N_i} = \frac{-PZVC_P}{C_VN_i[Z - P(\partial P/\partial V)_T]} \tag{P1.2-22}$$

Equation (P1.2-22) represents the speed of sound in a fluid characterized by the compressibility factor Z. For the ideal gas $Z = 1$. Let us assume that air is an ideal gas with the molar mass $M = 29$ (kg/kmol). The pressure is $P = 101$ (kPa), molar volume $V = 22.4$ (m^3), heat capacity $C_V = 21$ (J/mol) and $C_P = 29.3$ (J/mol). Substituting the data into Eq. (P1.2-22) we get

$$w^2 = \frac{-PVC_P}{C_VN_i} = \frac{101 \times 10^3\ (\text{Pa})22.4\ (\text{m}^3)29.3\ (\text{J/mol})}{21\ (\text{J/mol})29\ (\text{kg})} = 108,\ 848 \tag{P1.2-23}$$

$$w = \sqrt{108,848} = 330 \text{ (m/s)} \quad \blacksquare \tag{P1.2-24}$$

When using the Bridgeman table it is necessary to remember that it can be applied only to constant composition systems. The relations (1.2-132)–(1.2-139) and (1.2-143)–(1.2-145) have a general significance and can be used for systems consisting of many phases of variable composition.

Example 1.2-7 Let us consider the Joule–Thomson effect. The diagram of the Joule–Thomson effect is shown in Figure 1.2-2. Gas is flowing from tube A to tube B through a porous plug C (or a similar device that offers resistance to the flow; in the original experiment this was cotton-wool). The process is carried out without heat exchange with the surroundings, i.e., at constant enthalpy. This process is called isenthalpic expansion, or throttling process. Gas in tube A is in the state of equilibrium, as is the gas in tube B. It was found in the Joule–Thomson experiment that the temperature in tube A was not equal to the temperature in tube B ($T_A \neq T_B$). Let us compute the temperature difference as a function of the pressure drop between A and B.

SOLUTION The differential of temperature is

$$dT = \left(\frac{\partial T}{\partial P}\right)_H dP \tag{P1.2-25}$$

The composition of the gas flowing through C does not change, the number of moles is constant and enthalpy can be treated as a function of pressure and composition only. For a function of two independent variables relation (1.2-145) can be used. This leads to

$$dT = -\frac{(\partial H/\partial P)_T}{(\partial H/\partial T)_P} dP \tag{P1.2-26}$$

By differentiating Eq. (P1.2-14) (representing the differential of enthalpy versus pressure and temperature) at a constant number of moles, Eq. (P1.2-26) is rearranged as follows:

$$dT = -\frac{T(\partial S/\partial P)_T + V}{T(\partial S/\partial T)_P} dP \tag{P1.2-27}$$

The Maxwell relation (1.2-141)

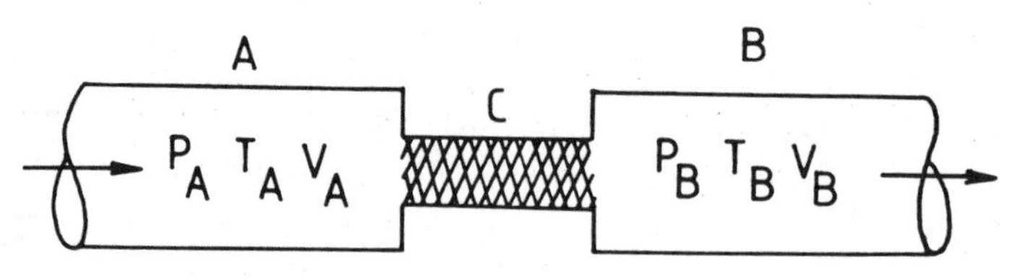

Figure 1.2-2
The Joule–Thomson experiment.

$$-\left(\frac{\partial S}{\partial P}\right)_{T,\mathbf{N}} = \left(\frac{\partial V}{\partial T}\right)_{P,\mathbf{N}} \tag{1.2-141}$$

and definitions (1.2-146)–(1.2-148) lead to

$$dT = -\frac{-TV\alpha + V}{C_P}\,dP = \frac{V}{C_P}(\alpha T - 1)\,dP \tag{P1.2-28}$$

The Joule–Thomson coefficient is defined by

$$\mu = \left(\frac{\partial T}{\partial P}\right)_H = -\frac{-TV\alpha + V}{C_P} \quad \blacksquare \tag{P1.2-29}$$

For an ideal gas the Joule–Thomson coefficient is equal to zero ($\alpha = 1/T$) and, consequently, the differential of temperature $(dT)_{\mathrm{id}}$ is equal to zero. This leads to the feature of the Joule–Thomson coefficient that its deviation from unity can be treated as a measure of the nonideality of a gas.

Example 1.2-8 Let us compute the difference between the heat capacity at constant pressure (C_P) and at constant volume (C_V) for a constant composition fluid.

SOLUTION Application of the Jacobian method (cf. Appendix B) to definition (1.2-19) yields

$$C_V = T\left(\frac{\partial S}{\partial T}\right)_V = T\,\frac{\partial(V,S)/\partial(T,P)}{\partial(V,T)/\partial(T,P)} = T\,\frac{\partial(S,V)/\partial(T,P)}{(\partial V/\partial T)_T} \tag{P1.2-30}$$

Substituting (1.2-145), using the Maxwell relations (1.2-132) and (1.2-138) and computing the Jacobi determinant we get

$$\begin{aligned} C_V &= -\frac{T}{V\kappa}\,\frac{\partial(S,V)}{\partial(T,P)} = -\frac{T}{V\kappa}\left[\left(\frac{\partial S}{\partial T}\right)_P\left(\frac{\partial V}{\partial P}\right)_T - \left(\frac{\partial S}{\partial P}\right)_T\left(\frac{\partial V}{\partial T}\right)_P\right] \\ &= -\frac{T}{V\kappa}\left[-\frac{C_P V\kappa}{T} + \left(\frac{\partial V}{\partial T}\right)_P^2\right] \end{aligned} \tag{P1.2-31}$$

Rearranging and substituting the definition (1.2-147), we obtain

$$C_P - C_V = T\,\frac{V\alpha^2}{\kappa} \quad \blacksquare \tag{P1.2-32}$$

The obtained relation is important in phase equilibrium thermodynamics. Another important problem is the sign of the heat capacity and its relation to the phase stability. As an example, the heat capacity at constant volume will be considered.

Example 1.2-9 Let us consider the conditions that determine the sign of heat capacity at constant volume (C_V) of a stable phase.

SOLUTION The condition of stability of a phase (1.2-106) can be written in the energy representation in the form (P1.2-33) on the assumption that only volumetric work is taken into account:

$$(\delta'' U(S, V, \mathbf{N}))_S > 0 \qquad \text{(P1.2-33)}$$

For a first-order phase transition at constant volume and constant number of moles, relation (P1.2-33) reduces to

$$\left(\frac{\partial^2 U}{\partial S^2}\right)_{V,\mathbf{N}} > 0 \qquad \text{(P1.2-34)}$$

According to the fundamental equation (1.2-31), we obtain

$$\left(\frac{\partial^2 U}{\partial S^2}\right)_{V,\mathbf{N}} = \left(\frac{\partial T}{\partial S}\right)_{V,\mathbf{N}} \qquad \text{(P1.2-35)}$$

From the general rules of partial differentiation we get

$$\left(\frac{\partial T}{\partial S}\right)_{V,\mathbf{N}} = \left(\frac{\partial S}{\partial T}\right)_{V,\mathbf{N}}^{-1} \qquad \text{(P1.2-36)}$$

The condition of stability (P1.2-34) can be rewritten using (P1.2-36) as

$$\left(\frac{\partial S}{\partial T}\right)_{V,\mathbf{N}} > 0 \qquad \text{(P1.2-37)}$$

The definition of heat capacity (P1.2-19) is

$$C_V = T\left(\frac{\partial S}{\partial T}\right)_V \qquad \text{(P1.2-38)}$$

The temperature is always positive and the derivative of entropy versus temperature is also always positive for a stable phase. Therefore, the molar heat capacity at constant volume is always positive for a stable phase

$$C_V > 0 \quad \blacksquare \qquad \text{(P1.2-39)}$$

The obtained result (P1.2-39) is called the *thermal stability condition* and can be formulated as follows: The heat capacity at constant volume is always positive for a stable phase. A similar deduction can be made for the heat capacity at constant pressure.

Example 1.2-10 Let us determine the sign of isentropic compressibility and heat capacity at constant pressure (C_P).

SOLUTION Another condition arising from (P1.2-33) for a stable phase is

$$\left(\frac{\partial^2 U}{\partial V^2}\right)_{S,\mathbf{N}} > 0 \qquad \text{(P1.2-40)}$$

A treatment analogous to that from Example (1.2-9) gives

$$-\left(\frac{\partial V}{\partial P}\right)_{S,\mathbf{N}} > 0 \tag{P1.2-41}$$

Analyzing the definition (1.2-148) of the coefficient of adiabatic (isentropic) compressibility and taking into account that volume is always positive, we obtain a new condition

$$\kappa_S > 0 \tag{P1.2-42}$$

Now, the formula (P1.2-32) determining the difference $(C_P - C_V)$ can be analyzed. The quantity α^2 is always positive as a square power and, as proven before, the left sides of (P1.2-41) and (P1.2-42) are also positive. Therefore, we obtain for a stable phase that

$$C_P > C_V > 0 \quad \blacksquare \tag{P1.2-43}$$

Condition (P1.2-42) is called the *mechanical stability condition*. The two conditions of thermal (P1.2-39) and mechanical (P1.2-42) stability can be transformed using the Legendre transformation into other coordinates. The most frequently used conditions are formulated on the basis of the Gibbs energy ($G(P, T, \mathbf{N})$) and require that the isothermal compressibility coefficient and the heat capacity at constant pressure be always positive for a stable phase.

1.2.11. Equations of State and Partial Molar Properties

In the previous section thermodynamic relations were given for a constant number of moles. For a phase equilibrium a change of the phase is usually connected with a change of composition. This leads to the necessity of introducing a method for treating the composition of a mixture as independent variables.

In the analysis of the properties of the Gibbs fundamental equation it was found that the intensive parameters can be treated as functions of all independent variables appearing in Eq. (1.2-40).

In the entropy representation the following equations of state are possible

$$\frac{1}{T} = \frac{1}{T}(U, V, \mathbf{N}) \tag{1.2-156}$$

$$\frac{P}{T} = \frac{P}{T}(U, V, \mathbf{N}) \tag{1.2-157}$$

$$\frac{\mu_i}{T} = \frac{\mu_i}{T}(U, V, \mathbf{N}) \tag{1.2-158}$$

Equation (1.2-156) can be simply rearranged to the form

$$U = U(T, V, \mathbf{N}) \tag{1.2-159}$$

which is traditionally called the *caloric equation of state*. Equation (1.2-157) can be rearranged to the form

$$P = P(U, V, \mathbf{N}) \tag{1.2-160}$$

traditionally called the *thermal equation of state*. The division of equations of state into "thermal" and "caloric" equations has a purely historical significance. In fact, both types of equations are practically equivalent due to the fact that they are defined in the phase space using a complete set of independent variables. Historically, different units and, subsequently, different values of the universal gas constant were used for both types (cf. Table 2.2-1). These differences are irrelevant at present when a unified unit system (SI) is used.

In a similar way equations of state can be developed for the energetic representation. According to the previous discussion the equations of state are, as derivatives of the fundamental equations, homogeneous functions of the zeroth order of extensive parameters. According to the definition of a homogeneous function they satisfy the relation

$$\mathcal{P}_{I,i}(\alpha\mathcal{P}_{E,1}, \ldots, \alpha\mathcal{P}_{E,r}) = \mathcal{P}_{I,i}(\mathcal{P}_{E,1}, \ldots, \mathcal{P}_{E,r}) \tag{1.2-161}$$

Now, a molar property will be defined. An extensive property divided by the number of moles of the system under consideration is called a *molar property*. In addition, a *density* of this property is obtained by dividing the extensive property by the volume of the system. Both molar properties, and densities are important for the thermodynamics of phase equilibria. For a system of constant number of moles the dimension of the phase space is $r = 3$. Coefficient α in Eq. (1.2-161) can take any value according to definition of the homogeneous function. On the assumption that $\alpha = N^{-1}$ in Eq. (1.2-161) three equations of state of the entropic representation are obtained in terms of the molar internal energy ($U_{\mathbf{N}} = U/N$) and molar volume ($V_{\mathbf{N}} = V/N$)

$$\mathcal{P}_{I,i} = \mathcal{P}_{I,i}\left(\frac{U}{N}, \frac{V}{N}, 1\right) = \mathcal{P}_{I,i}(U_{\mathbf{N}}, V_{\mathbf{N}}) \tag{1.2-162}$$

Similarly, three equations in terms of densities $\rho_U = U/V$ and $\rho_{\mathbf{N}} = N/V$ can be obtained

$$\mathcal{P}_{I,i} = \mathcal{P}_{I,i}\left(\frac{U}{V}, \frac{N}{V}, 1\right) = \mathcal{P}_{I,i}(\rho_U, \rho_{\mathbf{N}}) \tag{1.2-163}$$

For the energetic representation the corresponding molar properties are $S_{\mathbf{N}} = S/N$ and $V_{\mathbf{N}} = V/N$ and the densities are $\rho_S = S/V$ and $\rho_{\mathbf{N}} = N/V$.

The molar properties and densities are independent of the size of the system and are no more extensive properties. They are not intensive properties either as they do not fulfill the general relation (1.2-22) stating that the intensive parameters are constant in the state of equilibrium in all phases of which the system consists.

The use of two equations of state is sufficient for a complete description of a one-component system in the state of equilibrium due to the fact that they are homogeneous functions of the zeroth order of extensive parameters. The simultaneous use of all three equations is equivalent to the use of the Gibbs fundamental equation for this system.

The method of equations of state can be generalized for the representation of two- and more component systems. This will be described in further parts of this work. A less complex representation can be obtained by means of characteristic functions. The simplest and most popular function used for this purpose is the Gibbs energy (1.2-60). Similarly, as the fundamental equation, the Gibbs energy is a homogeneous function of the first order of extensive parameters. Applying this property, it can be written for an n-component mixture in the form

$$\alpha G = G(P, T, \alpha N_1, \ldots, \alpha N_n) \tag{1.2-164}$$

Assuming that

$$\alpha = \frac{1}{\sum_{i=1}^{n} N_i} \tag{1.2-165}$$

the following equation can be written for the molar Gibbs energy:

$$G_m = G_m(P, T, x_1, \ldots, x_{n-1}) \tag{1.2-166}$$

The concentration of the mixture is represented by *mole fractions*

$$x_i = \frac{N_i}{\sum_{i=1}^{n} N_i} \tag{1.2-167}$$

which, according to (1.2-165), always fulfill the relation

$$\sum_{i=1}^{n} x_i = 1 \tag{1.2-168}$$

The molar Gibbs energy (G_m) is a function of $n-1$ mole fractions as independent variables. It is always related to one mole of mixture. In Eq. (1.2-53) enthalpy was written as a function of the number of moles, pressure and temperature. Entropy (S) and molar volume (V) can also be represented as functions of these variables. A function of state can be represented by an arbitrarily chosen, complete and linearly independent set of variables of state. This leads to the molar enthalpy, entropy and volume. Therefore, any function of state $\mathscr{L}$ with P, T and $\mathbf{N}$ as independent variables can be represented as a molar function

$$\mathscr{L}_m(P, T, x_1, \ldots, x_{n-1}) = \frac{\mathscr{L}(P, T, N_1, \ldots, N_n)}{\sum_{i=1}^{n} N_i} \tag{1.2-169}$$

Lewis (1907) introduced the concept of the *partial molar property of component i* (e_i) and defined it as a derivative of an extensive property

$$e_i = \left(\frac{\partial E}{\partial N_i}\right)_{P,T,N_{j \neq i}} \tag{1.2-170}$$

Comparison of Eq. (1.2-170) and the last statement of (1.2-63) leads to a conclusion that the chemical potential (μ_i) is a partial molar property of the Gibbs energy. This is, however, a restricted statement. A more general definition is that the chemical potential is an intensive parameter of the Gibbs fundamental equation. The relations between molar properties are analogous to the relations between complete thermodynamic functions. The molar properties will be denoted by lowercase letters. For the molar Gibbs energy the following relations are obtained:

$$\left(\frac{\partial g}{\partial T}\right)_{P,\mathbf{x}} = -s \qquad \left(\frac{\partial g}{\partial P}\right)_{T,\mathbf{x}} = v \tag{1.2-171}$$

where $\mathbf{x}$ denotes the vector of mole fractions.

Equation (1.2-62) for the molar Gibbs energy is

$$dg = -s\, dT + v\, dP + \sum_{i=1}^{n} \left(\frac{\partial g}{\partial N_i}\right)_{P,T,N_{j \neq i,n}} dN_i \tag{1.2-172}$$

The derivatives of molar properties are not always equivalent to chemical potentials, and it is necessary to define their precise meaning. For each extensive variable of state E the following differential can be written:

$$dE = \left(\frac{\partial E}{\partial T}\right)_{P,\mathbf{N}} dT + \left(\frac{\partial E}{\partial P}\right)_{T,\mathbf{N}} dP + \sum_{i=1}^{n} \left(\frac{\partial E}{\partial N_i}\right)_{P,T,N_{j \neq i}} dN_i \tag{1.2-173}$$

According to the definition of an extensive property, its differential is a homogeneous function of the first order of the amount of substance (number of moles). On the basis of the Euler theorem and definition (1.2-172) of a partial molar property (e_i), the following equation can be written:

$$E = \sum_{i=1}^{n} e_i N_i \tag{1.2-174}$$

By differentiating (1.2-174) and substituting it into (1.2-173) the following equation is obtained

$$\sum_{i=1}^{n} N_i\, de_i - \left(\frac{\partial E}{\partial T}\right)_{P,\mathbf{N}} dT - \left(\frac{\partial E}{\partial P}\right)_{T,\mathbf{N}} dP = 0 \tag{1.2-175}$$

Equation (1.2-175) represents the general form of the *Gibbs–Duhem equation*. This equation is a direct consequence of the fact that the extensive properties appearing in the fundamental equation and those obtained by the Legendre transformation of the fundamental equation are homogeneous functions of the first order of the amount of substance (number of moles). The Gibbs–Duhem equation can be represented with the use of the definitions of the mole fraction (1.2-167) and the partial molar property (1.2-170)

$$\sum_{i=1}^{n} x_i \, de_i - \left(\frac{\partial e}{\partial T}\right)_{P,\mathbf{x}} dT - \left(\frac{\partial e}{\partial P}\right)_{T,\mathbf{x}} dP = 0 \tag{1.2-176}$$

The Gibbs–Duhem equation in the last form is frequently used in the phase equilibrium thermodynamics.

According to the general definition (1.2-170), partial molar properties can be defined for the Gibbs energy (G):

$$\left(\frac{\partial G}{\partial N_i}\right)_{P,T,N_{j\neq i}} = \mu_i \tag{1.2-177}$$

for entropy (S):

$$\left(\frac{\partial S}{\partial N_i}\right)_{P,T,N_{j\neq 1}} = s_i \tag{1.2-178}$$

for enthalpy (H):

$$\left(\frac{\partial H}{\partial N_i}\right)_{P,T,N_{j\neq i}} = h_i \tag{1.2-179}$$

and for volume (V):

$$\left(\frac{\partial V}{\partial N_i}\right)_{P,T,N_{j\neq i}} = v_i \tag{1.2-180}$$

The following derivatives can be obtained from the equation defining the differential of the Gibbs energy (1.2-62):

$$\left(\frac{\partial \mu_i}{\partial T}\right)_{P,\mathbf{N}} = -s_i \tag{1.2-181}$$

$$\left(\frac{\partial \mu_i}{\partial P}\right)_{T,\mathbf{N}} = v_i \tag{1.2-182}$$

An equation relating the partial molar properties is obtained from the definition of the Gibbs energy (1.2-61):

$$\mu_i = h_i - Ts_i \tag{1.2-183}$$

Substitution of (1.2-181) into (1.2-183) leads to an equation analogous to the Gibbs–Helmholtz equation

$$\mu_i = h_i - T\left(\frac{\partial \mu_i}{\partial T}\right)_{P,\mathbf{N}} \tag{1.2-184}$$

1.2.12. Properties of Mixtures

The concept of the chemical potential is very useful for the representation of mixture properties. In the thermodynamics of mixtures it is advantageous to establish some reference hypothetical mixtures whose properties can be computed from pure compound properties and subsequently used for the description of real mixtures as convenient standards of normal behavior. The most popular reference mixture is called an ideal mixture.

The ideal mixture was defined in thermodynamics in various ways. In this work the following definition will be used:

A mixture for which the chemical potential (μ_i) of a component i is described by the equation

$$\mu_i(P, T, \mathbf{x}) = \mu_i^0(P, T) + RT \ln x_i \tag{1.2-185}$$

is called an ideal mixture, where μ_i^0 denotes the chemical potential determined for a pure component i at the same conditions of pressure and temperature as the mixture.

This definition is applicable to systems containing gaseous, liquid and solid substances. For mixtures of two or more liquids the vapor pressure is usually not the same for pure components and the same components in the mixture. Therefore, the definition (1.2-185) is customarily used in conjunction with a correction for the difference between the pure-component equilibrium pressure P_i^0 and the actual pressure P. Many precise vapor–liquid equilibrium measurements have been carried out not far from the normal boiling point (at pressure of 0.1013 MPa) and can be compared with an ideal mixture. This definition is not used for those components of a two-phase system that form only one pure phase at the temperature of interest. More specific methods will be described later in this work.

The definition of the ideal mixture and the relations between thermodynamic functions described before enable us to write formulas for the computation of thermodynamic properties of an ideal mixture consisting of n substances:

$$G = \sum_{i=1}^{n} N_i \mu_i^0 + RT \sum_{i=1}^{n} N_i \ln x_i \tag{1.2-186}$$

$$S = \sum_{i=1}^{n} N_i s_i^0 - R \sum_{i=1}^{n} N_i \ln x_i \tag{1.2-187}$$

$$H = \sum_{i=1}^{n} N_i h_i^0 \tag{1.2-188}$$

$$V = \sum_{i=1}^{n} N_i v_i^0 \tag{1.2-189}$$

$$C_P = \sum_{i=1}^{n} N_i c_{Pi}^0 \tag{1.2-190}$$

where N_i denotes the number of moles of component i in the mixture and μ_i^0, s_i^0, h_i^0, v_i^0 and c_{Pi}^0 are the molar properties of the pure components at the same pressure and temperature as those of the mixture.

It can be seen from Eqs. (1.2-186)–(1.2-190) that enthalpy, heat capacity and volume are linear functions of composition while the Gibbs energy and entropy are not. Therefore it is convenient to introduce the *thermodynamic function* of mixing (F^m) and to define it as the difference between the value of any function in a mixture ($F^{(\text{MIXTURE})}$) and the sum of its values ($f_i^{(\text{PURE})}$) for the same amount of unmixed components at the same pressure and temperature.

$$F^m = F^{(\text{MIXTURE})} - \sum_{i=1}^{n} N_i f_i^{(\text{PURE})} \tag{1.2-191}$$

It is evident from Eqs. (1.2-186) and (1.2-187) that for an ideal mixture the Gibbs energy of mixing is always negative and the entropy of mixing is always positive. This result is in agreement with the stability conditions and is equivalent to the statement that the formation of an ideal mixture from its components is a spontaneous irreversible process, whether carried out isothermally or adiabatically.

Experimental measurements of physico-chemical properties provide information about changes in thermodynamic quantities rather than their absolute values. The third law of thermodynamics makes it possible to arrive at absolute values as it gives the value of entropy at absolute zero temperature. However, in practical calculations of thermodynamic properties, it is necessary to define a baseline for substances and their mixtures, to which the effect of changes may be referred. Such baselines are called *reference* states. Some of the reference states are used as *standard* states due to their generality and widespread popularity.

The widely used standard states for pure substances are, depending on their aggregation states, as follows:

1. For pure gases: a substance as an ideal gas at the standard state pressure, customarily 101,325 Pa.
2. For pure liquids and solids: a pure liquid or solid under the standard state pressure, usually 101,325 Pa.

These definitions are of importance for thermochemical calculations when tabulated values of thermodynamic functions referred to the standard states are compiled or used. For phase equilibrium computations different reference states can be used.

The situation is much more complicated in the case of mixtures for which the choice or reference states depends not only on the state of aggregation of components but also on the particular method used for computation.

Definitions of the reference states for mixtures are best developed from expressions for the chemical potential of some idealized mixture. Such mixtures may not exist in reality, but their properties must be readily calculated when the composition of the mixture is known. The composition of a mixture as well as pressure must be defined for application of the reference state concept.

If we adopt the mole fraction scale as a measure of concentration, we can use the ideal mixture defined by Eq. (1.2-185) as a reference state. For a real mixture the deviation from the ideal mixture behavior is accounted for by the activity coefficient defined as

$$\mu_i(P, T, \mathbf{x}) = \mu_i^0(P, T) + RT \ln x_i + RT \ln \gamma_i(P, T, \mathbf{x}) \tag{1.2-192}$$

The concept of activity was first introduced by Lewis (1901, 1907). It should be noted that the reference state for the mixture is the ideal mixture at system pressure. It is reasonable to use this reference state only if all pure components under consideration are in the same (usually liquid) state. The activity coefficients of all components tend to 1 as the mole fraction of a component approaches 1.

$$\lim_{x_i \to 1} \gamma_i = 1 \quad \text{for } i = 1, \ldots, n \tag{1.2-193}$$

Therefore, the use of the ideal mixture as a reference state is called the *symmetrical normalization*. The activity coefficients obeying relation (1.2-193) are called symmetrically normalized. Their use will be described in detail in Chapter 3 devoted to mixtures.

The above reference state was defined for a *mixture*, i.e., a system containing more than one component for which none of the components is distinguished. If we treat a system as a *solution*, i.e., when a solute and a solvent are distinguished, we can introduce *unsymmetrical normalization* of activity coefficients. In this case the reference state for the solvent is defined by the ideal mixture expression (1.2-185) and the solvent activity coefficient is defined by Eq. (1.2-192). On the other hand, the standard state of the solute is chosen to be a hypothetical infinitely dilute solution. It should be noted that such a state does not occur in reality; it is merely an extrapolation of the properties of dilute solutions. The chemical potential of an ideal solution can be then expressed as

$$\mu_i(P, T, \mathbf{x}) = \mu_i^*(P, T) + RT \ln x_i \tag{1.2-194}$$

where the infinite-dilution reference state is denoted by an asterisk (*). A corresponding equation defining the activity coefficient of the solute is

$$\mu_i(P, T, \mathbf{x}) = \mu_i^*(P, T) + RT \ln x_i + RT \ln \gamma_i^*(P, T, \mathbf{x}) \tag{1.2-195}$$

The activity coefficient approaches unity as the concentration of the solute approaches zero

$$\lim_{x_i \to 0} \gamma_i^* = 1 \quad \text{for } i = 1, \ldots, n \text{ (solutes)}$$
$$\lim_{x_k \to 1} \gamma_k^* = 1 \quad \text{for } k = 1, \ldots, m \text{ (solvents)} \tag{1.2-196}$$

Therefore, the activity coefficient defined by Eq. (1.2-195) is unsymmetrically normalized. It is reasonable to use the unsymmetrical normalization when the pure solute is not in the same state of aggregation as the solution. This is often the case for liquid solutions of electrolytes and noncondensable gases.

Example 1.2-11 Relation between the symmetrically and unsymmetrically normalized activity coefficients.

SOLUTION As the chemical potential is independent of the normalization system, the right sides of Eqs. (1.2-192) and (1.2-195) are equal. Hence, the difference of reference state chemical potentials is

$$\mu_i^0 - \mu_i^* = RT \ln \frac{\gamma_i^*}{\gamma_i} \tag{P1.2-44}$$

According to both definitions (1.2-192) and (1.2-195) the infinite-dilution limit of the chemical potential is

$$\lim_{x_i \to 0} \mu_i = \mu_i^0 + RT \ln x_i + RT \ln(\lim_{x_i \to 0} \gamma_i) = \mu_i^* + RT \ln x_i + RT \ln 1 \tag{P1.2-45}$$

Rearranging Eq. (P1.2-45) to obtain $(\mu_i^0 - \mu_i^*)$ and substituting the result into eq. (P1.2-44) we obtain an equation interrelating γ_i and γ_i^*:

$$\gamma_i^* = \frac{\gamma_i}{\gamma_i^\infty} \tag{P1.2-46}$$

where

$$\gamma_i^\infty = \lim_{x_i \to 0} \gamma_i \tag{P1.2-47}$$

The quantity γ_i^∞ is called the infinite-dilution activity coefficient and plays an important role in applied thermodynamics. Its applications will be discussed in detail in this book.

The second equation interrelating γ_i and γ_i^* can be obtained from the pure-component limit of the chemical potential:

$$\lim_{x_i \to 1} \mu_i = \mu_i^0 + RT \ln 1 + RT \ln 1 = \mu_i^* + RT \ln 1 + RT \ln(\lim_{x_i \to 1} \gamma_i^*) \tag{P1.2-48}$$

Rearranging Eq. (P1.2-48) to obtain $(\mu_i^0 - \mu_i^*)$ and substituting the result into Eq. (P1.2-44) we get

$$\gamma_i = \frac{\gamma_i^*}{\lim\limits_{x_i \to 1} \gamma_i^*} \tag{P1.2-49}$$

Although Eqs. (P1.2-46) and (P1.2-49) are conceptually straightforward, their practical application requires extrapolating γ_i or γ_i^* to $x_i = 0$ or $x_i = 1$, respectively. This extrapolation is unavoidably burdened with much uncertainty and precludes the interchangeable use of symmetrical and unsymmetrical normalizations. ■

For a solution a different concentration scale such as molality m (number of moles per 1 kg of solvent) can also be introduced. In this case the following defining equation for the chemical potential of a solute i in an ideal solution can be used:

$$\mu_i(P, T, \mathbf{x}) = \mu_i^*(P, T) + RT \ln\left(\frac{m_i}{m^0}\right) \tag{1.2-197}$$

Hence, the reference state of a solute in a solution is the hypothetical state of an ideal solution of standard molality m^0, customarily 1 mol · kg^{-1} at the system pressure. The activity coefficient can be defined analogously as in Eq. (1.2-195). In this book we shall not use definitions of this type. All systems dealt with hereinafter will be treated as mixtures rather than solutions.

Alternatively, we may use the ideal gas as a reference state. This definition is most natural for gaseous mixtures but may also be used for liquids. The chemical potential of an ideal gas mixture is given by

$$\mu_i(P, T, \mathbf{x}) = \mu_i^0(T) + RT \ln \frac{x_i P}{P^0} \tag{1.2-198}$$

Here, the superscript zero is used to denote the reference state that is a pure ideal gas at the reference pressure P^0 (usually 101,325 Pa) rather than the pure-liquid reference state. An important feature of this definition is the independence of μ_i^0 on pressure. This is a natural consequence of using the ideal gas (for which the equation of state is known a priori) as the reference state.

To account for the deviations of the real fluid behavior from Eq. (1.2-198), Lewis and Randall (1921) introduced the concept of fugacity f_i of component i in a mixture, defined in its most general form by

$$\mu_i(P, T, \mathbf{x}) = \mu_i^0(T) + RT \ln \frac{f_i(P, T, \mathbf{x})}{f_i^0(T)} \tag{1.2-199}$$

where, again, the superscript zero denotes the ideal gas reference state. An important feature of this definition is the asymptotic behavior of fugacity at zero pressure for which all substances approach the ideal gas state:

$$\lim_{P \to 0} \frac{f_i}{x_i P} = 1 \tag{1.2-200}$$

According to its definition, fugacity is an intensive property and is equal for each component i in coexisting phases (e.g., α and β) in equilibrium:

$$f_i^{\alpha} = f_i^{\beta} \tag{1.2-201}$$

The methods of calculating fugacity from PVT relationships using P and T or V and T as independent variables will be described later in this book. In this chapter we confine ourselves to illustrating the interrelation between fugacities and activities and the simplest applications of these concepts: Raoult's and Henry's laws.

Example 1.2-12 Relation between the fugacity and the symmetrically normalized activity coefficient. Raoult's law.

SOLUTION Writing the definition of fugacity for the component i of a liquid mixture and for pure liquid i and substituting them into the definition of activity coefficient (1.2-192), we get

$$f_i = f_i^{\text{pure}} x_i \gamma_i \tag{P1.2-50}$$

This is a general result provided that the pure component i and the mixture are in the same aggregation state at the temperature in question. Equation (P1.2-50) can be inserted into the general equilibrium condition (1.2-201) to represent the fugacity of the liquid phase in equilibrium with the gas phase. Then, the equilibrium condition becomes

$$f_i^{v} = f_i^{\text{pure}} x_i \gamma_i \tag{P1.2-50a}$$

The left-hand side of Eq. (P1.2-50a) refers to the vapor phase in equilibrium and the right-hand side to the liquid. Equation (P1.2-20) can be then simplified by assuming that

1. $f_i^{v} = P y_i$ (the vapor phase is an ideal gas).
2. $f_i^{\text{pure}} = P_i^{\text{sat}}$ (the fugacity of a pure liquid is equal to its saturation pressure).
3. $\gamma_i = 1$ (the liquid phase is ideal).

In this case we get

$$P y_i = P_i^{\text{sat}} x_i \tag{P1.2-51}$$

This is the well-known Raoult law. By substituting $\Sigma\, y_i = 1$, we obtain

$$P = \sum P_i^{\text{sat}} x_i \tag{P1.2-51a}$$

The total pressure is then a linear function of mixture composition and saturation pressures of components or, equivalently, it is a scalar product of the vectors of components' mole fractions and vapor pressures (cf. Appendix B).

The assumptions used to derive Raoult's law are extremely crude and can be approximately satisfied only for a very limited class of mixtures (e.g., mixtures of two isomers or isotopes at low pressure). ■

Example 1.2-13 Relation between the fugacity and the unsymmetrically normalized activity coefficient. Henry's law.

SOLUTION Writing the definition of fugacity for the solute i in a liquid solution and for the hypothetical infinite-dilution reference state with $x_i^* = 1$ and substituting the resulting expressions into Eq. (1.2-201), we get

$$f_i^v = f_i^* x_i \gamma_i^* = H_i x_i \gamma_i^* \tag{P1.2-52}$$

The extrapolated infinite-dilution reference fugacity f_i^* is called the Henry constant H_i. The value of H_i can be obtained according to the limiting formula

$$H_i = \lim_{x_i \to 0} \frac{f_i}{x_i} \tag{P1.2-53}$$

When applied to two-phase gas–liquid systems, Eq. (P1.2-52) is the general formulation of Henry's law. A simplified version of the law can be obtained by assuming that

1. $f_i = P_i$ (the vapor phase is ideal and the fugacity is equal to partial pressure).
2. $\gamma_i^* = 1$ (the solution is sufficiently dilute to assume that it is ideal with respect to the unsymmetric normalization).

Then, we obtain the simple Henry gas solubility law

$$P_i = H_i x_i \tag{P1.2-54}$$

This equation, while simplified, is obeyed in reality by very dilute solutions of gases at low pressures. Equation (P1.2-54) is equivalent to the statement that the pressure is a linear function of composition in the immediate vicinity of the pure solvent. ■

The above examples showed the ultrasimplified cases for which analytical correlations such as Raoult's and Henry's laws can be derived.

It must be emphasized that the choice of reference states is entirely at our discretion. We should be guided by the convenience of calculations. This will be illustrated in the text example, which shows the application of a pure subcooled liquid as a reference state. This is an important example as it provides the thermodynamic framework for the calculation of solid–liquid equilibria.

Example 1.2-14 Pure subcooled liquid as a reference state. An equation for the solubility of solids.

SOLUTION Let us assume that a pure solid dissolves in a solvent at temperature T. We choose the reference state as the pure, subcooled liquid at temperature T under its own saturation pressure. The quantity of interest is then the difference between the chemical potential of the pure solid solute μ_i^{0S} and the pure subcooled liquid μ_i^{0L}. Under isothermal-isobaric conditions the difference $(\mu_i^{0L} - \mu_i^{0S})$ is equal to the change of the molar Gibbs energy upon melting $\Delta g^{S,L}$. This Gibbs energy change is related to the corresponding enthalpy and entropy changes by

$$\Delta g^{S,L} = \Delta h^{S,L} - T\,\Delta s^{S,L} = \mu_i^{0L} - \mu_i^{0S} = \Delta \mu_i^{S,L} \qquad \text{(P1.2-55)}$$

The melting process to produce a subcooled liquid can be visualized by means of a thermodynamic cycle consisting of three steps:

1. Heating of the solid to the triple-point temperature:

$$\Delta h_{(1)} = \int_T^{T_t} C_p^S \, dT \qquad \text{(P1.2-56)}$$

$$\Delta s_{(1)} = \int_T^{T_t} \left(\frac{C_p^S}{T} \right) dT \qquad \text{(P1.2-57)}$$

where C_p^S is the molar heat capacity at constant pressure of the solid and T_t is the triple-point temperature.

2. Melting at the triple-point temperature:

$$\Delta h_{(2)} = \Delta h_f \qquad \text{(P1.2-58)}$$

$$\Delta s_{(2)} = \frac{\Delta h_f}{T_t} \qquad \text{(P1.2-59)}$$

where Δh_f is the measurable enthalpy of fusion.

3. Cooling from the triple-point temperature to the temperature of interest:

$$\Delta h_{(3)} = \int_{T_t}^{T} C_p^L \, dT \qquad \text{(P1.2-60)}$$

$$\Delta s_{(3)} = \int_{T_t}^{T} \left(\frac{C_p^L}{T} \right) dT \qquad \text{(P1.2-61)}$$

where C_p^L is the molar heat capacity of the subcooled liquid. Summing the contributions [(P1.2-56–(P1.2-61)], substituting the result into Eq. (P1.2-55) and integrating on the assumption that $\Delta C_p = C_p^L - C_p^S$ is independent of temperature, we obtain

$$\frac{\Delta \mu_i^{S,L}}{RT} = \frac{\Delta h_f}{RT_t} \left(\frac{T_t}{T} - 1 \right) - \frac{\Delta C_p}{R} \left(\frac{T_t}{T} - 1 \right) + \frac{\Delta C_p}{R} \ln \frac{T_t}{T} \qquad \text{(P1.2-62)}$$

This is the solubility equation. In the state of equilibrium, μ_i^{0S} is equal to the chemical potential of the solute in saturated solution μ_i

$$\mu_i^{0S} = \mu_i \tag{P1.2-63}$$

Equation (P1.2-62) can be simplified by introducing the easy-to-measure melting temperature T_m instead of the triple-point temperature T_t because both temperatures have very close values for most substances. Moreover, the second and third term on the right-hand side of Eq. (P1.2-62) can be neglected because they have a tendency to cancel each other especially if T and T_t are not far apart.

The thermodynamic framework developed in this chapter will be used in Chapters 3 and 4 to solve real problems of phase equilibria.

1.3 OUTLINE OF SOME USEFUL THEORETICAL CONCEPTS

Theoretical representation of fluid properties is still one of the most complicated and challenging tasks of chemical physics. An exact solution to the problem would necessitate answering the following questions:

1. What are the intermolecular forces between molecules?
2. What is the effect of intermolecular interactions on the spatial arrangement of molecules and, subsequently, their thermodynamic properties?

The first problem belongs to the realm of quantum mechanics and can be solved exactly only for very simple cases. The second problem can be, in principle, answered by statistical mechanics if a solution to the first problem is known.

The purpose of a statistical mechanical treatment is to arrive at the partition function Z. The definition and evaluation of partition functions is discussed in detail in standard textbooks of statistical mechanics (e.g., Hill, 1956; McQuarrie, 1973).

If a system has discrete quantum states 1, 2, 3, . . . with energies E_1, E_2, E_3, . . . , then the canonical partition function is defined as the sum over all states i of the Boltzmann factor $\exp(-E_i/kT)$:

$$Z = \sum_i \exp\left(\frac{-E_i}{kT}\right) \tag{1.3-1}$$

where k is the Boltzmann constant. The Helmholtz energy is related to the canonical partition function by the equation

$$A = -kT \ln Z \tag{1.3-2}$$

In most thermodynamic applications temperature is sufficiently high so that classical statistical mechanics can be used and the sum over all states can be replaced by an integral over all coordinates and momenta of the system. For a system of N molecules each of mass m in a volume V, the integration over momenta can be performed immediately, leading to the result:

$$Z = \left(\frac{2\pi mkT}{h^2}\right)^{3N/2} Qq_{r,v,e} = \left(\frac{1}{\Lambda^3}\right)^N Qq_{r,v,e} \tag{1.3-3}$$

where h is the Planck constant, Λ is frequently called the thermal de Broglie wavelength, Q is the configurational integral and $q_{r,v,e}$ is the rotational, vibrational and electronic contribution to the partition function. The configurational integral is defined by

$$Q(N, V, T) = \frac{1}{N!}\int \cdots \int \exp\left(\frac{-U}{kT}\right) dx_1 \ldots dx_N \tag{1.3-4}$$

where U is the potential energy of intermolecular interactions in a system of N particles. Once A is known as a function of N, V and T, all the thermodynamic properties of a system can be derived as explained before. The Helmholtz energy and other functions are frequently expressed in terms of the configurational Helmholtz energy, defined as

$$A^{\text{conf}} = -kT \ln Q \tag{1.3-5}$$

which differs from the Helmholtz energy by an amount that is independent of density, depending on temperature alone. Most calculations of the properties of fluids assume that the interactions between different pairs of molecules are additive, so that the potential energy of an assembly of molecules is the sum of the potential energies of interaction of all the pairs of molecules. In the simplest case when the potential energy of interaction of molecules i and j $[u(R_{ij})]$ depends only on their separation R_{ij}, we get

$$U = \sum_{i<j} u(R_{ij}) \tag{1.3-6}$$

CORRESPONDING STATES PRINCIPLE

If a class of substances have a common potential energy function apart from scale factors of interaction energy and distance, the substances can be shown to obey the principle of corresponding states. If the potential energy function for the ith substance is given by

$$u_i(R) = \epsilon_i u_0\left(\frac{R}{\sigma_i}\right) \tag{1.3-7}$$

where ϵ_i, σ_i are constants characteristic of the substance i and u_0 is the same function for all the substances, then the configurational Helmholtz energy A^{conf} for substance i can be expressed as

$$A^{\text{conf}} - A^{\text{conf},ig} = N\epsilon_i f\left(\frac{V}{N\sigma_i^3}, \frac{kT}{\epsilon_i}\right) \tag{1.3-8}$$

where the superscript *ig* refers to the ideal gas state and f is the same function for all the substances, being determined solely by the function u_0. Since all configurational thermodynamic properties can be derived from the Helmholtz energy by differentiation, it follows that the substances of the class have identical configurational thermodynamic functions provided that energies, volumes, temperatures and pressures are scaled in units of ϵ_i, σ_i^3, ϵ_i/k and ϵ_i/σ_i^3, respectively. This is the principle of corresponding states. Its more usual statement in terms of reduced parameters related to critical temperature, pressure and volume follows if it is recognized that these must be proportional to ϵ_i/k, ϵ_i/σ_i^3 and σ_i^3, respectively. The practical applications and empirical extensions of the corresponding-states principle will be described in detail later in this book.

LATTICE MODELS

The partition function or configurational integral can only be evaluated explicitly when the system under consideration can be broken down into units that are in some sense independent, or almost so. For example, in an approximate theory of solids the normal lattice vibrations can be treated as independent units. In the theory of ideal gases the independent units are the actual molecules. The theory of real fluids is much more complex because no straightforward method exists for dividing the fluid into a set of independent or almost independent systems.

Most theories of fluids can be divided roughly into two basic groups. The first group includes theories that derive the relationship between intermolecular forces and the structure of the system in terms of distribution functions. The problem of finding these relationships is complicated so that it can be solved only in stages that subsequently encompass an increasingly wider range of systems (i.e., monatomic fluids, diatomic fluids, polyatomic fluids, polar molecules etc.). This is an extremely broad field of current research in theoretical chemical physics. Reviews of these methods can be found, for example, in books by Gray and Gubbins (1984), Boublík et al. (1980), Fisher (1964) and Münster (1969, 1975). At present, these first-principle methods are not used in practically oriented calculations of phase equilibria.

The second group of methods includes lattice theories that start from an initially assumed structure of the system and neglect the effect of molecular properties on this structure. A priori assumptions about the arrangement of molecules make it possible to introduce simplifying assumptions into the calculation of the partition function that are analogous to those employed in the theories of crystals. The lattice theories have been reviewed by Barker (1963). They were, in the 1930s to 1950s, the most frequently used methods for the theoretical interpretation of liquid phase behavior. Now, they are no longer a subject of research in theoretical chemical physics. However, they influenced, due to their simplicity, some of the currently used methods of practical phase equilibrium modelling, especially some models used for the Gibbs energy and other thermodynamic functions. Therefore, their main assumptions will be briefly summarized here.

The lattice model was introduced by Guggenheim and Fowler (1935), Guggenheim (1952) and further refined by Prigogine (1957) and Barker (1963). The initial version of the lattice model, i.e., the rigid lattice model, assumed that all sites on a quasi-crystalline lattice were occupied by mixture components of similar size and spherical symmetry. The lattice was characterized by a fixed coordination number. The distance between molecules, the coordination number and the type of the lattice structure were not affected by the mixing process. The interactions of molecules on a lattice were represented using temperature-independent energetic interaction parameters. The energy and volume were additive for all components. These assumptions made it possible to formulate the canonical partition function (1.3-1). The rigid lattice model was later refined by Miller (1948) to account for the differences in molecular size. Tompa (1953) and Barker (1952, 1953) introduced the orientation dependence of interactions between polar and polarizable molecules. All these methods were based on a mathematical approximation method developed by Bethe (1935) for the calculation of the partition function. Other modifications of the lattice model included the introduction of the "hole" theories, which use similar models with some of the lattice sites unoccupied, and the "tunnel" theories, based on a partially disordered lattice structure (Barker, 1963).

A detailed review of lattice models would be beyond the scope of this work as they cannot be recommended in their original forms for practically oriented phase equilibrium computations. However, some elements of lattice theories served as an inspiration for some practical models. Therefore, the basic relations of lattice models will be outlined here using the Barker (1953) method as an illustrative example.

According to Barker, each molecule can occupy more than one lattice site, depending on its volume. The number of lattice sites occupied by a molecule is treated as a number of its segments r. All lattice sites are occupied. The geometrical parameters of the model are the coordination number z, number of segments r_i and number of contacts Q_i. The numbers z, r_i and the sum of contacts $\Sigma\, Q_i$ are not changed upon mixing. The geometrical parameters are interrelated by

$$\sum Q_i = r_i z - 2r_i + 2 \tag{1.3-9}$$

There may be several types of contacts between various types of molecules. For example, the interaction energy between contacts of type μ in molecule i and contacts of type ν in molecule j is $E_{ij}^{\mu\nu+}$. The interactions between contacts are treated by a quasi-chemical reaction

$$(i - i) + (j - j) = 2(i - j) \tag{1.3-10}$$

where $i-j$ denotes a pair of molecules i and j. The numbers of pairs of molecules i and j interacting through contacts μ or ν are interrelated by the quasi-chemical expression

$$\frac{(N_{ij}^{\mu\nu})^2}{N_{ii}^{\mu\mu}N_{jj}^{\nu\nu}} = 4(\eta_{ij}^{\mu\nu})^2 \tag{1.3-11}$$

where

$$\eta_{ij}^{\mu\nu} = \exp\left(\frac{-E_{ij}^{\mu\nu}}{kT}\right) \tag{1.3-12}$$

which bears a resemblance to the well-known mass action law. The energetic parameter $E_{ij}^{\mu\nu}$ is defined as a deviation of the cross-interaction energy from the arithmetic mean of interaction energies of like molecules:

$$E_{ij}^{\mu\nu} = E_{ij}^{\mu\nu+} - \frac{E_{ii}^{\mu\nu+} + E_{jj}^{\mu\nu+})}{2} \tag{1.3-13}$$

The pair numbers satisfy a balance equation

$$2N_{ii}^{\mu\mu} + \sum_{\nu,j} N_{ij}^{\mu\nu} = Q_i^{\mu} N_i \tag{1.3-14}$$

The contact pair numbers can be reexpressed in terms of contact parameters X_i^{μ} and X_i^{ν}:

$$N_{ii}^{\mu\mu} = (X_i^{\mu})^2 N \tag{1.3-15}$$

$$N_{jj}^{\nu\nu} = (X_j^{\nu})^2 N \tag{1.3-16}$$

$$N_{ij}^{\mu\nu} = 2X_i^{\mu} X_j^{\nu} \eta_{ij}^{\mu\nu} N \tag{1.3-17}$$

Therefore, we obtain a system of equations that have to be solved numerically:

$$X_i^{\mu} \sum_{\nu,j} \eta_{ij}^{\mu\nu} X_j^{\nu} = \left(\frac{Q_i^{\mu}}{2}\right) x_i \tag{1.3-18}$$

The number of these equations is equal to the number of possible contact types. After solving these equations for the pair number parameters, the partition function can be computed from

$$Q = \sum_{\{N_{ij}^{\mu\nu}\}} g(N_{ij}^{\mu\nu}, N_i, N_j) \exp\left(-\frac{\Pi\, N_{ij}^{\mu\nu} E_{ij}^{\mu\nu}}{kT}\right) \tag{1.3-19}$$

It should be noted that the quasi-chemical approximation has been used in several practically oriented models that will be described in Chapter 3 (e.g., UNIQUAC, DISQUAC).

n-FLUID THEORIES

Finally, we should mention in this short introduction the methods of combining mixture properties with the properties of pure components. In other words, we are interested if mixing rules for thermodynamic functions can be established without resorting to some particular models of fluids. Scott (1956) analyzed some feasible approaches, which he called one-fluid, two-fluid and three-fluid models.

According to the one-fluid model, a mixture is treated as a pseudopure fluid. The Gibbs energy of the pseudopure fluid G_x differs from that of the mixture by the ideal Gibbs energy of mixing

$$G_m(T, P, \mathbf{x}) = G_x(T, P, \mathbf{x}) + RT \sum_i x_i \ln x_i \qquad (1.3\text{-}20)$$

Similar expressions hold for other thermodynamic functions. The parameters (macroscopic or microscopic) of the pseudopure fluid are obtained from the pure component parameters through suitable mixing rules. For nonpolar fluids, these mixing rules are usually quadratic in mole fraction. The one-fluid model is also known as the random mixing model (Brown, 1957).

In the case of the two-fluid model each component of the mixture is replaced by a pseudocomponent. The name *two-fluid model* follows from the fact that for a binary mixture two pseudopure fluids are considered. The pseudopure components are assumed to mix ideally:

$$G_m(T, P, \mathbf{x}) = \sum_i x_i G_i(T, P, x_i) + RT \sum_i x_i \ln x_i \qquad (1.3\text{-}21)$$

In this case the mixture is no longer random. Each component changes its properties because of interactions with the remaining components. Therefore, the microscopic or macroscopic parameters of the pseudocomponents are different from the parameters of the actual pure components. They have to be determined in principle, from mixture data. This model has been called by Brown (1957) the *semi-random mixing approximation*.

For a three-fluid model, the interactions between like molecules are assumed to be the same as in actual pure components. The interactions between unlike molecules are accounted for by another hypothetical fluid. For example, the properties of a binary mixture of components 1 and 2 are represented by a combination of pure fluids 1 and 2 with intermolecular interactions of the types 1–1 and 2–2, respectively, and by a third fluid with interactions of the 1–2 type only. Therefore, the Gibbs energy of a mixture takes the form

$$G_m(T, P, \mathbf{x}) = \sum_i \sum_j x_i x_j G_{ij}(T, P) + RT \sum_i x_i \ln x_i \qquad (1.3\text{-}22)$$

In this case it is not necessary to use mixing rules for the model parameters. The mixture behavior is characterized completely by the parameters of the G_{ij}

($i \neq j$) contribution. The applicability of the one-, two- and three-fluid models was studied by Rowlinson and Swinton (1982), Leland et al. (1968, 1969) and Quitzsch et al. (1981). In this work their applicability in conjunction with the corresponding-states principle will be described in Section 4.3.2.

Throughout this book more specific methods for the modelling of phase equilibria will be studied. The general methods described briefly in this chapter will be applied to the construction of practically oriented techniques.

REFERENCES

Barker, J. A., 1952. *J. Chem. Phys.* 20: 1526.

Barker, J. A., 1953. *J. Chem. Phys.* 21: 1391.

Barker, J. A., 1963. *Lattice Theories of the Liquid State*, Pergamon Press, Oxford.

Bethe, M. A., 1935. *Proc. Royal Soc.* A150: 552.

Boublík, T., I. Nezbeda and K. Hlavaty, 1980. *Statistical Thermodynamics of Simple Liquids and Their Mixtures*, Academia, Praha.

Bridgeman, P. W., 1926. *Condensed Collection of Thermodynamic Formulas*, Cambridge, MA.

Brown, W. B., 1957. *Phil. Trans.* 250: 157, 221.

Caratheòdory, C., 1909. *Math. Ann.* 67: 355.

Clausius, R., 1870. *Ann. Phys.*, 141: 124; *Phil. Mag.* 40: 122.

Fisher, I. Z., 1964. *Statistical Theory of Liquids*, University of Chicago Press, Chicago.

Gibbs, J. W., 1957. *Collected Works*. Vol. I. Yale University Press, New Haven.

Goranson, W. R., 1930. *Thermodynamic Relations in Multicomponent Systems*, Carnegie Inst. Washington Pub., Washington, D.C.

Gray, C.G. and K.E. Gubbins, 1984. *Theory of Molecular Fluids*, Clarendon Press, Oxford.

Guggenheim, E. A., 1952. *Mixtures*, Clarendon Press, Oxford.

Guggenheim, E.A. and R. H. Fowler, 1935. *Proc. Royal Soc.* 148: 304.

Gumiński, K., 1982. *Termodynamika* (in Polish), PWN–Polish Scientific Publishers, Warszawa.

Hill, T. L., 1956. *Statistical Mechanics. Principles and Selected Applications*, McGraw-Hill, New York.

Leland, T. W., J. S. Rowlinson and G. A. Sather, 1968. *Trans. Faraday Soc.* 64: 1447.

Leland, T. W., J. S. Rowlinson, G. A. Sather and J. O. Watson, 1969. *Trans. Faraday Soc.* 65: 2034.

Lewis, G. N., 1901. *Z. Phys. Chem.* 38: 205.

Lewis, G. N., 1907. *Z. Phys. Chem.* 61: 129.

Lewis, G. N. and M. Randall, 1921. *J. Am. Chem. Soc.* 43: 233.

Lewis, G. N., M. Randall, K. S. Pitzer and L. Brewer, 1961. *Thermodynamics*, 2nd ed., McGraw-Hill, New York.

Margenau, H. and G. M. Murphy, 1956. *The Mathematics in Physics and Chemistry*, Van Nostrand, New York.
McQuarrie, D. A., 1973. *Statistical Thermodynamics*, Harper & Row, New York.
Miller, A. R., 1948. *Theory of Solutions of High Polymers*, Clarendon Press, Oxford.
Model, M. and R. C. Reid, 1974. *Thermodynamics and its Applications*, Prentice-Hall, Englewood Cliffs, NJ.
Münster, A., 1969 and 1975. *Statistical Thermodynamics*, Vols. I and II, Springer, Berlin.
Münster, A. 1970. *Classical Thermodynamics*, Wiley-Interscience, London.
Prausnitz, J. M., 1969. *Molecular Thermodynamics of Fluid Phase Equilibria*, Prentice-Hall, Englewood Cliffs, NJ.
Prigogine, I., 1957. *Molecular Theories of Solutions*, North-Holland, Amsterdam.
Quitzsch, K., U. Messow, R. Pfestorf and K. Sühnel, 1981. In: Bittrich, H. J. (Ed.), *Modellierung von Phasengleichgewichten als Grundlage von Stofftrennprozessen*, Akademie Verlag, Berlin.
Rowlinson, J. S. and F. Swinton, 1982. *Liquids and Liquid Mixtures*, 3rd. ed., Butterworth, London.
Scott, R. L., 1956. *J. Chem. Phys.* 25: 193.
Shaw, A. N., 1935. *Phil. Trans. Roy. Soc.* (*London*) A234: 299.
Thielsen, 1885. *Ann. Phys.*, 24: 467.
Tompa, H., 1953. *J. Chem. Phys.*, 21: 250.

2

Pure-Component Properties Relevant to Phase Equilibrium Calculations

The accurate representation of pure-component properties is a fundamental task in the modelling of phase equilibria of multicomponent mixtures. Thermodynamic functions of pure components are necessary to establish the reference states required for phase equilibrium modelling. Vapor pressure, liquid and vapor volumes are the properties of primary importance.

2.1 COEXISTENCE OF TWO PHASES AND VAPOR PRESSURE OF A PURE COMPONENT

The accurate knowledge of pure-component vapor pressures is crucial in the modelling of vapor–liquid or vapor–liquid–liquid equilibria. For one set of isothermal data the vapor pressure of each component can be represented by a single value. When the existing data should be extended to different temperatures or nonisothermal data are to be correlated, an equation representing the vapor pressure of pure components as a function of temperature has to be introduced.

The phase behavior of a pure substance is illustrated in Figure 2.1-1 as a function of temperature, pressure and volume. The figure shows regions of the phase diagram corresponding to the solid (S), liquid (L) and gas (G) phases separated by surfaces corresponding to the coexistence of the solid and liquid (S–L) or liquid and gaseous (L–G) or solid and gaseous (S–G) phases. There are three characteristic points: the triple point T where three phases coexist (S–L–G) and the critical point where the liquid and gas phases converge into one fluid phase. Projections of the three-phase diagram on the P–V and P–T planes are also shown. In the P–T projection three coexistence curves exist.

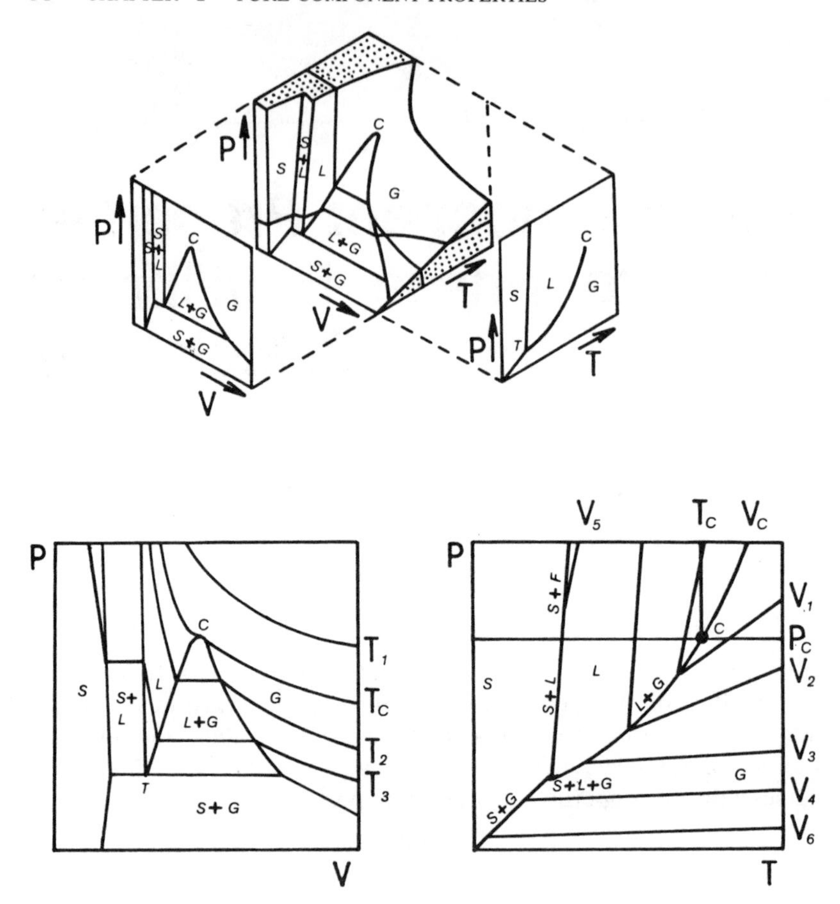

Figure 2.1-1
The pressure–volume–temperature phase diagram of a pure substance. The solid, liquid and gas phases are denoted by S, L and G, respectively. The critical temperature and pressure are denoted by T_c and P_c, respectively.

They represent the coexistence regions between solid and gas (S–G), solid and liquid (S–L) and liquid and gas (L–G). T_c and P_c are the critical temperature and pressure above which only one fluid state exists. According to the phase rules (1.2–124) the number of degrees of freedom is $\phi = 2$ in the one-phase regions. Two parameters—pressure and temperature—can be changed without affecting the equilibrium state. Along the coexistence curve only one variable, pressure or temperature, can be changed ($\phi = 1$). In the triple point $\phi = 0$ and all variables have to be fixed to maintain equilibrium.

Let us consider the coexistence of two phases as a function of temperature and pressure. If two phases are in equilibrium, the values of all intensive parameters should be equal in both phases. To maintain equilibrium, a change of an intensive parameter in phase α must entail an analogous change in phase β. For the chemical potential we get

$$d\mu^{\alpha} = d\mu^{b} \tag{2.1-1}$$

In a one-component system a partial molar quantity is equal to a molar quantity. Therefore, the molar chemical potential is equal to the molar Gibbs energy (1.2-172), and Eq. (2.1-1) can be rearranged, following (1.2-183), as

$$v^{\alpha}\, dP - s^{\alpha}\, dT = v^{\beta}\, dP - s^{\beta}\, dT \tag{2.1-2}$$

Taking into account that $\mu^{\alpha} = \mu^{\beta}$ in equilibrium and

$$\mu = h - Ts \tag{1.2-183}$$

Eq. (2.1-2) can be rearranged as

$$\frac{dP}{dT} = \frac{s^{\alpha} - s^{\beta}}{v^{\alpha} - v^{\beta}} = \frac{h^{\alpha} - h^{\beta}}{T(v^{\alpha} - v^{\beta})} = \frac{\Delta h}{T\,\Delta v} \tag{2.1-3}$$

Equation (2.1-3) is a differential equation determining the equilibrium in a two-phase one-component system. It was first formulated by Clapeyron in 1834. The present form of the equation was given by Clausius in 1850.

The quantities $\Delta h = h^{\alpha} - h^{\beta}$ and $\Delta v = v^{\alpha} - v^{\beta}$ are the molar enthalpy and volume change accompanying the phase transition, respectively. Their values depend strongly on the kind of phase transition. If both phases are condensed, the quantity Δv is small and the value of the derivative dP/dT is large. If one of the phases is gaseous, Δv is large and dP/dT is small. The following two examples are given to illustrate the orders of magnitude and relative errors of the quantities involved.

Example 2.1-1 What is the effect of pressure on melting and vaporization?

SOLUTION The molar volume of ice, liquid water and gaseous water at 273.15 K are $19.6 \times 10^{-6}\ m^3/mol$, $18.0 \times 10^{-6}\ m^3/mol$ and $3.71\ m^3/mol$, respectively. The heat of melting of ice is 6 kJ/mol and the heat of vaporization of ice is 45 kJ/mol. Therefore, the following values are obtained for the derivative of equilibrium pressure against temperature.

$$(dP/dT)_{\text{melting at 273.15 K}} = -14\ \text{MPa/K}$$

$$(dP/dT)_{\text{vaporization at 273.15 K}} = 44.5\ \text{Pa/K}$$

The obtained values differ by 6 orders of magnitude. Therefore, the effect of pressure on the melting of ice is very small whereas the effect on vaporization is large. ■

Example 2.1-2 Let us estimate the errors of quantities that can be computed from the Clausius–Clapeyron equation in the case of an equilibrium between liquid and vapor.

SOLUTION The volume of the vapor phase can be computed from the virial equation of state truncated after the second term (for a detailed discussion of the virial equation see Section 2.2):

$$v^g = \frac{RT}{P} + B \tag{P2.1-1}$$

Therefore, the virial coefficient B can be computed from the measured values of liquid molar volume v^1, vaporization enthalpy Δh, pressure P and its derivative versus temperature:

$$B = -\frac{T}{P}\left(R - \frac{\Delta h\, P}{T^2}\frac{dT}{dP}\right) + v^1 \tag{P2.1-2}$$

For benzene the liquid molar volume at the normal boiling point ($T =$ 353.25 K, $P =$ 101,325 Pa) is 0.000096 m^3, the vaporization enthalpy measured in the National Bureau of Standards is 30,780 J/mol and $dP/dT =$ 0.0003205 Pa/K. The virial coefficient is then

$$B = 0.000096 - \frac{353.25}{101325}\left(8.3144 - 0.0003205\,\frac{30{,}780 \times 101{,}325}{353.25^2}\right) = -964\ \text{cm}^3/\text{mol}$$

Assuming very small errors of pressure and temperature measurements, i.e., 15 Pa (0.015 percent) and 1 mK, respectively, the error in the calculated value of B is 180 cm^3/mol (18.5 percent) whereas the error in the dP/dT derivative is only 0.5 percent. A value of B derived from a direct measurement is -945 ± 20 cm^3/mol and accidentally agrees with the computed value. The error of enthalpy measurements affects the calculations in a similar way. It is, on average, 0.1–0.4 percent. ■

Example 2.1-3 What is the sensitivity of vaporization enthalpy calculated from the Clapeyron equation to gas-phase nonideality?

SOLUTION This time we shall back-calculate the vaporization enthalpy of benzene using the data of Example 2.1-3. The molar volume of benzene vapor is

$$v^g = \frac{RT}{P} + B = \frac{8.3144 \times 353.25}{101{,}325} - 0.000945 = 0.028042\ \text{m}^3/\text{mol}$$

According to the Clapeyron equation the vaporization enthalpy is

$$\Delta h = T(v^g - v^1)\frac{dP}{dT} = 353.25(0.028042 - 0.000096)3120 = 30{,}800\ \text{J/mol}$$

If the vapor phase nonideality is neglected, i.e., v^g is calculated from the ideal gas law ($v^g = RT/P$), the vaporization enthalpy is estimated at

$$\Delta h = \frac{RT^2}{P}\frac{dP}{dT} = \frac{8.3144 \times 353.25}{101{,}325} \times 3121 = 31{,}957\ \text{J/mol}$$

The error of the vaporization enthalpy calculated from the Clapeyron equation is only 0.1 percent and falls within the experimental uncertainty range. If the vapor phase nonideality is neglected, the error increases to 3.7 percent. ■

There are two methods for correlating vapor pressure as a function of temperature. The first one is based on integration of the Clapeyron equation [Eq. (2.1-3)]. The second method is based on the equation-of-state technique. This method can be used to correlate saturation pressure of pure compounds in a wide temperature range, from the triple to the critical point, as well as volumetric properties in the one- and two-phase regions. A detailed description of the equation-of-state techniques is deferred to Chapter 4.

The first group of methods is based on the Clapeyron differential equation (2.1-3), which provides an exact relation between the properties of vapor and liquid phases. An exact solution of the Clapeyron equation would necessitate the use of functions representing the change of enthalpy and volume on vaporization. Such functions are not known in a general form. Therefore, approximate relations for the enthalpy (ΔH) and volume change (ΔV) accompanying a phase transition are necessary for this purpose.

The simplest solution to Eq. (2.1-3) can be obtained on the assumption that Δh is independent of temperature and Δv is the ideal gas volume ($\Delta v = RT/P$). The simplest vapor pressure equation based on this assumption was proposed by Wrede in 1984 (Partington, 1955) and is sometimes called the Clapeyron equation

$$\ln p_i^0 = A_i - \frac{B_i}{T} \tag{2.1-4}$$

where A_i and B_i are adjustable parameters.

The equation is able to represent saturation pressures within the experimental accuracy over small temperature intervals not exceeding 20 K and is widely used for smoothing experimental data for pure compounds and fluids of constant composition.

The more elaborate assumptions are that the heat capacity of both phases is independent of temperature (i.e., Δh is a linear function of temperature):

$$\Delta h = h^0 - a(T - T^0) \tag{2.1-5}$$

and similarly as in Eq. (2.1-4), the volume of the liquid phase is negligible in comparison with that of the vapor phase, which is treated as an ideal gas:

$$v^g = \frac{RT}{P} \tag{2.1-6}$$

Equation (2.1-3) is then rearranged to the form

$$\frac{dp_i^0}{dT} = \frac{p_i^0(h_i^0 - a_i T)}{RT^2} \tag{2.1-7}$$

where h_i^0 is the enthalpy of phase change at a reference state and a_i is the heat capacity at phase change. Integration gives

$$\ln p_i^0 = A_i - \frac{B_i}{T} - C_i \ln T \tag{2.1-8}$$

where A_i is the integration constant, and $B = h_i^0/R$ and $C = a_i/R$.

This equation was proposed by Rankine in 1849 and by Kirchhoff in 1858. Although the equation is empirically not very effective and is now rarely used, it served as a basis for the development of more elaborate equations (Partington, 1955).

A very interesting empirical modification of (2.1-4) was proposed in 1888 by Antoine. The equation is able to represent the vapor pressure of nonassociating organic compounds in the pressure range from 5 to 200 kPa with the accuracy achieved in direct measurements of this property. For associating compounds the pressure range in which such accuracy can be achieved is much smaller, e.g., for alcohols it reduces to 5 to 80 kPa. The equation can be written in two explicit forms: the first one for pressure and the second one for temperature:

$$\log_{10} p_i^0 = A_i - \frac{B_i}{C_i + t} \tag{2.1-9}$$

$$t = \frac{B_i}{A_i - \log_{10} p_i^0} - C_i \tag{2.1-10}$$

The adjustable parameters of the Antoine equation are only empirical and can be obtained by smoothing experimental vapor pressure data. Due to its accuracy the equation can be used for the calculation of derivatives dP/dT or dT/dP in an analytical way, which is very important for numerical computations. All the above properties have rendered the Antoine equation very popular among experimentalists measuring low-pressure vapor–liquid equilibrium data.

The constants of the Antoine equation are tabulated for more than 8000 various organic compounds (Boublik et al., 1987; Dykyj and Repas, 1979; Dykyj et al., 1984; Ohe, 1976; Lencka et al., 1984; Reid et al., 1977; Stephenson and Malanowski, 1987; and TRC Thermodynamic Tables). Only the Antoine parameters accompanied by estimated errors can be treated with confidence. In some collections (Boublik et al., 1987; Lencka et al., 1984) the parameters are accompanied by original experimental data. In such cases the user's own estimation can be made. This is important because significant progress has been recently achieved in the numerical estimation methods (Gregorowicz et al., 1987). The values of constants depend on the units used. Traditionally, millimeters of mercury and °C were used with decimal logarithms. With the introduction of SI units each data collection has been used with its own set of units. The most popular units are kilopascal, kelvin and decimal logarithms. Each set of A, B and C values should be accompanied by its range of applicability. The constants should not be used for extrapolation outside the recommended ranges of pressure or temperature. Constants from one set should never be mixed with those from a different one. When the extrapolation is unavoidable, it should be made only with great care and only in the close vicinity of the recommended range. Extrapolation beyond these limits may produce absurd results. Special attention should be paid to this problem as sometimes unreasonable intervals are given even in data compilations (Ohe, 1976).

The Antoine equation can be used for the computation of the enthalpy of vaporization according to the formula

$$\Delta H = T\,\Delta V\,\frac{dP}{dT} = T\,\Delta V\,\frac{PB\ln(10)}{(C+t)^2} \tag{2.1-11}$$

On the assumption that the volume of the vapor phase can be represented by the virial equation of state truncated after the second term (β), equation (2.1-7) can be rearranged as

$$\Delta H = T\left(\frac{RT}{P} - \beta - V^L\right)\frac{PB\ln(10)}{(C+t-273.15)^2} \tag{2.1-12}$$

The computation of the enthalpy of vaporization enables the thermodynamic consistency of the measured vapor pressure to be checked. The enthalpy computed from vapor pressure data should be compared with that obtained by direct calorimetric determination. For very good measurements, the difference between the measured and computed values is of the order of 0.1 to 0.4 percent of the measured enthalpy. The tables of enthalpy of vaporization published by Majer and Svoboda (1985) are a useful source of data for this purpose.

The use of the Antoine equation constants for high-precision correlations needs special caution. Some data collections (TRC Thermodynamic Tables) were prepared on the assumption that the values of the B and C constants are monotonic functions of the number of carbon atoms in homologous series or boiling points (Kręglewski and Zwoliński, 1961; Wilhoit, 1976; Zwoliński and Wilhoit, 1966; Wilhoit and Zwoliński, 1971). This is very dangerous in the modelling of vapor–liquid equilibria for mixtures when the excess Gibbs energy is computed from isobaric data. The constants are consistent with the enthalpy data published in the same tables but do not represent the vapor pressure with the accuracy of the original experiment (Willingham et al., 1945).

There are other three-parameter equations of similar accuracy like that proposed by Miller (Reid et al., 1977)

$$\ln\frac{p_i^0}{P_c} = -\frac{A_i T}{T_c}\left[1-\left(\frac{T}{T_c}\right)^2 + B_i\left(3+\frac{T}{T_c}\right)\left(1-\frac{T}{T_c}\right)^2\right] \tag{2.1-13}$$

where A_i, B_i and P_c are adjustable parameters and T_c is the real critical temperature. The P_c parameter has the dimension of critical pressure but is generated by the shape of the vapor pressure curve and, due to large deviations from the real critical pressure, it cannot be used for its prediction. Equation (2.1-9) is not popular due to its complicated shape and implicit form against temperature.

The three-parameter equations are generally incapable of representing a wide range of temperature from the triple to the critical point. There were several attempts to improve this situation by adding more terms. The most popular one was due to Frost and Kalkwarf (1953). The Frost–Kalkwarf equation was derived on the assumption that the vaporization enthalpy is a linear function of temperature and Δv can be estimated from the van der Waals equation of state (see Chapter 4). Parameters for this equation are available in the third and fourth editions of the book by Reid et al. (1977, 1987) and in the paper by Harlacher and Braun (1975):

$$\ln p_i^0 = A_i + \frac{B_i}{T} + C_i \ln T + D_i \frac{p_i^0}{T^2} \tag{2.1-14}$$

Another four-parameter modification was proposed by Plank and Riedel (1948):

$$\ln p_i^0 = A_i + \frac{B_i}{T} + \frac{C_i}{\ln T} + D_i T^6 \tag{2.1-15}$$

Due to the insignificant improvement in the correlation results, a tendency to increase even more the number of adjustable parameters was observed. If very accurate data are available, orthogonal polynomials can be used for their correlation as proposed by Ambrose and his co-workers from the National Physical Laboratory (Ambrose et al., 1970). The proposed equation is

$$T \ln P = \frac{A_0 E_0(x)}{2} + \sum_{i=1}^{n} A_i E_i(x) \tag{2.1-16}$$

where n is the number of terms and x is a function of temperature defined as

$$x = \frac{2T - (T_{\max} + T_{\min})}{T_{\max} - T_{\min}} \tag{2-1-17}$$

The consecutive terms are given by

$$E_0(x) = 1 \tag{2.1-18}$$

$$E_1(x) = x \tag{2.1-19}$$

$$E_{i \neq 1}(x) - 2xE_i(x) + E_{i-1}(x) = 0 \tag{2.1-20}$$

When using the orthogonal polynomials, the number of terms must be carefully selected. Otherwise, it is possible to arrive at smaller deviations than the experimental error. Moreover, the orthogonal polynomials cannot be used to calculate analytically the derivative dP/dT. The derivative can be obtained numerically using sufficiently wide temperature intervals to avoid any accidental oscillations of the correlating function. Power series and Chebyshev polynomials were also used for saturation pressure correlation (Angus, 1983). In general, the obtained improvement in the representation of the experimental data obtained by increasing the number of adjustable parameters is not sufficient to compensate for the loss in accuracy and predictive ability.

A significant improvement in the correlation of vapor pressures has been obtained with the Wagner (1974) equation:

$$\ln\left(\frac{P}{P_c}\right) = [a_1\theta + a_2\theta^{3/2} + a_3\theta^{5/2} + f(\theta)](1 - \theta)^{-1} \tag{2.1-21}$$

where $\theta = 1 - T/T_c$, $f(\theta)$ contains one or two further terms raised to integral powers of θ, a_i are adjustable parameters and P_c is the real critical temperature.

The Wagner equation is very efficient, and it is the only equation able to accurately describe the data with few constants and with the reasonable behavior from $T_r = T/T_c = 0.5$ to the critical point. When applying the equation to the correlation of vapor pressure data, one should be aware that the representation of the data can be quite precise and yet violate the experimentally observed behavior of the saturation curve. Therefore, both factors should be balanced. Observed behavior of the saturation curve has been generalized and may be expressed by a number of quantitative parameters.

Qualitatively, a constrained correlation strategy taking into account some reasonable physical conditions should be superior to unconstrained correlation of $T - P$ data, especially when the experimental data points are burdened with much uncertainty. This may be the case when the investigated compound partly decomposes at higher temperatures, thus making it impossible to obtain high-quality data. Correlations that do not match the observed characteristics of the saturation curve may be particularly misleading when derived quantities are desired (e.g., dP/dT values for estimating vaporization enthalpy). Chase (1987) presented an overview of techniques used to obtain physically meaningful parameters of correlation equations.

Among the many observations of the correct shape of the P versus T curve, Thodos (1950) detailed experimental confirmation of the general definition of the saturation curve (shallow s on a $\ln P$ vs. $1/T$ plot). Waring (1954) stated general requirements for the analytical representation of the saturation curve, based on the consideration of the $(\Delta H/\Delta z)$ versus T plot where ΔH is the enthalpy of vaporization and Δz is the difference between the compressibility factors of vapor and liquid. King and Al-Najjar (1974) and King (1976) detailed advantages of using thermal data when evaluating the coefficients of $P - T$ correlations. Ambrose and co-workers (Ambrose et al., 1975, 1978; Ambrose and Davies, 1980; Ambrose and Ghiassee, 1987) further defined the characteristics of the saturation curve. McGarry (1983) and Chase (1984) amplified the criteria for correlating vapor pressure data.

The following three criteria have a sound physical basis and can be used in constrained optimization techniques:

1. The derived $(\Delta H/\Delta z)$ versus T curve has a minimum within a specified range of T_R values (Waring's criterion). Ambrose and co-workers used this criterion in conjunction with the following version of the Wagner equation:

$$\ln\left(\frac{P}{P_c}\right) = [a_1\theta + a_2\theta^{3/2} + a_3\theta^3 + a_4\theta^6](1-\theta)^{-1} \tag{2.1-22}$$

Since

$$\frac{\Delta H}{\Delta z} = RT^2 \frac{d \ln p}{dT}\ , \tag{2.1-23}$$

this minimum is revealed by differentiation of the vapor pressure curve. Study of many vapor pressure equations has shown that for nonassociating substances there is a minimum in $\Delta H/\Delta z$ at $T_R \cong 0.85$ whereas for associating substances except water the minimum occurs much closer to the critical temperature. Differentiation of Eq. (2.1-23) gives

$$\frac{d(\Delta H/\Delta z)}{dT} = \frac{RT(0.75a_2\theta^{-0.5} + 3.75a_3\theta^{0.5} + 20a_4\theta^3)}{T_c} \tag{2.1-24}$$

Since the minimum is expected to be in the range $0.8 < T_R\,(\text{min}) < 1.0$, where the effect of the last term in Wagner's equation is negligible,

$$0.75a_2[\theta(\text{min})]^{-0.5} + 3.75\text{a}_3[\theta(\text{min})]^{0.5} \cong 0 \tag{2.1-25}$$

2. The second constraint requires that the quantity $\ln\,(P/P')_{T_R=0.95}$ falls within a specified range (Ambrose's criterion):

$$-0.010 < \ln\left(\frac{P}{P'}\right)_{T_r=0.95} < -0.002 \tag{2.1-26}$$

where P' is given by

$$\ln\left(\frac{P}{P'}\right) = -B\left(1 - \frac{1}{T_R}\right) \tag{2.1-27}$$

and B is calculated from the critical pressure and the value $P' = P$ at $T_R = 0.7$. This constraint quantifies the experimentally observed shallow s shape of the $\ln P$ against $1/T$ curve.

3. The derived $\Delta H/\Delta z$ values should approximate the enthalpy of vaporization at low reduced temperatures.

It is also possible to relate the shape of the curve to parameters used in corresponding-states schemes. Such criterion can be inferred from the curves of α against T_R where $\alpha = d(\ln P_R)/d(\ln T_R)$. Since at $T_R = 1$, $\Delta H = 0$, we get $d\alpha/dT_R = 0$ at $T_R = 1$. Thus, the vapor pressure is related for any compound to a single factor (Riedel, 1954). For the special case of Eq. (2.1-23)

$$a_1 = -\left(\frac{d \ln P}{d \ln T}\right)_{T_R=1} = \alpha_c \tag{2.1-28}$$

Chase (1987) showed that the Riedel criterion is at least qualitatively representative of the Waring and Ambrose criteria. The Riedel criterion may be useful to ensure the correctness of the correlation in the high reduced temperature region.

In general, there can be some degree of degradation of correlation deviations if the constrained parameter optimization is applied. Thus, phenomenological correctness must be balanced against the accuracy of correlation.

The predictive ability of the equations based on the Clapeyron equation (2.1-3) is, in general, very poor. There are some techniques developed specially for prediction, or more strictly speaking the extrapolation of experimental data.

The best ones are based on the principle of corresponding states. Among them the Pitzer perturbation expansion in terms of the acentric factor ω is the most popular approach:

$$\ln p_i^0 = f_0 + \omega f_1 \tag{2.1-29}$$

where ω is the acentric factor (Pitzer, 1955) and f_0, f_1 are functions of the reduced temperature $T_R = T/T_c$. The functions f_0 and f_1 have been expressed in various forms (Reid et al., 1977). Good results can be obtained with the use of those proposed by Lee and Kesler (1975):

$$f_0 = 5.92714 - \frac{6.09648}{T_R} - 1.28862 \ln T_R + 0.169347 T_R^6 \tag{2.1-30}$$

$$f_1 = 15.2518 - \frac{15.6875}{T_R} - 13.4721 \ln T_R + 0.43577 T_R^6 \tag{2.1-31}$$

The acentric factor is defined as

$$\omega = -\log_{10} P_R - 1.000 \tag{2.1-32}$$

where $P_R = p_i^0/P_c$ should be calculated at reduced temperature $T_R = T/T_c = 0.7$.

The factor ω represents the geometric shape of the molecule. For monatomic gases it is almost equal to zero. It rises with the size and polarity of a molecule. The values of acentric factor are tabulated in various sources (Reid et al., 1977, 1987; TRC Thermodynamic Tables).

The acentric factor used un the Lee–Kesler correlation should be calculated by means of a formula derived from Eqs. (2.1-29)–(2.1-31):

$$\omega = \frac{-\ln P_c - 5.92714 + 6.09648\theta^{-1} + 1.28862 \ln \theta - 0.169347\theta^6}{15.2518 - 15.6875\theta^{-1} - 13.4721 \ln \theta + 0.43577\theta^6} \quad (2.1\text{-}33)$$

where $\theta = T_b/T_c$, and T_b is normal boiling temperature K.

Another vapor pressure estimation method was proposed by Gomez–Nieto and Thodos (1977a, b, 1978). Its theoretical background is less straightforward than that of the Lee–Kesler method, but the claimed accuracy is higher. The Gomez–Thodos equation is as follows:

$$\ln \frac{P_i^c}{P_c} = \beta\left[\left(\frac{T_c}{T}\right)^m - 1\right] + \gamma\left[\left(\frac{T}{T_c}\right)^7 - 1\right] \quad (2.1\text{-}34)$$

where

$$\gamma = ah + b\beta \quad (2.1\text{-}35)$$

is related to the normal boiling point by

$$a = \frac{1 - 1/\theta}{\theta^7 - 1} \quad (2.1\text{-}36)$$

$$b = \frac{1 - 1/\theta^m}{\theta^7 - 1} \quad (2.1\text{-}37)$$

$$h = \theta \frac{\ln(P_c/1.01325)}{1 - \theta} \quad (2.1\text{-}38)$$

For nonpolar compounds β and m are computed from the equation

$$\beta = -4.267 - \frac{221.79}{h^{2.5} \exp(0.0384h^{2.5})} + \frac{3.8126}{\exp(2272.44/h^3)} + d \quad (2.1\text{-}39)$$

where $d = 0$ for all nonpolar compounds except for helium ($d = 0.41815$), hydrogen ($d = 0.19904$) and neon ($d = 0.02319$).

$$m = 0.78425 \exp(0.089315h) - \frac{8.5217}{\exp(0.74826h)} \quad (2.1\text{-}40)$$

For polar compounds that do not hydrogen bond

$$m = 0.466T_c^{0.166} \quad (2.1\text{-}41)$$

$$\gamma = 0.0854 \exp(7.462 \times 10^{-4}T_c) \quad (2.1\text{-}42)$$

$$\beta = \frac{\gamma}{b} - \frac{ah}{b} \quad (2.1\text{-}43)$$

Table 2.1-1
COMPARISON OF VARIOUS METHODS FOR THE CORRELATION AND PREDICTION OF VAPOR PRESSURE (P) FOR ACETONE

Method	Range T_r	AAD(P)[a]
Antoine	0.51–0.93	0.675
Antoine	0.51–0.69	0.017
Wagner	0.51–0.93	0.112
Lee–Kesler	0.51–0.93	2.888
Gomez and Thodos	0.51–0.93	0.437

[a] $\mathrm{AAD}(P) = (100/n) \sum_{I=1}^{n} |(P_{\mathrm{calc}} - P_{\mathrm{exp}})/P_{\mathrm{exp}}|$.

For hydrogen bonding compounds β is computed from (2.1-43) and the remaining values from

$$m = 0.0052 M^{0.29} T_c^{0.72} \tag{2.1-44}$$

$$\gamma = \frac{2.464}{T_c} \exp(9.8 \times 10^{-6} M T_c) \tag{2.1-45}$$

where M is the molecular weight.

An example of the performance of various equations is given in Table 2.1-1. The high accuracy vapor pressure data measured for acetone at the National Physical Laboratory (U.K) (Ambrose et al., 1974) for reduced temperatures between $T_r = 0.51$ and $T_r = 0.69$, complemented with some high-pressure data (Reid et al., 1987) were used.

The results obtained illustrate the applicability range of various methods. The Antoine equation is best for the correlation of experimental vapor pressure data between 5 and 200 kPa when the information obtained by accurate measurements should not be lost by a poor correlation. For strongly polar compounds the pressure range should be smaller. For higher and wider pressure ranges the Wagner equation with the number of parameters adjusted to the compound type and the pressure range can be recommended. The Pitzer method with the Lee–Kesler parameters is recommended for the prediction of vapor pressures at reduced temperatures 0.8 and lower, while the Gomez–Thodos method should be used for temperatures between 0.5 and 1 (Reid et al., 1987). At present the experience with the last method is limited.

2.2 VAPOR VOLUME

For the determination of the properties of the liquid phase by measuring the vapor–liquid equilibrium, the knowledge of the thermodynamic properties of the vapor phase determined in an independent way is necessary. Fugacity coefficients are generally used for this purpose. For the computation of these coefficients the knowledge of the volume of the vapor phase as a function of pressure and temperature (i.e., the *PVT* properties) is necessary.

Therefore, an equation of state for the vapor phase is required:

$$P = P(T, V, N) \tag{2.2-1}$$

The equation can also be expressed in terms of the compressibility factor (1.2-153):

$$Z \equiv \frac{PV}{RT} \tag{2.2-2}$$

Here, we are interested in equations of state representing only the *PVT* (pressure–volume–temperature) properties of the vapor phase. Chapter 4 will be devoted to the much more complicated problem of equations of state for *both* gas and liquid phases.

In the low and moderate pressure range the virial equation of state can be used. The equation was formulated by Thiesen in 1885 as an expansion of pressure in a McLaurin series in terms of density (reciprocal volume). The name *virial equation* stems from the Latin word *vis* denoting *power* (for a power series). This expansion was later called the Leiden version of the virial equation of state. The equation written for the compressibility factor becomes

$$Z = \frac{PV}{RT} = Z_0^{(0)} + Z_0^{(1)} \frac{1}{V} + \frac{Z_0^{(2)}}{2} \left(\frac{1}{V}\right)^2 + \cdots \tag{2.2-3}$$

where $Z_0^{(n)}$ is the nth derivative of Z in the reference point 0. The consecutive terms on the right side of Eq. (2.2-3) are called the first, second, third, etc. virial coefficients. First, the compressibility factor in the reference state has to be defined. It is convenient to use the ideal gas as the reference state:

$$Z = Z_0 = \frac{PV}{RT} = 1 \tag{2.2-4}$$

It is interesting to note that the ideal gas law was formulated for the first time by Horstman in the same year (1873) when van der Waals published his celebrated equation for *real* fluids. The gas constant appearing in (2.2-4) was introduced much earlier by Clapeyron (1834) who assumed that for each gas a characteristic constant exists equal to the ratio of specific volume and absolute temperature (273.1 K) at 1 atm pressure. The constant defined in this way is equal to the contemporary gas constant for one mole of an ideal gas. Equation (2.2-4) is not satisfied by any real gas but defines the limiting value of the compressibility factor at zero density. The gas constant can be determined by extrapolating the properties of real gases to zero density. The values of the gas constant in various units are given in Table 2.2-1.

Table 2.1-2
THE GAS CONSTANT IN VARIOUS UNITS

Unit System	R
SI	8.31433 J/mol · K
cgs	83.1434 cm^3 · bar/mol · K
Metric	82.05 cm^3 · atm/mol · K = 1.9872 cal/mol · K
British	10.73 psia · ft^3/lbmol · R = 1545 ft · lb/lbmol · R = 1.986 Btu/lbmol · R

To completely define the virial equation, the derivatives of the compressibility factor appearing in Eq. (2.2-3) have to be determined. As the virial equation is an expansion of pressure in terms of density, these coefficients can be functions of temperature only. The virial equation takes then the following form:

$$Z = \frac{PV}{RT} = 1 + \frac{B(T)}{V} + \frac{C(T)}{V^2} + \frac{D(T)}{V^3} + \cdots \tag{2.2-5}$$

where the parameters B, C, D, ... are called the second, third, fourth, etc. virial coefficients.

The virial equation is sometimes written as a power series in pressure (the so-called Berlin form)

$$Z = \frac{PV}{RT} = 1 + B'P + C'P^2 + D'P^3 + \cdots \tag{2.2-6}$$

The coefficients B', C', D', ... are once more dependent only on temperature. Both types of coefficients can be evaluated according to

$$B = \lim_{\rho \to 0} \left(\frac{\partial Z}{\partial \rho} \right)_T \tag{2.2-7}$$

$$C = \frac{1}{2!} \lim_{\rho \to 0} \left(\frac{\partial^2 Z}{\partial \rho^2} \right)_T \tag{2.2-8}$$

$$D = \frac{1}{3!} \lim_{\rho \to 0} \left(\frac{\partial^3 Z}{\partial \rho^3} \right)_T \quad \text{etc.} \tag{2.2-9}$$

where ρ is the density ($\rho = 1/V$).

$$B' = \lim_{P \to 0} \left(\frac{\partial Z}{\partial P} \right)_T \tag{2.2-10}$$

$$C' = \frac{1}{2!} \lim_{P \to 0} \left(\frac{\partial^2 Z}{\partial P^2} \right)_T \tag{2.2-11}$$

$$D' = \frac{1}{3!} \lim_{P \to 0} \left(\frac{\partial^3 Z}{\partial P^3} \right)_T \quad \text{etc.} \tag{2.2-12}$$

The two sets of coefficients A, B, C,.. and A', B', C',... can be interrelated. The derivative of the compressibility factor against pressure can be expressed as

$$\left(\frac{\partial Z}{\partial P}\right)_{T,\mathbf{x}} = \left[\frac{\partial Z}{\partial(1/V)}\right]_{T,\mathbf{x}} \left[\frac{\partial(1/V)}{\partial P}\right]_{T,\mathbf{x}} \tag{2.2-13}$$

Taking the derivative of the compressibility factor we get

$$\left[\frac{\partial(1/V)}{\partial P}\right]_{T,\mathbf{x}} = \frac{1}{ZRT}\left[1 - \frac{P}{Z}\left(\frac{\partial Z}{\partial P}\right)_{T,\mathbf{x}}\right] \tag{2.2-14}$$

Rearrangement of Eqs. (2.2-13) and (2.2-14) yields

$$\left(\frac{\partial Z}{\partial P}\right)_{T,\mathbf{x}} = \frac{\left[\dfrac{\partial Z}{\partial(1/V)}\right]}{ZRT + \dfrac{P}{Z}\left[\dfrac{\partial Z}{\partial(1/V)}\right]_{T,\mathbf{x}}} \tag{2.2-15}$$

Since $\lim_{\rho \to 0} P = 0$ and $\lim_{P \to 0} \rho = 0$, Eq. (2.2-15) reduces to

$$\left(\frac{\partial Z}{\partial P}\right)_{T,\mathbf{x},1/V=0} = \frac{1}{RT}\left[\frac{\partial Z}{\partial(1/V)}\right]_{T,\mathbf{x},P=0} \tag{2.2-16}$$

Finally, differentiation of both forms of the virial equation gives

$$B' = \frac{B}{RT} \tag{2.2-17}$$

$$C' = \frac{C - B^2}{(RT)^2} \tag{2.2-18}$$

$$D' = \frac{D - 3BC + 2B^2}{(RT)^3} \tag{2.2-19}$$

Relations (2.2-17)–(2.2-19) hold exactly only for expansions (2.2-5) and (2.2-6) as infinite Taylor series. In practice, virial coefficients calculated by Eqs. (2.2-7)–(2.2-9) may differ even for very accurate measurements by 5% from those computed from (2.2-10)–(2.2-12) and recalculated by Eqs. (2.2-17)–(2.2-19) (Scott and Dunlap, 1962). There are two reasons: the inaccuracy of experimental data very near zero pressure and the inability of the truncated virial equation to represent the *PVT* properties of real gases.

When applying the virial equation of state, one should remember that a function that is to be expanded into a McLaurin series must be differentiable at least as many times as the number of terms in the expansion.

Figures 2.2-1 and 2.2-2 show the isotherms of the compressibility factor of water as a function of pressure. A figure for constant densities is analogous. Water is a good example because its properties are very well known in the vapor as well as liquid phase.

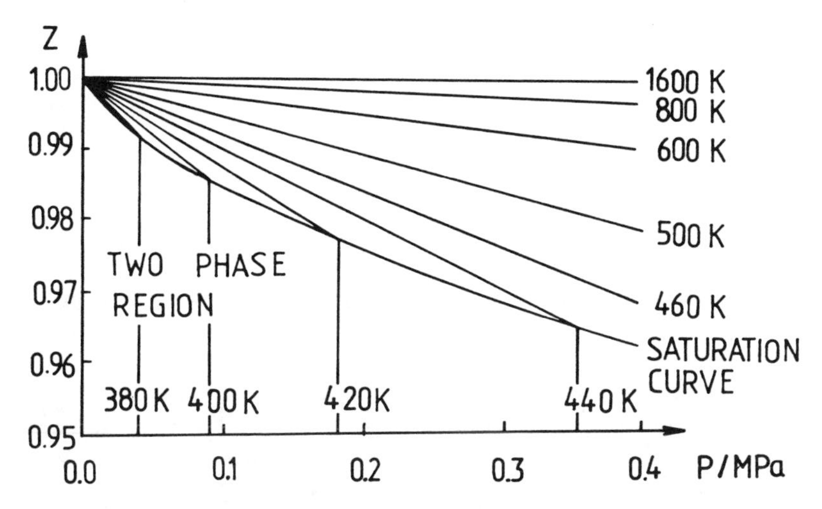

Figure 2.2-1
Phase diagram of water in the low-pressure region (Keenan et al., 1969).

In the low-pressure region the isotherms can be approximated by straight lines. Therefore, they can be represented by the virial equation of state. At higher pressures a more marked curvature of isotherms is observed. The curvature can be reproduced by additional terms of the virial equation. However, the compressibility factor is no longer differentiable at the border line between the one-phase and two-phase regions. This means that the virial expansion can be used only in the one-phase region. In fact, it is widely used for the vapor phase. Equations of state, which are applicable to one-phase as well as two-phase regions, will be described in Chapter 4.

In practice the truncated forms of virial equation are used. In this case the accuracy can be lower, but the simplicity of the equation compensates for it. For subatmospheric pressures the two-term equation is used:

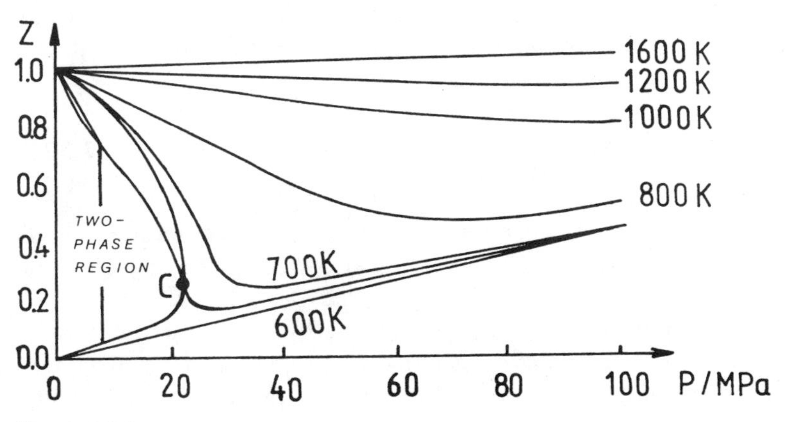

Figure 2.2-2
Phase diagram of water (Keenan et al., 1969).

$$\frac{PV}{RT} = 1 + \frac{B}{V} \tag{2.2-20}$$

For higher pressure the three term equation is used:

$$\frac{PV}{RT} = 1 + \frac{B}{V} + \frac{C}{V^2} \tag{2.2-21}$$

The second virial coefficients used in these equations should be computed from the data determined well below atmospheric pressure, while third coefficients from the data well below the critical point. Equations (2.2-20) and (2.2-21), representing the truncated forms of the virial equation, should be treated only as an approximation. They are rigorous only close to the zero pressure.

The experimental *PVT* data are relatively scarce. Comprehensive collections of second virial coefficient data have been prepared by Dymond and Smith (1980) and Choliński et al. (1986). There is no general collection of high-pressure *PVT* data. For several low molecular weight compounds the IUPAC Thermodynamic Tables are the most reliable source.

The limited volume of available virial coefficients data has led to the research on prediction methods. Many papers have been published on this subject. As the virial equation is used for the computation of fugacity coefficients of gases, the values of virial coefficients should be reliable and not introduce additional errors to the modelling of phase equilibria. It is much better to assume that the vapor phase is ideal than to obtain systematic errors. Due to this some earlier and more complex methods are no longer in use.

The early methods were based on the application of statistical-mechanical expressions derived for assumed intermolecular potentials (Hirschfelder et al., 1954). However, the results obtained were not satisfactory for the applications to phase equilibria modelling.

More reliable results were obtained by applying the principle of corresponding states. Pitzer and Curl (1957) have found that the reduced second virial coefficient ($B_R = B/V_c$ or $B_R = BP_c/RT_c$) is a linear function of Pitzer's acentric factor ω at constant reduced temperature:

$$B_R = \frac{BP_c}{RT_c} = B^{(0)} + \omega B^{(1)} \tag{2.2-22}$$

where $B^{(0)}$ represents the temperature dependence of the reduced virial coefficient for simple fluids ($\omega = 0$) and $B^{(1)}$ is a correction for the effect of acentricity.

For polar fluids, Eq. (2.2-22) can be extended by adding one more term ($B^{(2)}$) representing the influence of polarity as proposed by Tsonopoulos (1974, 1975). Instead of using the original Pitzer–Curl expression for $B^{(0)}$ and $B^{(1)}$, those proposed by Abbot (Smith and Van Ness, 1975) are recommended as much simpler. Finally we obtain

$$B_R = \frac{BP_c}{RT_c} = B^{(0)} + \omega B^{(1)} + B^{(2)} \tag{2.2-23}$$

where

$$B^{(0)} = 0.083 - \frac{0.422}{T_R^{1.6}} \tag{2.2-24}$$

$$B^{(1)} = 0.139 - \frac{0.172}{T_R^{4.2}} \tag{2.2-25}$$

$$B^{(2)} = \frac{a}{T_R^6} - \frac{b}{T_R^8} \tag{2.2-26}$$

where μ_R is the reduced dipole moment defined as

$$\mu_R = \frac{10^5 \mu^2 P_c}{T_c^2} \tag{2.2-27}$$

where μ is the dipole moment in Debyes.

Equation (2.2-23), due to its simplicity, can also be used for direct correlation of second virial coefficient data with two or three adjustable parameters.

The values of the constants *a* and *b* proposed for various compounds by Tsonopoulos are given in Table 2.2-2. The values computed for those compounds for which experimental data exist are given in Choliński et al. (1986). The average accuracy of prediction for all types of compounds is about $50\ cm^3/mol$, which is equivalent to about 5 to 10 percent of the actual value of the viral coefficient.

Other correlations for the second virial coefficients based on the corresponding-states principle and allowing for polarity effects have been developed by Black (1958), O'Connell and Prausnitz (1967), Hayden and O'Connell (1975) and Vetere (Reid et al., 1977). Another group of correlations is based on the chemical theory of gas phase imperfections (Nothnagel et al., 1973,

Table 2.2-2
VALUES OF *a* AND *b* for EQ. (2.2-26)

Compound Class	*a*	*b*
Ketones, aldehydes, nitriles, ethers, esters, NH3, H2S, HCN	$-2.112 \times 10^{-4}\mu_R - 3.877 \times 10^{-21}\mu_R^8$	0
Mercaptans	0	0
Monoalkilhalides	$2.078 \times 10^{-4}\mu_R - 7.048 \times 10^{-21}\mu_R^8$	0
Alcohols	0.0878	0.04–0.06
Phenol	−0.0136	0

From Tsonopoulos 1974, 1975.

Książczak and Anderko, 1987a,b, 1988). The latter authors developed a method for predicting second virial coefficients of polar and associating compounds that is based on vapor pressure data for the polar compounds and their homomorphs and does not require any specific adjustable parameters. The compilation of Choliński et al. (1986) compares the accuracy of some of the above methods (those of Tsonopoulos, O'Connell–Prausnitz, Hayden–O'Connell, Black, Vetere and Nothnagel et al.)

To avoid the necessity of introducing specific adjustable parameters for polar and associating compounds, some authors developed group contribution methods (McCann and Danner, 1984, Brandani et al., 1988, Abusleme and Vera, 1989). Among these methods, the method of McCann and Danner is once more based on the principle of corresponding states and was developed for reduced temperatures from 0.5 to 5.0. The Benson and Buss (1958) second-order additivity scheme was applied to split organic compounds into groups. The second virial coefficient B is represented as a sum over an overall number of groups of the primary $B^{(c1)}$ and, additionally for some groups, secondary contributions $B^{(c2)}$ according to the formulas

$$B_i = \sum_{i=1}^{n} m_i B_i^{(c1)} + \sum_{i=1}^{n} (m_i - 1)^2 B^{(c2)} \tag{2.2-28}$$

where each contribution is represented as a function of reduced temperature (T_R)

$$B_i^{(c)} = a_i + \frac{b_i}{T_R} + \frac{c_i}{T_R^3} + \frac{d_i}{T_R^7} + \frac{e_i}{T_R^9} \tag{2.2-29}$$

The table of parameters a, b, c, d and e for groups and for the ring or trans isomers is given by McCann and Danner (1984). The authors claim that the accuracy is similar to that obtained from the Tsonopoulos method.

A more comprehensive group contribution method for the prediction of second virial coefficients (including cross coefficients) has been proposed by Abusleme and Vera (1989). The authors represented the virial coefficient as a product of a nonpolar contribution B_0 and a polar one g^*:

$$B = B_0 g^* \tag{2.2-30}$$

The nonpolar contribution is calculated from the Pitzer–Curl–Tsonopoulos correlation. In the case of polar compounds a "nonpolar" acentric factor is used instead of the original acentric factor:

$$W = 0.006R_H + 0.02087R_H^2 - 0.00136R_H^3 \tag{2.2-31}$$

where R_H is the mean radius of gyration [used previously by Hayden and O'Connell (1975)]. The polar contribution is calculated from group contributions:

$$g^* = \exp(-\epsilon/T) \tag{2.2-32}$$

$$-\epsilon = \sum_{k}^{NG} \sum_{l=k}^{NG} \theta_k \theta_l \epsilon_{kl}^{G} \tag{2.2-33}$$

where θ_k is the group surface area fraction of group k, ϵ_{kl}^{G} is the $k-l$ group interaction parameter and NG is the number of groups composing a molecule. The group surface area is given by

$$\theta_k = \frac{\nu_k(zQ_k)}{\sum \nu_1(zQ_1)} \tag{2.2-34}$$

where

$$zQ_k = 0.4228V_k^* + (2 - J_k) \tag{2.2-35}$$

and J_k is the number of groups bounded to group k and V_k^* is the van der Waals volume. The temperature dependence of the ϵ_{kl}^{G} parameters is defined using two constants a_{kl} and b_{kl}, tabulated for 20 groups:

$$\epsilon_{kl}^{G} = \frac{298.15}{T^{0.5}} \left\{ a_{kl} + b_{kl}\left[\left(\frac{298.15}{T}\right)^{0.5} - 1\right]\right\} \tag{2.2-36}$$

The results obtained with this group contribution method compare favorably with those of correlations of Tsonopoulos (1974) and Hayden and O'Connell (1975). A comparison of the three methods is given in Table 2.2-3. For the Tsonopoulos and, in particular, for the Hayden–O'Connell methods, specific

Table 2.2-3
AVERAGE ABSOLUTE DERIVATIONS $\sigma(B)$ OF SECOND VIRIAL COEFFICIENTS PREDICTED FOR SEVERAL CLASSES OF COMPOUNDS BY MEANS OF THE TSONOPOULOS (1974), HAYDEN–O'CONNELL (1975) AND ABUSLEME–VERA (1989) CORRELATIONS

	$\sigma(B)^b$ (cm^3/mol)		
Class[a]	**Tsonopoulos**	**Hayden–O'Connell**	**Abusleme–Vera**
Alcohols, phenols (87, 9)	106	259	97
Ketones, ethers (77, 5)	46	66	41
Esters (96, 6)	57	36	67
Halogenated compounds (168, 10)	51	63	33
Sulfides, thiols	357	59	57
Amines (63, 5)	–	21	15
Pyridines (24, 4)	346	60	57
Others (111, 9)	187	119	22
Overall	167	94	50

[a] Numbers in parentheses indicate number of experimental points and number of sets, respectively.
[b] $\sigma(B) = [(1/n)\,\Sigma\,(B_1^{\text{cal}} - B_i^{\text{exp}})^2]^{1/2}$.

values of polar parameters have been used for several compounds instead of the generalized values (cf. Table 2.2-2). This increases the accuracy while reducing the predictive capability.

The Abusleme–Vera correlation can be viewed as a method to extend the predictive capability of the Pitzer–Curl–Tsonopoulos method to polar compounds. In comparison with the McCann–Dammer correlation the Abusleme–Vera method have fewer groups and group constants and is readily applicable to cross coefficients.

Prediction of third virial coefficients is much more difficult due to the limited database, which could have been used to establish reliable correlations. The correlation of Orbey and Vera (1983) can be used for this purpose.

2.3 LIQUID VOLUME

Another property used is the modelling of vapor–liquid equilibrium at low and moderate pressure is the volume of the saturated liquid phase. This property is applied in some equations representing the excess Gibbs energy of the liquid phase. For this purpose there is no necessity to have high accuracy.

For the correlation of experimental data resonable results can be obtained with the equation

$$V^L = a + bT + cT^2 \tag{2.3-1}$$

where a, b and c are adjustable parameters

For small temperature intervals or when few experimental points are available, the third term of Eq. (2.3-1) can be omitted.

There are various methods for the prediction of liquid molar volumes. Some of them can be used for the prediction of the densities of compressed liquids. They are reviewed in the two last editions of Reid et al. (1977, 1987). The methods proposed by Hankinson and Thomson (1979), Thomson et al. (1982), Rackett (1970), Spencer and Danner (1972) and Campbell and Thodos (1984, 1985) can be used for prediction of saturated molar volumes for polar as well as nonpolar compounds. The methods usually require the same input as the methods used for the prediction of second virial coefficients, i.e., the critical properties, the normal boiling point, etc.

Apart from the liquid volume, it is worthwhile to mention a simple relationship between the liquid and vapor volumes along the saturation curve. In 1886 Cailletet and Mathias formulated an empirical rule, known also as the rectilinear diameter rule, for the densities of coexisting phases:

$$\tfrac{1}{2}(\rho^L + \rho^G) = \rho_0 + aT \tag{2.3-2}$$

Equation (2.3-2) states that the saturated densities can be approximated by a parabolic curve in relation to temperature.

The Cailletet–Mathias rule is purely empirical but if frequently quite accurate. However, for larger temperature intervals a polynomial equation in terms of temperature is needed:

$$\tfrac{1}{2}(\rho^L + \rho^G) = \rho_0 + aT + bT^2 + cT^3 \tag{2.3-3}$$

This rule is convenient for the representation of saturated densities as a function of temperature and for the calculation of critical volumes. Its accuracy is quite high and even some extrapolation is feasible. However, it should be noted that relationships of this type are of very limited generality and cannot serve as a substitute for more elaborate equations of state.

REFERENCES

Abusleme, J. A. and J. H. Vera, 1989. *AIChE J.* 35: 481.

Ambrose, D. and R. H. Davies, 1980. *J. Chem. Thermodyn.* 12: 871.

Ambrose, D, and N. B. Ghiassee, 1987. *J. Chem. Thermodyn.* 19: 505.

Ambrose, D., J. F. Counsell and A. J. Davenport, 1970. *J. Chem. Thermodyn.* 2: 283.

Ambrose, D., C. H. S. Sprake and R. Townsend, 1974. *J. Chem. Thermodyn.* 6: 693.

Ambrose, D., C. H. S. Sprake and R. Townsend, 1975. *J. Chem. Thermodyn.* 7: 185.

Ambrose, D., J. F. Counsell and C. P. Hicks, 1978. *J. Chem Thermodyn.* 10: 771.

Angus, S., Ed., *International Thermodynamic Tables of Fluid State*, Vol. 1: *Argon* (1971), Vol. 2: *Ethylene* (1972), Vol. 3: *Carbon Dioxide* (1976), Vol. 4: *Helium*-4 (1977), Vol. 5: *Methane* (1978), Vol. 6: *Nitrogen* (1979), Vol. 7: *Propylene* (1980), Vol. 8: *Chlorine* (1985), Vol. 9: *Oxygen* (1987), Pergamon Press, Oxford, New York (continuous publication).

Angus, S., 1983. *Guide to the Preparation of Thermodynamic Tables and Correlation of Fluid State*, CODATA Bull., No. 51.

Antoine, C., 1888. *Compt. Rend. Acad. Sci., Paris* 107: 681, 836, 1143.

Benson, W. W. and J. H. Buss, 1958, *J. Chem. Phys.* 29: 546.

Black, C., 1958. *Ind. Eng. Chem.* 50: 391.

Boublik, T., V. Fried and E. Hala, 1973, 1987. *The Vapour Pressures of Pure Substances*, 2nd ed., Elsevier, Amsterdam.

Brandani, V., G. Di Giacomo and V. Mucciante, 1988. *Chem. Biochem. Eng. Q.*, 2: 69.

Cailletet, L. and E. Mathias, 1886. *J. Phys.* 5: 549.

Campbell, S. W. and G. Thodos, 1984. *Ind. Eng. Chem., Fundam.* 23: 500.

Campbell, S. W. and G. Thodos, 1985. *J. Chem. Eng. Data* 30: 102.

Chase, J. D., 1984. *Chem. Eng. Prog.* 80: 63.

Chase, J. D., 1987. *Ind. Eng. Chem. Res.* 26: 107.

Choliński, J., A. Szafranski and D. Wyrzykowska–Stankiewicz, 1986. *Com-*

puter-Aided Second Virial Coefficient Data for Organic Individual Compounds and Binary Systems, PWN–Polish Scientific Publishers, Warszawa.

Clapeyron, 1834. *J. l'Ecole Polytechn.* 14: 153.

Dykyj, J. and M. Repas, 1979. *Tlak Nasytenej Pary Organickych Zlucenin*, VEDA, Vydavatelstvo Slovenskej Akademie Ved, Bratislava CSRS.

Dykyj, J., M. Repas and J. Svoboda, 1984. *Tlak Nasytenej Pary Organickych Zlucenin*, Suppl., VEDA, Vydavatelstvo Slovenskej Akademie Ved, Bratislava CSRS.

Dymond, J. H. and E. B. Smith, 1980. *The Virial Coefficients of Gases and Gaseous Mixtures*, Clarendon Press, Oxford.

Frost, A. A., and D. R. Kalkwarf, 1953. *J. Chem. Phys.* 21: 264.

Gomez-Nieto, M. and G. Thodos, 1977. *Can. J. Chem. Eng.* 55: 445.

Gomez-Nieto, M. and G. Thodos, 1977. *Ind. Eng. Chem., Fundam.* 16: 254.

Gomez-Nieto, M. and G. Thodos, 1978. *Ind. Eng. Chem., Fundam.* 17: 45.

Gregorowicz, J., K. Kiciak and S. Malanowski, 1987. *Fluid Phase Equilibria* 38: 97.

Hankinson, R. W. and G. H. Thomson, 1979. *AIChE J.*, 25: 653.

Harlacher, E. A. and W. G. Braun, 1975, *Ind. Eng. Chem., Process Des. Develop.* 9: 479.

Hayden, J. G. and J. P., O'Connell, 1975. *Ind. Eng. Chem., Process Des. Develop.* 14: 209.

Hirschfelder, J. O., C. F. Curtiss and R. B. Bird, 1954. *Molecular Theory of Gases and Liquids*, Wiley, New York.

Horstmann, 1873. *Ann.* 170: 192.

Keenan, J. H., F. Keyes, P. Hill and J. Moore, 1969. *Steam Tables: Thermodynamic Properties of Water*, Wiley, New York.

King, M. B., 1976. *Trans. Inst. Chem. Eng.* 54: 54.

King, M. B. and H. Al-Najjar, 1974. *Chem. Eng. Sci.* 29: 1003.

Kręglewski, A., and B. J. Zwoliński, 1961. Roczniki Chem.; 35, 1041.

Książczak, A. and A. Anderko, 1987a. *Bull. Pol. Ac.: Chem.* 35: 61.

Książczak, A. and A. Anderko, 1987b. *Ber. Bunsenges. Phys. Chem.* 91: 1048.

Książczak, A. and A. Anderko, 1988. *Ber. Bunsenges. Phys. Chem.* 92: 496.

Lee, B. I., and M. G. Kesler, 1975. *AIChE J.* 21: 510.

Lencka, M., A. Szafrański and A. Mączyński, 1984. *Verified Vapor Pressure Data*, Vol. 1. *Organic Compounds Containing Nitrogen*, PWN–Polish Scientific Publishers, Warszawa.

Majer, V., and V. Svoboda, 1985. *Ethalpies of Vaporization of Organic Compounds, a Critical Review and Data Compilation*, Blackwell Scientific, Oxford.

McCann, D. W. and R. P. Danner, 1984, *Ind. Eng. Chem., Process Des. Dev.* 23: 529.

McGarry, J., 1983. *Ind. Eng. Chem., Process Des. Dev.* 22: 313.

Nothnagel, K. H. D. S. Abrams and J.M. Prausnitz, 1973. *Ind. Eng. Chem., Process Des. Dev.* 12: 25.

O'Connell, J. P. and J. M. Prausnitz, 1967. *Ind. Eng. Chem. Process Des. Dev.*, 6: 245.

Ohe, S., 1976. *Computer Aided Data Book of Vapor Pressure*, Data Book Publishing Company, Tokyo.
Orbey, H. and J. H. Vera, 1983. *AIChE J.*, 29: 107.
Partington, J. R., 1955. *An Advanced Treatise on Physical Chemistry* Vol. 2. *The Properties of Liquids*, Longmans, London.
Pitzer, K. S., 1955. *J. Am. Chem. Soc.* 77: 3427.
Pitzer, K. S. and R.F. Curl, 1957. *J. Am. Chem. Soc.* 79: 2369.
Plank, R. and L. Riedel, 1948. *Ing. Arch.* 16: 255.
Rackett, H. G., 1970. *J. Chem. Eng. Data* 15: 514.
Reid, R. C., J. M. Prausnitz and B.E. Poling, 1987. *The Properties of Gases and Liquids*, 4th ed., McGraw-Hill, New York.
Reid, R. C., J. M. Prausnitz and T.K. Sherwood, 1977. *The Properties of Gases and Liquids*, 3rd ed., McGraw-Hill, New York.
Riedel, L., 1954. *Chem. Ing. Technol.* 26: 83.
Scott, R. L. and R. D. Dunlap, 1962, *J. Phys. Chem.* 66: 639.
Smith, J. M. and H. C. Van Ness, 1975. *Introduction to Chemical Engineering Thermodynamics*, 3rd ed., McGraw-Hill, New York.
Smith, J. M. and H. C. Van Ness, 1988. *Introduction to Chemical Engineering Thermodynamics*, 4th ed., McGraw-Hill, New York.
Spencer, C. F. and R. P. Danner, 1972. *J. Chem. Eng. Data* 18: 230.
Stephenson, R. M. and S. Malanowski, 1987. *Handbook of Thermodynamics of Organic Compounds*, Elsevier, New York.
Thiesen, 1885. *Ann. Phys.* 24: 467.
Thodos, G., 1950. *Ind. Eng. Chem.* 42: 1514.
Thomson, G. H., K. R. Brobst, and R.W. Hankinson, 1982. *AIChE J.* 28; 671.
TRC, *Thermodynamic Tables, Hydrocarbons*, Engineering Experiment Station, Texas A&M University, College Station TX, updated half-yearly.
TRC, *Thermodynamic Tables, Non-Hydrocarbons*, Engineering Experiment Station, Texas A&M University, College Station TX, updated half-yearly.
Tsonopoulos, C., 1974. *AIChE J.* 20: 263.
Tsonopoulos, C., 1975. *AIChE J.* 21: 827.
Wagner, W., 1974. *Eine Neue Korrelationsmethode fur thermodynamische Daten angewendet auf die Dampfdruckkurve von Argon, Stickstoff und Wasser*, Fortschr. Ber. VDI-Z, Reiche 3, Nr. 39.
Waring, W., 1954. *Ind. Eng. Chem.* 46: 762.
Wilhoit R. C., 1976. *TRC Current Data News* 4(No. 6).
Wilhoit, R. C. and B. J. Zwolinski, 1971. *Handbook of Vapor Pressure and Heats of Vaporization of Hydrocarbons and Related Compounds*, Texas A & M University, TRC, College Station, TX.
Willingham C. J., W. J. Taylor, J. M. Pignocco and F. D. Rossini, 1945. *J. Res. Natl. Bur. Standards* 35: 219.
Zwolinski, B. J. and R. C. Wilhoit, 1966. *Proceedings Division of Refining (A.P.I.)* 46: 125.

3

The Gamma-Phi Method

3.1 THERMODYNAMIC FRAMEWORK OF VAPOR–LIQUID, LIQUID–LIQUID AND SOLID–LIQUID EQUILIBRIA

The gamma–phi method consists of calculating separately the activity coefficient of liquid phase (gamma) and the fugacity coefficient of the vapor phase (phi). This method was first developed for vapor–liquid equilibrium computations (Margules, 1895) but is general and can be used for the modelling of various types of phase equilibria.

To introduce the gamma–phi method of calculating phase equilibria, we should first define the concept of thermodynamic excess functions and relate them to the activity and fugacity coefficients.

For the representation of real mixtures the concept of an *excess property* proposed by Scatchard in 1936 is very useful. The thermodynamic excess properties, sometimes called the *excess functions of mixing*, are defined as the differences between the values of actual properties of a mixture and their values calculated for an ideal mixture of the same composition at the same pressure and temperature. Thus, by definition, we get for any property M:

$$M^E = M - M^{\mathrm{id}} \tag{3.1-1}$$

The excess functions (M^E) are a convenient measure of the deviations of any given mixture from the properties of the ideal mixture (M^{id}). The functional relations between excess functions are identical to those for corresponding thermodynamic properties of mixtures. For this reason they are widely used. The most widely used excess properties can be obtained by differentiation of the excess Gibbs energy. For computations it is convenient to use properties defined for one mole of a mixture (molar functions):

$$g^E = \frac{G^E}{\sum_{i=1}^{n} N_i} = u^E - Ts^E + Pv^E = h^E - Ts^E \tag{3.1-2}$$

$$s^E = -\left(\frac{\partial G^E}{\partial T}\right)_{\mathbf{x},P} \tag{3.1-3}$$

$$h^E = -T^2\left[\frac{\partial(g^E/T)}{\partial T}\right]_{\mathbf{x},P} \tag{3.1-4}$$

$$v^E = \left(\frac{\partial G^E}{\partial P}\right)_{\mathbf{x},T} \tag{3.1-5}$$

$$c_p^E = -T\left(\frac{\partial^2 G^E}{\partial T^2}\right)_{\mathbf{x},P} \tag{3.1-6}$$

For the purposes of modelling the thermodynamic properties of mixtures, it is convenient to use the activity coefficient γ_j of the substance j in a real n-component mixture. The equation defining the chemical potential is then

$$\mu_j(P, T, x_1, \ldots, x_{n-1}) = \mu_j^0(P, T) + RT \ln x_j\gamma_j(P, T, x_1, \ldots, x_{n-1}) \tag{3.1-7}$$

The function γ_j can be easily related to the excess chemical potential of component j, which is a partial molar property:

$$\mu_j^E = \left(\frac{\partial g^E}{\partial N_j}\right)_{N_{i\neq j},P,T} = RT \ln \gamma_j \tag{3.1-8}$$

Other partial derivatives of g^E are computed according to general thermodynamic relations:

$$s^E = \left(\frac{\partial g^E}{\partial T}\right)_{\mathbf{x},P} = -R\sum_{i=1}^{n} x_i \ln x_i - RT\sum_{i=1}^{n} x_i\left(\frac{\partial \ln \gamma_i}{\partial T}\right)_{\mathbf{x},P} \tag{3.1-9}$$

$$h^E = \left[\frac{\partial(g^E + T_S^E)}{\partial T}\right]_{\mathbf{x},P} = -RT^2\sum_{i=1}^{n} x_i\left(\frac{\partial \ln \gamma}{\partial T}\right)_{\mathbf{x},P} \tag{3.1-10}$$

$$c_p^E = \left(\frac{\partial h^E}{\partial T}\right)_{\mathbf{x},P} = -2RT\sum_{i=1}^{n} x_i\left(\frac{\partial \ln \gamma_i}{\partial T}\right)_{\mathbf{x},P} - RT^2\sum_{i=1}^{n} x_i\left(\frac{\partial^2 \ln \gamma_i}{\partial T^2}\right)_{\mathbf{x},P} \tag{3.1-11}$$

$$v^E = \left(\frac{\partial g^E}{\partial P}\right)_{\mathbf{x},T} = RT\sum_{i=1}^{n} x_i\left(\frac{\partial \ln \gamma_i}{\partial P}\right)_{\mathbf{x},T} \tag{3.1-12}$$

The equivalence between the excess molar Gibbs energy and activity coefficients is given by

$$g^E = \frac{G^E}{\sum_{i=1}^{n} N_i} = RT\sum_{i=1}^{n} x_i \ln \gamma_i \tag{3.1-13}$$

A converse transformation is given by

$$\ln \gamma_j = g^E + \left(\frac{\partial g^E}{\partial x_j}\right)_{x_{k\neq j}} - \sum_{i=1}^{n} x_i\left(\frac{\partial g^E}{\partial x_i}\right)_{x_{k\neq i}} \tag{3.1-14}$$

Equation (3.1-14) makes it possible to obtain the activity coefficients by differentiating g^E with respect to mole fractions rather than numbers of moles as in the definition (3.1-8).

The definition of the activity coefficient of a component in a mixture used in this book is analogous to that proposed by IUPAC (1979) with one significant difference. IUPAC defines activity coefficient as a number, we have defined it as a function. The treatment of activity coefficient as a function is very important for modelling the thermodynamic properties of real mixtures. It enables the empirical models to represent the dependence of activity coefficients not only on concentration but also on pressure and temperature.

The activity coefficient has been defined by Eq. (1.2-192) using the ideal liquid mixture as a reference state. On the other hand, the fugacity is introduced using the ideal gas as a reference state. According to Eq. (1.2-199) fugacity is given by

$$\mu_i = \mu_i^0 + RT \ln \frac{f_i}{f_i^0} \tag{3.1-15}$$

If we specify the reference state as the pure ideal gas state at temperature T, the reference fugacity becomes

$$f_i^0 = Px_i \tag{3.1-16}$$

As the properties of any real mixture approach the ideal gas properties in the limit of zero pressure, we obtain

$$\lim_{P \to 0} \frac{f_i}{Px_i} = 1 \tag{3.1-17}$$

Equation (3.1-17) serves as an inspiration for the definition of the dimensionless quantity called *fugacity coefficient*:

$$\phi_i = \frac{f_i}{Px_i} \tag{3.1-18}$$

The fugacity coefficient is a measure of the deviation of the real fluid behavior from the properties of an ideal gas.

In the gamma–phi method the definition of the fugacity coefficient (ϕ) is used to describe the properties of the gas phase whereas the definition of the activity coefficient (γ) is used for the liquid phase(s). The name of the method is derived from the γ and ϕ symbols.

VAPOR–LIQUID EQUILIBRIUM

To derive a general relation for the computation of vapor–liquid equilibria using the γ–ϕ approach, we can utilize the condition of equality of intensive properties in the coexisting phases. This means that the chemical potentials of any component should be equal in all coexisting phases. This condition can be written in terms of fugacities according to eq. (3.1-15):

$$f_i^L = f_i^v \tag{3.1-19}$$

where the superscripts v and L denote the vapor and liquid phases, respectively. The fugacity of the vapor can be expressed using the definition of the fugacity coefficient (3.1-18):

$$f_i^v = Py_i\phi_i \tag{3.1-20}$$

To evaluate the liquid phase fugacity f_i^L we can adopt as reference state the pure liquid i at equilibrium temperature T and pressure P. According to Eq. (P1.2-50) the fugacity can be related to the activity coefficient γ_i by

$$f_i^L = f_i^0 x_i \gamma_i \tag{3.1-21}$$

where f_i^0 is the fugacity of the pure substance at pressure P. This quantity should be expressed using the fugacity of the pure substance $f_i^0(P_i^0)$ at saturation pressure P_i^0, which is a more easily available variable:

$$f_i^0 = f_i^0(P) = P_i^0\left[\frac{f_i^0(P_i^0)}{P_i^0}\right]\left[\frac{f_i^0(P)}{f_i^0(P_i^0)}\right] \tag{3.1-22}$$

The ratio $f_i^0(P_i^0)/P_i^0$ is according to Eq. (3.1-18), equal to the fugacity coefficient of the pure substance ϕ_i^0 (either vapor or liquid as both the phases are in equilibrium at temperature T and pressure P_i^0). The ratio $f_i^0(P)/f_i^0(P_i^0)$ can be evaluated taking into account that

$$\left(\frac{\partial \ln f_i^0}{\partial P}\right)_T = \frac{v_i}{RT} \tag{3.1-23}$$

which is merely a consequence of the general thermodynamic relation $(\partial\mu_i/\partial P)_T = v_i$ and the definition of fugacity (3.1-16). Integrating Eq. (3.1-23) from P_i^0 to P and substituting the result into Eq. (3.1-22) we get

$$f_i^0(P) = P_i^0\phi_i^0 \exp\left(-\int_{P_i^0}^{P} \frac{v_i}{RT}\,dP\right) \tag{3.1-24}$$

The exponential factor is sometimes called the Poynting correction. Inserting Eq. (3.1-24) into (3.1-21) and, subsequently, into (3.1-19), a general equilibrium condition is obtained:

$$y_i P\phi_i = x_i\gamma_i P_i^0\phi_i^0 \exp\left(-\int_{P_i^0}^{P} \frac{v_i}{RT}\,dP\right) \tag{3.1-25}$$

or, in simplified notation,

$$y_i P\Phi_i = x_i\gamma_i P_i^0 \tag{3.1-26}$$

where Φ_i is a correction factor for vapor phase nonideality and the difference between the equilibrium pressure P and the pure component vapor pressure P_i^0:

$$\Phi_i = \frac{\phi_i}{\phi_i^0} \exp\left(-\int_{P_i^0}^{P} \frac{v_i}{RT}\, dP\right) \tag{3.1-27}$$

Taking into account that the sum of vapor phase mole fractions is equal to one:

$$\sum_j y_i = 1 \tag{3.1-28}$$

and substituting the values of y_j calculated from Eq. (3.1-26) into (3.1-28), we obtain an expression for the total pressure:

$$P = \sum_j \frac{x_j \gamma_j P_j^0}{\Phi_j} \tag{3.1-29}$$

The equation obtained can be used for modelling vapor–liquid equilibrium at low and moderate pressures. At high pressures the symmetrical definition of activity coefficients is not applicable to supercritical components and the unsymmetrical definition with hypothetical standard states should be employed. Usually, Eq. (3.1-29) is used in conjunction with the virial equation of state (truncated after the second term) for the vapor phase, which becomes impractical at high pressures. Moreover, this heterogeneous treatment (i.e., with different models for the liquid and vapor phases) cannot be expected to yield good results in the vicinity of the liquid–vapor critical region.

Equation (3.1-29) interrelates four thermodynamic variables: pressure P, temperature T, liquid phase composition vector $\mathbf{x}$ and vapor phase composition vector $\mathbf{y}$. If any two of these variables are known, the remaining two can be calculated by solving Eqs. (3.1-29) and (3.1-26). The possible choices of computation schemes are described in Table 3.1-1.

Table 3.1-1
POSSIBLE TYPES OF COMPUTATION FOR VAPOR–LIQUID EQUILIBRIUM MODELLING

Computation Type	Independent Variable	Dependent Variable
Bubble point, isothermal	T, $\mathbf{x}$	P, $\mathbf{y}$
Bubble point, isobaric	P, $\mathbf{x}$	T, $\mathbf{y}$
Dew point, isothermal	T, $\mathbf{y}$	P, $\mathbf{x}$
Dew point, isobaric	P, $\mathbf{y}$	T, $\mathbf{x}$
P–T flash	P, T, q[a]	$\mathbf{x}$, $\mathbf{y}$

[a] q = feed composition.

LIQUID–LIQUID EQUILIBRIA

In the case of liquid–liquid equilibrium, the chemical potentials in both coexisting liquid phases α and β are equal:

$$\mu_j^\alpha = \mu_j^\beta \tag{3.1-30}$$

Application of Eq. (3.1-7) to both phases leads to the equilibrium condition

$$x_j^\alpha \gamma_j^\alpha = x_j^\beta \gamma_j^\beta \tag{3.1-31}$$

for all components of the mixture.

SOLID–LIQUID EQUILIBRIA

To apply the general solubility equation (P1.2-62) to practical modelling tasks, the difference between the chemical potential of a pure subcooled liquid solute i and that of the solute in saturated solution $(\mu_i^{0L} - \mu_i^L)$ should be expressed through the activity coefficient. As the pure subcooled liquid has been chosen as the reference state (cf. Example 1.2-14), we obtain

$$\mu_i^{0L} - \mu_i^L = -RT \ln x_i \gamma_i \tag{3.1-32}$$

Therefore, the solubility equation becomes

$$-\ln x_i \gamma_i = \frac{\Delta h_f}{RT_m}\left(\frac{T_m}{T} - 1\right) - \frac{\Delta C_p}{R}\left(\frac{T_m}{T} - 1\right) + \frac{\Delta C_p}{R} \ln \frac{T_m}{T} \tag{3.1-33}$$

where Δh_f is the enthalpy of fusion at the melting temperature T_m and ΔC_p is the heat capacity of fusion at temperature T. As explained in Example 1.2-14 the terms including ΔC_p can be, to a good approximation, neglected. When applying Eq. (3.1-33), the temperature dependence of the activity coefficient γ_i is of prime importance.

It should be noted that Eq. (3.1-32) assumes that there are no transitions in the solid phase between the temperatures T and T_m. Modifications of Eq. (3.1-32) that are necessary to account for such transitions have been discussed by Preston et al. (1971) and Choi and McLaughlin (1983).

To apply Eqs. (3.1-29), (3.1-31) and (3.1-33) to practical modelling problems, it is only necessary to establish models describing with sufficient accuracy the saturation pressure of pure components as a function of temperature and the activity or fugacity coefficients of the vapor and liquid phases as functions of pressure, temperature and composition. Many models applying the gamma–phi method were proposed and are in use. In the following sections the most useful models will be described along with the criteria for their selection.

3.2 MODELLING THE VAPOR PHASE OF A MULTICOMPONENT MIXTURE

The fugacity coefficient of the vapor phase can be computed for a gaseous mixture of a known volume as a function of pressure, temperature and composition. For this purpose it is necessary to differentiate the Gibbs energy taking into account the relationship between the chemical potential and fugacity coefficient. An equation representing the chemical potential of component j in a mixture is then obtained:

$$\mu_j = \int_0^P \left(v_j - \frac{RT}{P} \right) dP + RT \ln y_j P + h_j^0 - Ts_j^0 \tag{3.2-1}$$

where v_j is the partial molar volume of component j in a mixture, h_j^0 and s_j^0 are the molar enthalpy and entropy of pure component j in the ideal gas state.

An equation representing the fugacity coefficient (ϕ_j) of component j in an n-component gaseous mixture as a function of pressure and temperature is obtained by rearranging Eq. (3.2-1)

$$\phi_j = \exp\left[\frac{1}{RT} \int_0^P \left(v_j - \frac{RT}{P} \right) dP \right] \tag{3.2-2}$$

This equation is convenient if the PVT relationship is represented by an equation explicit in volume, i.e., with P and T as independent variables. The volume-explicit virial equation of state is an example. In an analogous way an equation can be derived for calculating the fugacity coefficient with the use of volume and temperature as independent parameters. The latter case will be elaborated on in Chapter 4. A detailed description of the methods for the calculation of these quantities is given for both cases by Beattie (1949). The integration of (3.2-2) can be performed with the use of experimental *PVT* data for a gaseous mixture or by means of an equation of state representing with sufficient accuracy the thermal properties of a gaseous mixture. If fugacity coefficients are to be applied to vapor–liquid equilibrium calculations at low and moderate pressures, the virial equation of state truncated after the second term can be used.

The second virial coefficient $[B(T)]$ of an n-component gaseous mixture can be computed according to the formula

$$B(T) = \sum_{i=1}^{n} \sum_{k=1}^{n} y_i y_k B_{i,k}(T) \tag{3.2-3}$$

The function integrated in (3.2-2) for this case will be

$$v_j - \frac{RT}{P} = \frac{1}{RT} \left(B_{i,j} + \sum_{i=1}^{n} y_i d_{i,j} - \frac{1}{2} \sum_{i=1}^{n} \sum_{k=1}^{n} y_i y_k d_{i,k} \right) \tag{3.2-4}$$

where $d_{i,k} = 2B_{i,k} - B_{i,1} - B_{k,k}$ and $B_{i,k} = B_{k,i}$.

Integration of (3.2-4) according to formula (3.2-2) yields the fugacity coefficient (ϕ_j) of component j in an n-component gaseous mixture:

$$\phi_j = \exp\left[\frac{P}{RT}(B_{j,j} - D_j^E)\right] \tag{3.2-5}$$

The function D_j^E represents the deviation from linearity of the mixed virial coefficients as a function of composition.

$$D_j^E = \sum_{i=1}^{n} y_i B_{i,j} - \frac{1}{2}\sum_{i=1}^{n}\sum_{k=1}^{n} y_i y_k B_{i,k} \tag{3.2-6}$$

For the computation of vapor–liquid equilibrium not only the volume of the vapor but also the volume (V^L) and compressibility κ [eq. (1.2-148)] of the condensed phase should be considered. The last two quantities are necessary to calculate the integral appearing in Eq. (3.1-27). Substituting Eqs. (3.2-4)–(3.2-6) into Eq. (3.1-27) and integrating, we obtain an expression for the correction term Φ_j:

$$\Phi_j = \exp\left[\frac{(B_{j,j} - v^{0,L})(P - P_j^0)}{RT} + \frac{D_j^E P}{RT} + \frac{\kappa_j(P^2 - P_j^{0\,2})}{2RT}\right] \tag{3.2-7}$$

This is a basic quantity in the computation of vapor–liquid equilibrium by the gamma–phi method. For the low-pressure region where this method is normally used, the last term representing the properties of the liquid phase can be omitted as a small one. The first term corresponds to an ideal mixture of nonideal gases. A detailed discussion of various cases can be found in the paper by Beattie (1949). The experimental information about gaseous mixtures is rather scarce. Collections of second cross virial coefficient data (B_{ij} for $i \neq j$) can be found in Warowny and Stecki (1979), Dymond and Smith (1980) and Choliński et al. (1986).

Correlations for the estimation of cross second virial coefficients are identical to those used for the virial coefficients of pure gases (cf. Section 2.2) but require combining rules for the characteristic parameters.

For the Tsonopoulos (1974, 1975) correlation the pseudocritical constants for each binary pair take the form:

$$T_{c_{ij}} = (T_{c_i} T_{c_j})^{1/2}(1 - k_{ij}) \tag{3.2-8}$$

$$P_{c_{ij}} = \frac{z_{c_{ij}} R T_{c_{ij}}}{(\frac{1}{8})(V_{c_i}^{1/3} + V_{c_j}^{1/3})^3} \tag{3.2-9}$$

$$\omega_{ij} = \frac{\omega_i + \omega_j}{2} \tag{3.2-10}$$

$$z_{c_{ij}} = \frac{z_{c_i} + z_{c_j}}{2} \tag{3.2-11}$$

where k_{ij} is a characteristic constant. In general, it can be derived from experimental cross virial coefficient data. For components of similar size and shape, $k_{ij} \cong 0$. For mixtures of polar and nonpolar components the polar parameters a and b are equal to zero. For mixtures of two polar components

$$a_{ij} = \frac{a_i + a_j}{2} \tag{3.2-12}$$

$$b_{ij} = \frac{b_i + b_j}{2} \tag{3.2-13}$$

In the case of the Abusleme–Vera group contribution method (1989[b]) the combining rules (3.2-8)–(3.2-10) are also used for the nonpolar contribution B_0. For the polar interaction parameter ϵ_{ij} the geometric mean rule is assumed:

$$\epsilon_{ij} = (\epsilon_{ij}\epsilon_{jj})^{0.5} \tag{3.2-14}$$

The term ϵ_{ij} vanishes if one of the components is nonpolar.

In general, the predictive capability of the above methods is limited by the necessity to determine a binary constant. Therefore, the accuracy of estimating second cross virial coefficients is not very high in the absence of experimental B_{ij} data. However, the effect of the B_{ij} value on the calculated fugacity coefficient is small. The first term on the right-hand side of Eq. (3.2-7) is usually sufficient for the calculation of ϕ_j with satisfactory accuracy.

3.3 MODELLING THE LIQUID PHASE OF A MULTICOMPONENT MIXTURE

3.3.1. Introduction

For the modelling of the thermodynamic properties of a multicomponent liquid phase by the gamma–phi method, it is necessary to establish equations describing the excess Gibbs energy (G^E) as a function of mixture composition, temperature and pressure. For this function an assumption can be made that g^E of a mixture is split into terms corresponding to binary ($g^E_{i,j}$), ternary ($g^E_{i,j,k}$) and more component systems formed by all possible components present in the mixture under consideration. As a result an equation is obtained:

$$\frac{G^E}{NRT} = g^E = \sum_{i=1}^{n-1} \sum_{j=i+1}^{n} g^E_{i,j} + \sum_{i=1}^{n-2} \sum_{j=i+1}^{n-1} \sum_{k=j+1}^{n} g^E_{i,j,k} + \cdots \tag{3.3-1}$$

Besides pure component properties it is necessary to have separate equations describing all binary, ternary and higher-order terms. These equations can be established either with the use of experimental data for binary, ternary and more component mixtures or on the basis of theoretical considerations. Such a complex description is unrealistic. The number of data increases for each term with the second power of the number of components in the mixture. For the description of a quaternary mixture it is necessary to have 6 binary, 4 ternary and 1 quaternary data sets. For a 10-component mixture the total number of systems involved is 1013 and for 20-component 1,048,555 separate data sets are necessary.

Consequently, reduction of the information involved is unavoidable. The best situation from the point of view of data handling would be when only information on pure component properties were satisfactory. This is impossible within the gamma–phi method when the pure component properties are used only to define a hypothetical ideal mixture. A possible way to represent G^E of a liquid mixture containing more than three components is to truncate Eq. (3.3-1) to

$$\frac{G^E}{NRT} = g^E = \sum_{i=1}^{n-1} \sum_{j=1+1}^{n} g_{i,j}^E \tag{3.3-2}$$

Values of the binary terms can be computed either from binary data or by a simultaneous reduction of binary and ternary or even more component data. Experimental phase equilibrium data are always subject to errors. Measurement of concentration is, in general, less accurate for a multicomponent mixture than for a binary one. Additional errors are introduced by the limited ability of a model equation to represent G^E as a function of mixture composition. Due to all these reasons better results are usually obtained by reduction of only binary data. It is necessary to remember that this is not a general rule. There are cases when better results can be obtained by reduction of multicomponent data.

There is no reliable collection of parameters for equations representing the excess Gibbs energy of a binary mixture. To obtain the parameters, it is necessary to perform a reduction of experimental data. There are several collections of experimental vapor–liquid equilibrium data. Some of them report data together with the results of some correlations (Gmehling et al., 1977–, Hella et al., 1968; Hirata et al., 1975; Mączyński et al., 1976–1988) and some only raw experimental data (Kogan et al., 1966; Landolt–Börnstein Tables, 1977; Timmermans, 1959). The *Vapor–Liquid Equilibrium Data Bibliography* by Wichterle et al. (1973), with supplements published each third year, is very useful. There are also data distributed on magnetic tape (the *Dortmund Data Bank* of Gmehling et al., 1977–) or floppy disks (the *Floppy Books* of Mączyński et al., 1988–). Such data can be directly transferred to a computer for further processing.

Any equation representing the Gibbs excess energy as a function of mixture composition and equilibrium temperature can be used for the binary term in (3.3-2). In principle, various equations can be used to represent different binary interaction terms $g_{i,j}$. There is only one condition, i.e., the proper handling of concentration variables and units. Special care should be taken when mole fractions are used as concentration variables. There is no evidence that the use of various equations leads to increased accuracy, and due to this it is much safer to represent all binary terms in (3.3-2) by one equation and to use the same number of adjustable parameters for all binaries. The widely used equations are, in principle, functions of concentration. Their dependence on temperature is not very well established and, therefore, they can be used only in a narrow temperature range. For a wide temperature range the equation-of-state technique is recommended. Equation (3.3-2) is written for the excess Gibbs energy. However, the corresponding partial molar properties like the chemical potential and its dimensionless equivalent (i.e. activity coefficient) are more useful for the purpose of modelling equilibrium data. They can be obtained by a Legendre transformation as explained in Chapter 1.

Not all equations representing $g_{i,j}^E$ as a function of temperature and composition are suitable for the use in Eq. (3.3-2). Some of them were proposed only for binary mixtures and their extension to multicomponent systems is not justified. It may be very difficult to choose an equation. The simplest criterion for such a selection is the comparison of the values of differences between the computed and measured pressure, temperature and phase composition. This can be sometimes misleading. The number of adjustable parameters should be in a realistic proposition to the number of experimental points. Another factor is the ability of an equation to represent the proper shape of the $g_{i,j}^E(x, T)$ curve. It may happen that an equation exhibits extrema or inflection points for which no experimental evidence exists. This is very dangerous and may lead to serious errors.

Here, general advice can be given. For the representation of a multicomponent mixture a model should be chosen that enables the pressure, temperature and composition of coexisting phases to be represented within the experimental errors. The errors should be estimated in a realistic way. It should be taken into account that authors usually tend to overestimate the accuracy of measurements. On the other hand, the correlation should be made in such a way that no information obtained from the measurements should be lost in the process of data reduction. It is almost impossible to formulate general criteria for phase equilibrium data reduction. It is an art and should be performed by a qualified person with some experience in thermodynamics.

Many equations have been proposed for the representation of $g_{i,j}^E$ for a binary mixture. A general review of such equations would be very large. Therefore, only the most important ones will be described in the following sections.

3.3.2. Polynomial Equations

No general theory exists that would adequately describe the composition dependence of the excess thermodynamic properties of liquid mixtures. Therefore, this functional dependence is commonly represented by various empirical equations. The oldest and still very popular ones are those proposed in 1895 by Margules and in 1910 by Van Laar. Scatchard and Hamer (1935) extended the number of adjustable parameters of the Margules equation, and Redlich and Kister (1948) proposed its most popular form for a binary mixture of components i and j:

$$g_{i,j}^E = x_i x_j \sum_{k=0}^{m} A_k z^k \tag{3.3-3}$$

where $z \equiv x_i - x_j$.

The Van Laar equation was generalized in 1964 by Van Ness (cf. Van Ness and Abbott, 1982)

$$g_{i,j}^E = \frac{x_i x_j}{\sum_{k=0}^{m} A_k z^k} \tag{3.3-4}$$

Orthogonal polynomials and spline functions have been also proposed for the representation of excess functions as they are very flexible and do not introduce model errors. This, however, has not been proven. As a matter of fact the Redlich–Kister equation is very similar to an orthogonal polynomial. There were many similar equations proposed by various authors. Some of them have been described in textbooks on the thermodynamics of phase equilibria (Hala et al., 1967; Van Ness and Abbott, 1982). In general, the systems that are well described by the Margules-type equation give poor results with the Van Laar equation and vice versa. All these equations can be reduced to a quotient of two polynomials (Malanowski, 1974; cf. Van Ness and Abbott, 1982):

$$g_{i,j}^E = x_i x_j \frac{\sum_{k=1}^{m} A_k z^k}{1 + \sum_{k=1}^{l} B_k z^k} \tag{3.3-5}$$

where A, B = adjustable parameters, and m, l = variable numbers of polynomial terms.

The main shortcoming of higher-order equations of this type is their tendency to describe experimental random errors and show unjustified oscillations, even inflections points, with small root-mean-square deviations at the experimental points. These oscillations propagate when the equation is extended to multicomponent mixtures by means of Eq. (3.3-2). Therefore, only one-parameter equations of the (3.3-5) type can be applied to represent multicomponent mixtures. Malesiński (1965) proved that an equation of the (3.3-5) type can be used for the accurate representation of multicomponent azeotropic mixtures with only one adjustable parameter. There is no general relation describing the dependence of adjustable parameters as a function of temperature. The polynomial equations are incapable of representing polythermal data. This type of equation is very useful for smoothing experimental data and analyzing the thermodynamic consistency of isothermal vapor–liquid equilibrium data, especially in the cases when the experimental errors are well known.

When using the polynomial equations for smoothing experimental excess function data, the number of adjustable parameters has to be judiciously

Table 3.3-1
CORRELATION OF TWO SETS OF VAPOR–LIQUID EQUILIBRIUM DATA FOR THE CHLOROBENZENE–HEXANE SYSTEM AT 338.20 K BY MEANS OF THE REDLICH–KISTER EXPANSION [EQ. (3.3-3)] WITH THREE OR FOUR ADJUSTABLE CONSTANTS

	Redlich–Kister Constants				
Data	A_1	A_2	A_3	A_4	$\Delta P \times 10^5$/kPa
Brown (1952)	0.5768	0.0275	−0.0150	—	357
	0.5752	0.2910	−0.0178	−0.0231	319
Rogalski and	0.5782	0.0323	−0.0195	—	73
Malanowski (1980)	0.5882	0.0367	0.013	0.0394	49

chosen to reflect the accuracy of the data. This is illustrated in Table 3.3-1, which compares the accuracy of correlating two sets of isothermal vapor–liquid equilibrium data for the chlorobenzene–hexane system measured by Brown (1952) and Rogalski and Malanowski (1980). In both cases the Redlich–Kister expansion [Eq. (3.3-3)] was used, and the number of adjustable constants was varied to obtain the best fit.

As shown in Table 3.3-1, the accuracy of the fit is markedly different for both data sets. In the case of Brown's (1952) data introduction of the fourth adjustable constants practically does not improve the correlation. Three parameters are sufficient for these data. On the other hand, the more precise data of Rogalski and Malanowski are better reproduced by a four-term polynomial (the fourth term reduces the deviation by 33 percent). In this case the use of four terms is more justified. In general, a polynomial equation with a properly chosen number of coefficients helps estimate the precision of measurements.

However, there is always a danger in using polynomial equations arising from increasing the number of adjustable parameters. Namely, too small deviations of the fit can be easily obtained. The error of the fit should always be greater or equal to the properly estimated experimental error.

3.3.3. Local Composition Models

In 1964 Wilson proposed a new equation for $g^E_{i,j}$. In analogy to volume fractions he introduced the *local mole fractions* of component i in a mixture of components i and j:

$$\xi_i = \frac{v_i x_i \exp(-\lambda_{i,i}/RT)}{v_i x_i \exp(\lambda_{i,i}/RT) + v_j x_j \exp(-\lambda_{j,i}/RT)} \tag{3.3-6}$$

where $-\lambda_{i,j}$ are adjustable parameters related to the potential energy of a pair of molecules introduced instead of overall segment fractions to the Flory (1941, 1942) and Huggins (1941, 1942) equation.

$$g^E = \sum_{i=1}^{n} x_i \ln \frac{\xi_i}{x_i} \tag{3.3-7}$$

The Flory–Huggins equation was derived originally from a lattice model to approximate the behavior of athermal polymer mixtures.

To simplify notation a new parameter was defined:

$$\Lambda_{i,j} = \frac{v_i}{v_j} \exp\left(-\frac{\lambda_{i,j} - \lambda_{i,i}}{RT}\right) \tag{3.3-8}$$

where $\Lambda_{i,i} = 1$.

The Wilson equation for an n-component mixture finally becomes

$$g^E = -\sum_{i=1}^{n} x_i \ln\left(\sum_{j=1}^{n} x_j \Lambda_{i,j}\right) \tag{3.3-9}$$

An equation for the activity coefficient γ_i can be obtained by a transformation of (3.3-9) according to (3.1-14):

$$\ln \gamma_i = -\ln\left(\sum_{j=1}^{n} x_j \Lambda_{i,j}\right) + 1 - \sum_{k=1}^{n} \frac{x_k \Lambda_{k,i}}{\sum_{l=1}^{n} x_l \Lambda_{k,l}} \tag{3.3-10}$$

Numerical values of the parameters $\lambda_{i,j} - \lambda_{i,i}$ can be found only by the reduction of experimental phase equilibrium data. The theoretical background of the equation is too weak to establish the values of the parameters from molecular considerations.

The Wilson equation has two advantages. First, the parameters $(\lambda_{i,j} - \lambda_{i,i})/RT$ can be treated as temperature independent for narrow ranges of temperature. This enables γ_i to be computed not only from isothermal but also isobaric data or even those taken at various pressures and temperatures. Second, this is a very flexible equation and is able to represent various mixtures, even those exhibiting strong deviations from ideality and an asymmetry of the g^E curve. The error introduced by the Wilson equation is of the order of 0.5 to 2 percent except for systems exhibiting extrema on the activity coefficient versus composition curve, which cannot be reproduced.

The Wilson equation is also unable to represent the phase splitting and therefore should be used only for liquid systems that are completely miscible.

To overcome this difficulty Renon and Prausnitz proposed in 1968 the nonrandom two-liquid (NRTL) equation. This model is based on a "local composition" concept similar to that used by Wilson (1964) and the "two-liquid" model proposed by Scott (1956). Once more the theoretical background is too weak to establish the values of parameters on the basis of molecular considerations. There are three adjustable parameters for each binary system $g_{j,i} - g_{i,i}$ and $\alpha_{j,i}$. The numerical values of parameters can be established only by the reduction of experimental phase equilibrium data.

For an n-component mixture the NRTL equation is

$$g^E = \sum_{i=1}^{n} x_i \frac{\sum_{j=1}^{n} \tau_{j,i} G_{j,i} x_j}{\sum_{l=1}^{n} G_{l,i} x_l} \tag{3.3-11}$$

where

$$\tau_{j,i} = \frac{g_{j,i} - g_{i,i}}{RT} \tag{3.3-12}$$

$$G_{j,i} = \exp(-\alpha_{j,i}\tau_{j,i}) \qquad \alpha_{j,i} = \alpha_{i,j} \tag{3.3-13}$$

The activity coefficient is given by

$$\ln \gamma_i = \frac{\sum_{j=1}^{n} \tau_{j,i} G_{j,i} x_j}{\sum_{l=1}^{n} G_{l,i} x_l} + \sum_{j=1}^{n} \frac{x_j G_{i,j}}{\sum_{l=1}^{n} G_{l,i} x_l} \left(\tau_{i,j} - \frac{\sum_{k=1}^{n} \tau_{k,j} G_{k,j} x_k}{\sum_{l=1}^{n} G_{l,j} x_l} \right) \tag{3.3-14}$$

The physical significance of $g_{j,i}$ is similar to that of $\lambda_{j,i}$ in the Wilson equation. It is an adjustable parameter having the dimension of energy units and is characteristic for the $i - j$ interaction. The parameter $\alpha_{j,i}$ is related to the value of the $g^E_{j,i}(x)$ curve at maximum. Its typical values are between -1 and 0.5. The value of this parameter is often set arbitrarily, typically $\alpha_{j,i} = 0.3$ or -1. For $\alpha_{j,i} = 0$ the NRTL equation reduces to a one-term polynomial equation.

For moderately nonideal binary mixtures both Wilson and NRTL equations offer no advantages over the simpler polynomial equations. However, for strongly nonideal mixtures, and especially for partially miscible systems, the NRTL equation often provides a good representation of experimental data (Renon et al., 1971).

The parameters $(g_{j,i} - g_{i,i}/RT)$ are similar to those of the Wilson equation and can be treated as independent of temperature within narrow temperature ranges. For wider temperature ranges Renon et al. introduced a linear temperature dependence of the form

$$\tau_{j,i} = \tau^{(0)}_{j,i} + \tau^{(T)}_{j,i}(T - 273.15) \tag{3.3-15}$$

$$\alpha_{j,i} = \alpha^{(0)}_{j,i} + \alpha^{(T)}_{j,i}(T - 273.15) \tag{3.3-16}$$

The introduction of temperature-dependent parameters raises the number of empirical constants from 3 to 6 for each binary mixture. There is no theoretical justification for the above equations. It cannot be proved experimentally in a general way owing to the lack of polythermal data. Substitution of (3.3-15) into (3.3-11) and differentiation according to the Gibbs–Helmholtz formula (1.2-82) yields an equation for the excess enthalpy:

$$h^E = \sum_{i=1}^{n} \frac{x_i}{\sum_{k=1}^{n} x_k G_{k,i}} \sum_{j=1}^{n} x_j G_{j,i}$$

$$\left\{ (\tau_{j,i}^{(0)} - 273.15\tau_{j,i}^{(T)}) \left[1 - \alpha_{i,j} \left(\tau_{j,i} - \frac{\sum_{k=1}^{n} \tau_{k,i} G_{k,i} x_k}{\sum_{k=1}^{n} G_{k,i} x_k} \right) \right] \right.$$

$$\left. + RT^2 \alpha_{j,i}^{(T)} \tau_{j,i} \left(\tau_{j,i} - \frac{\sum_{k=1}^{n} \tau_{k,i} G_{k,i} x_k}{\sum_{k=1}^{n} G_{k,i} x_k} \right) \right\} \quad (3.3\text{-}17)$$

To combine the advantages of the Wilson and NRTL equations, Abrams and Prausnitz (1975) proposed a two-parameter equation that, in a sense, extends the quasi-chemical theory of Guggenheim (1952). Owing to this, the proposed equation was called UNIversal QUAasi-Chemical or, in short, UNIQUAC. This equation consists of two parts, a combinatorial (COM) part that attempts to describe the entropic contribution and a residual (RES) part that is to describe the energetic contribution. The UNIQUAC equation for the excess Gibbs energy is

$$g^E = g^{E(\text{COM})} + g^{E(\text{RES})} \quad (3.3\text{-}18)$$

The combinatorial part describes the interactions of molecules of the same type but different sizes and shapes, while the residual part is related to the interaction between molecules of a different chemical nature. The combinatorial part depends only on the pure component properties. The residual part depends on the properties of a binary mixture.

For a multicomponent mixture the UNIQUAC equation is given by the sum of the following terms:

$$g^{E(\text{COM})} = \sum_{i=1}^{n} x_i \ln \frac{\Phi_i}{x_i} + \frac{Z}{2} \sum_{i=1}^{n} q_i x_i \ln \frac{\theta_i}{\Phi_i} \quad (3.3\text{-}19)$$

$$g^{E(\text{RES})} = -\sum_{i=1}^{n} q_i x_i \ln\left(\sum_{j=1}^{n} \theta_j \tau_{j,i} \right) \quad (3.3\text{-}20)$$

where the coordination number, $Z = 10$, the energy parameter, $\tau_{j,i}$, the segment fraction Φ_i and area fraction θ_i are given by

$$\tau_{j,i} = \frac{u_{i,j} - u_{i,i}}{RT} \quad (3.3\text{-}21)$$

$$\Phi_i = \frac{q_i x_i}{\sum_{j=1}^{n} q_j x_j} \quad (3.3\text{-}22)$$

$$\theta_i = \frac{r_i x_i}{\sum_{j=1}^{n} r_j x_j} \tag{3.3-23}$$

The parameters r_i and q_i are calculated as the sum of the group volume (R) and area parameters (Q):

$$r_i = \sum_{j=1}^{k} n_j R_j \tag{3.3-24}$$

$$q_i = \sum_{j=1}^{k} n_j Q_j \tag{3.3-25}$$

where n is the number of groups of type j in the molecule of component i in which k various groups can be distinguished.

Group parameters R_j and Q_j are obtained from the van der Waals group volume and surface areas V_{wk} and A_{wk} given by Bondi (1968):

$$R_j = \frac{V_{wk}}{15.17} \tag{3.3-26}$$

$$Q_j = \frac{A_{wk}}{2.5 \times 10^9} \tag{3.3-27}$$

The values obtained are valid for $Z = 10$. The activity coefficients can be computed by

$$\ln \gamma_i = \ln \gamma_i^{\mathrm{COM}} + \ln \gamma_i^{\mathrm{RES}} \tag{3.3-28}$$

where γ_i^{COM} denotes the combinatorial and γ_i^{RES} the residual contribution to the activity coefficient of component i in an n-component mixture.

$$\ln \gamma_i^{\mathrm{COM}} = \ln \frac{\Phi_i}{x_i} + \frac{Z}{2} q_i \ln \frac{\theta_i}{\Phi_i} + l_i - \frac{\Phi_i}{x_i} \sum_{j=1}^{n} x_j l_j \tag{3.3-29}$$

$$\ln \gamma_i^{\mathrm{RES}} = q_i \left[1 - \ln\left(\sum_{j=1}^{n} \theta_j \tau_{j,i} \right) - \sum_{j=1}^{n} \frac{\theta_j \tau_{i,j}}{\sum_{k=1}^{n} \theta_k \tau_{k,j}} \right] \tag{3.3-30}$$

where $l = Z/2(r_i - q_i) - (r_i - 1)$.

The UNIQUAC equation is more complicated and slightly less accurate than the NRTL and Wilson equations. The advantage of the equation is its flexibility. It can be used to represent g^E of multicomponent mixtures formed by binary systems exhibiting different types of interactions between molecules.

Both NRTL and UNIQUAC can be used to describe liquid–liquid equilibrium (LLE). In general, the parameters generated from vapor–liquid equilibrium (VLE) data do not describe properly LLE and vice versa. The values of activity coefficients generated from LLE are always too high to represent VLE in the homogeneous region, and those computed from VLE lead to a much narrower immiscibility gap.

According to (1.2-82) differentiation of the UNIQUAC equation (3.3-18) with respect to temperature on the assumption that the energetic parameters $\lambda_{i,j} - \lambda_{i,i}$ are temperature independent leads to the UNIQUAC equation for the heat of mixing:

$$h^E = R \sum_{i=1}^{n} \left\{ \frac{q_i x_i}{\sum_{j=1}^{n} \theta_j \tau_{j,i}} \left[\sum_{j=1}^{n} \theta_j \tau_{j,i} (\lambda_{i,j} - \lambda_{i,i}) \right] \right\} \tag{3.3-31}$$

Equation (3.3-31) should be treated carefully. The assumption that the energetic parameters of the UNIQUAC equation are independent of temperature is not justified in many cases. The obtained values of G^E and H^E should be always checked for consistency with experiment. When consistent results are obtained, the above simple relations enabling H^E and G^E data to be treated simultaneously can be useful. When this is not the case temperature-dependent energetic parameters can be introduced:

$$\tau_{j,i}^{(T)} = \exp \frac{\tau_{j,i}^{(0)} + \tau_{j,i}^{(1)}/T}{T} \tag{3.3-32}$$

The corresponding equation for H^E is then

$$h^E = R \sum_{i=1}^{n} \left\{ \frac{q_i x_i}{\sum_{j=1}^{n} \theta_j \tau_{j,i}^{(T)}} \left[\sum_{j=1}^{n} \theta_j \tau_{j,i}^{(T)} \left(\frac{\tau_{i,j}^{(0)} - \tau_{j,i}^{(1)}/T}{T} \right) \right] \right\} \tag{3.3-33}$$

Frequently, it is necessary to use correlation equations with only one binary parameter because only one meaningful data point exists or the quality of data does not justify using more than one parameter. In such cases it is advantageous to use one-parameter simplified versions of the local composition equations. The Wilson equation can be simplified by assuming that $\lambda_{i,j} = \lambda_{j,i}$ in Eq. (3.3.3-3) and that

$$\lambda_{i,i} = -\beta(\Delta H_{vi} - RT) \tag{3.3-34}$$

where β is a proportionality factor and ΔH_{vi} is the enthalpy of vaporization of pure component i. The proportionality factor β is typically set equal to $2/Z$ where Z is the coordination number (Wong and Eckert, 1971). It is usually assumed that $Z = 10$. Ladurelli et al. (1975) have suggested that $\beta = 2/Z$ for component 2, having the smaller molar volume, while for component 1, having the larger molar volume, $\beta = (2/Z)(v_2^L/v_1^L)$. This suggestion was based on some considerations of the $\lambda_{i,i}$ parameters as interaction energies per segment.

When β is fixed, the only binary parameter is $\lambda_{i,j}$. Bruin and Prausnitz (1971) also used Eq. (3.3-34) for the $g_{i,i}$ parameters in the NRTL equation. Additionally, a fixed value was used for the $\alpha_{i,j}$ parameter.

Abrams and Prausnitz (1975) reduced the UNIQUAC equation to a one-parameter form by assuming that the energetic parameters $u_{i,i}$ in Eq. (3.3-21) are given by

$$u_{i,i} = \frac{-\Delta U_i}{q_i} \tag{3.3-35}$$

whereas the cross-interaction energy is expressed by

$$u_{i,j} = u_{j,i} = (u_{i,i}u_{j,j})^{1/2}(1 - c_{i,j}) \tag{3.3-36}$$

where ΔU_i can be approximated by $\Delta H_{vi} - RT$. The only adjustable parameter $c_{i,j}$ is small compared with unity in the case of nonpolar mixtures but may be large for mixtures containing polar molecules.

In the majority of cases the widely used, well-elaborated equations based on the local composition concept (Wilson, NRTL, UNIQUAC) provide reasonable results of simultaneous correlation of the excess Gibbs energy and the enthalpy of mixing of multicomponent mixtures of organic compounds. The more accurate polynomial-type equations with a larger number of adjustable parameters are suitable only for an individual correlation of isothermal binary data of either the excess Gibbs energy or the excess enthalpy.

3.3.4. The Sum of Symmetric Functions Equation

When the purpose of the correlation is the critical evaluation of data by checking the thermodynamic consistency or testing the apparatus used for experiment, it is advantageous to utilize the multiparameter polynomial equations. In practice, whenever a complicated form of the excess function should be represented, an adequate equation must be selected by a trial-and-error procedure. Such a procedure is described in the literature for the analysis of VLE and heat of mixing data (Abbott et al., 1975; Abbott and Van Ness, 1975; Morris et al., 1975).

In the case of highly nonideal multicomponent mixtures for which the complexity of excess functions is characteristic, local composition equations with two or three adjustable parameters are not sufficient for the representation of these functions within the experimental accuracy. The universality of the local composition equations is obtained at the cost of a partial loss of the information included in the primary data. This loss becomes greater if the measurements are very accurate.

For the correlation of heat of mixing h^E, excess volume V^E and excess Gibbs energy g^E, a special equation, called SSF sum of symmetric functions (SSF) has been proposed by Rogalski and Malanowski (1977). An excess function $F_{i,j}^{\mathrm{BIN}}$ of a binary system formed by components i and j is given by

$$F_{i,j}^{\text{BIN}} = \sum_{i=1}^{m} \frac{A_i x_i x_j (x_i + x_j)^2}{(x_i/a_l + x_j a_l)^2} \tag{3.3-37}$$

where A_l and $a_l = \sqrt{q_{i,1}/q_{l,2}}$ are adjustable parameters, $q_{l,i}$ are mole fractions of component i, corresponding to the maximum on the lth term versus composition curve.

The excess function for an n-component mixture is

$$F^E = \sum_{i=1}^{n-1} \sum_{j=i+1}^{n} F_{i,j}^{\text{BIN}} \tag{3.3-38}$$

For $F^E = G^E$ the activity coefficients of a binary mixture can be computed as follows:

$$\ln \gamma_1 = \sum_{l=1}^{m} x_2^2 A_l \frac{a_l + x_1(a_l - 1/a_l)}{(x_1/a_l + x_2 a_l)^3} \tag{3.3-39}$$

$$\ln \gamma_2 = \sum_{l=1}^{m} x_1^2 A_l \frac{a_l + x_2(a_l - 1/a_l)}{(x_1/a_l + x_2 a_l)^3} \tag{3.3-40}$$

Equations (3.3-39) and (3.3-40) yield formulas for the activity coefficients at infinite dilution:

$$\ln \gamma_1^\infty = \sum_{l=1}^{m} \frac{A_l}{a_1^2} \tag{3.3-41}$$

$$\ln \gamma_2^\infty = \sum_{l=1}^{m} A_l a_L^2 \tag{3.3-42}$$

For the correlation by means of the SSF equation the following derivatives are useful:

$$\frac{\partial F^E}{\partial A_l} = \frac{x_i x_j (x_i + x_j)^2}{(x_i/a_l + x_j a_l)^3} \tag{3.3-43}$$

$$\frac{\partial F^E}{\partial a_l} = \frac{2A_l x_i x_j (x_i + x_j)^2 (x_i/a_l^2 - x_j)}{(x_i/a_l + x_j a_l)^3} \tag{3.3-44}$$

$$\frac{\partial F^E}{\partial x_i} = \sum_{l=1}^{m} \frac{A_l x_j \left\{[2x_i(x_i + x_j) + (x_i + x_j)^2](x_i/a_l + x_j a_l) + \frac{2x_i}{a_l^2}(x_i + x_j)^2\right\}}{(x_i/a_l + x_j a_l)^3} \tag{3.3-45}$$

$$\frac{\partial F^E}{\partial x_j} = \sum_{l=1}^{m} \frac{A_l x_i \{[2x_j(x_i + x_j) + (x_i + x_j)^2](x_i/a_l + x_j a_l) - 2a_l x_j (x_i + x_j)^2\}}{(x_i/a_l + x_j a_l)^3} \tag{3.3-46}$$

The SSF equation is very useful for the correlation of H^E, V^E and G^E of strongly nonideal mixtures. Examples are shown in Tables 3.3-2 and 3.3-3.

Christensen et al. (1984) have shown that the best results of correlating H^E data are obtained by means of the Redlich–Kister equation in 85 percent of cases. For the remaining 15 percent cases, all strongly nonideal mixtures, this equation was unsuitable and the SSF equation was shown to yield the best results. They have found the SSF equation to be far superior to the Redlich–Kister equation for the correlation of binary data for which the H^E versus x_1 curve is highly skewed. On the other hand, they have found the Redlich–Kister equation to be better for systems that exhibit a simple composition dependence of h^E.

3.3.5. Comparison of Activity Coefficient Models

Before we show numerical examples of the application of the equations discussed above, we shall outline the computational procedure needed to arrive at practical results using the γ–ϕ method. For a more detailed account the reader is referred to the book by Van Ness and Abbott (1982). The simplest case is to perform isothermal bubble-point calculations, i.e., to compute P and $\mathbf{y}$ for known T and $\mathbf{x}$.

1. The vapor composition $\mathbf{y}$ appear only in the secondary functions Φ_i [Eq. 3.1-27)] and need not be known initially. The iterative process starts with $\Phi_i = 1$.
2. The activity coefficient γ_i can be calculated at the composition x_i and temperature T from one of the equations described above.
3. The pure component vapor pressure P_i^0 can be taken from the VLE data set under consideration or calculated from a suitable correlation equation as described in Chapter 2.
4. If Φ_i, γ_i and P_i^0 are known, the total pressure can be calculated from Eq. (3.1-29).
5. With the total pressure known, Eq. (3.1-26) yields the vapor phase mole fraction y_i. This makes it possible to calculate Φ_i.
6. With the Φ_i value refined, we can go back to step 4 and iterate until convergence of the calculated P and y_i values is achieved.
7. If the parameters of the correlation equation for γ_i were known a priori, the calculations are completed. If not, stage 1 to 6 constitute an inner loop in an optimization procedure for finding the parameters of the model for γ_i. The optimization is performed by minimizing the objective function

$$F = \sum_i (P_i^{\text{cal}} - P_i^{\text{exp}})^2$$

or other suitable function designed to adjust the model to experimental data. The choice of objective functions is important especially for the purpose of consistency testing.

Therefore, a thorough discussion of objective functions will be deferred to Chapter 5. The data reduction scheme outlined above was proposed for the

Table 3.3-2
CORRELATION OF VAPOR–LIQUID EQUILIBRIUM BY VARIOUS EQUATIONS

Equation No. of Adjustable Parameters		*SSF* *4*		*Redlich–Kister* *4*	
Absolute Average Deviation	T (K)	$\delta(P)$	$\delta(y)$	$\delta(P)$	$\delta(y)$
Binary mixture					
Nitromethane + CCl_4	318	0.07	0.12	0.12	0.20
Acetonitrile + CCL_4	318	0.11	0.18	0.29	0.33
Pyridine + water	243	0.08	0.35	0.29	0.68
Ethanol + isooctane	223	1.10	1.20	2.38	2.24
4-Picoline + water	263	0.53	0.34	1.13	1.26
3-Picoline + water	263	0.45	0.64	1.68	1.80
Ethanol + toluene	308	0.13	0.59	0.66	0.91
Ethanol + methylcyclohexane	308	0.31	0.92	0.91	1.29
Isopropanol + isooctane	318	0.32	0.59	1.19	1.30
Pyridine + nonane	369	0.34[a]	0.69[a]	0.44	0.78
Mean: 10 *systems*		0.34	0.56	0.91	1.08

Equation **No. of Adjustable Parameters**		*Wilson* *2*		*NRTL* *3*		*UNIQUAC* *2*	
Absolute Average Deviations	**T (K)**	**$\delta(P)$**	**$\delta(y)$**	**$\delta(P)$**	**$\delta(y)$**	**$\delta(P)$**	**$\delta(y)$**
Binary mixture							
Nitromethane + CCl_4	318	0.56	0.33	0.56	0.50	1.24	0.85
Acetonitrile + CCl_4	318	0.26	0.30	0.46	0.51	1.21	1.01
Pyridine + water	243	0.92	1.03	0.75	0.49	1.17	1.73
Ethanol + isooctane	223	1.21	0.82	2.11	2.39	3.99	3.12
4-Picoline + water	263	1.07	0.54	0.63	0.41	1.93	1.20
3-Picoline + water	263	1.54	0.67	0.73	0.48	1.49	1.27
Ethanol + toluene	308	0.50	0.29	0.75	0.39	1.92	1.05
Ethanol + methylcyclohexane	308	0.45	0.30	1.20	0.48	3.19	1.33
Isopropanol + isooctane	318	1.71	1.45	2.30	0.97	2.71	1.93
Pyridine + nonane	369	1.52	0.65	1.92	0.37	1.90	0.40
Mean: 10 systems		0.87	0.64	1.14	1.41	2.07	1.59

From Rogalski and Malanowski (1977).

[a] Only 2 parameters used:

$$\delta(P) = \left(\frac{100}{n}\right) \sum_{n=1}^{n} \left|\frac{P_{\text{calc}} - P_{\text{exp}}}{P_{\text{exp}}}\right|$$

$$\delta(P) = \left(\frac{100}{n}\right) \sum_{n=1}^{n} \left|\frac{y_{\text{calc}} - y_{\text{exp}}}{y_{\text{exp}}}\right|$$

first time by Barker (1953). It has been used in this chapter to produce the numerical results shown below.

The results of correlation obtained by means of various equations are given in Table 3.3-2. All systems chosen for comparison satisfied two different consistency tests and were measured in laboratories of good reputation. It is evident that SSF outperforms other equations in its capability of reproducing the experimental data. The local composition equations give very good results. Among them, the Wilson equation is best for homogeneous nonideal mixtures. The NRTL equation is very good and should be used for multicomponent systems containing some binaries close to liquid phase splitting. The poorest results are usually obtained with the UNIQUAC equation. This is, however, only a general remark and all system should be always treated individually with respect to the selection of the equation most suitable for this particular case.

The local composition equations are usually best for correlating literature data when there is no reliable information about experimental errors. This type of equations is not advantageous for the statistical analysis of one's own measurements. In Table 3.3-3 a comparison of the flexibility of some equations is shown. The heat of mixing data of alcohol–hydrocarbon systems have been chosen for this purpose. The results show that the SSF equation achieves the same accuracy of the representation of the data as the accuracy of the original experiment. The large number of experimental points justifies the use of six adjustable parameters in this case. The accuracy obtained with the Redlich–Kister expansion is much lower with the same number of adjustable parameters.

Table 3.3-3
CORRELATION OF HEAT OF MIXING BY REDLICH–KISTER AND SSF EQUATIONS

		RMSD[a] J/mol			
		4 parameters		6 parameters	
System	Temperature, T (K)	R–K	SSF	R–K	SSF
Methanol + hexane	298	34.2	15.9	17.6	0.1
Methanol + hexane	318	14.0	6.3	1.5	0.5
Methanol + heptane	303	54.0	15.5	21.1	0.8
Ethanol + heptane	283	38.6	16.8	25.3	2.3
Propanol + heptane	303	60.3	8.7	30.4	2.0
Isopropanol + isooctane	298	12.6	2.2	7.1	2.0
Mean: 6 systems		37.7	9.9	17.6	1.5

From Rogalski and Malanowski (1977).

[a] $$\mathrm{RMSD}(H^E) = 100\left[\frac{\sum_{n=1}^{n}(H^E_{\mathrm{calc}} - H^E_{\mathrm{exp}})^2}{n}\right]^{1/2}$$

Table 3.3-4.
PREDICTION OF ACTIVITY COEFFICIENTS AT INFINITE DILUTION BY VARIOUS EQUATIONS

System	T (K)	Activity Coefficient					
		Measured Value	Predicted by Equation				
			R–K4	Van Ness	Wilson	NRTL	SSF4
Prydine–water	263	23	16	23	28	20	23
3-Picoline–water	243	63	27	48		62	56
4-Picoline–water	243	59	25	44		57	58
2.6-Lutidine–water	243	136	103	83		126	139

From Rogalski and Malanowski (1977).

Another advantage of an equation for the representation of the excess Gibbs energy is its ability to represent the limiting values of activity coefficients. A comparison of the performance of various equations is made in Table 3.3-4. Both vapor–liquid equilibrium data and the limiting values of activity coefficients were measured in the same, very good laboratory (NPL–Teddington). The best results have been obtained with the SSF and NRTL equations.

3.3.6. Explicit Treatment of Association in Activity Coefficient Models

The activity coefficient models based on the local composition concept represent accurately nonidealities in most mixtures. However, vapor–liquid and liquid–liquid equilibria in some systems containing associating components cannot be correlated as satisfactorily as those in systems containing nonpolar or weakly polar compounds. This can be explained by allowing for specific chemical forces between molecules existing in a mixture. The chemical forces are usually due to hydrogen bonding or coulombic interactions between polar and quadrupolar molecules or charge transfer complexes. A concise introduction to intermolecular interactions from a thermodynamicist's point of view has been given by Prausnitz et al. (1986).

The specific chemical interactions lead to *self-association* when molecules of the same chemical species form aggregates or *cross-association* when the aggregates are formed from unlike molecules. Frequently, the cross-association is also called *solvation*.

Association of one or more components in a liquid mixture and the chemical forces due to coulombic interactions between an associating and an active compound influence strongly the excess properties of mixtures. In the case of some mixtures it is advantageous to treat chemical and physical interactions separately in liquid phase models.

If we assume the existence of specific associates, we can treat an associating substance as a mixture of several species in equilibrium. The only constraint imposed by phenomenological thermodynamics states that the chemical potential of the associating component μ_A is equal to the chemical potential of its monomer μ_{A_1} (Prigogine and Defay, 1954):

$$\mu_A = \mu_{A_1} \tag{3.3-47}$$

The application of the chemical theory to the interpretation of the properties of mixtures goes back to the work of Dolezalek (1908) who introduced the chemical reaction concept to the theory of solutions. Dolezalek treated an associated mixture as an ideal mixture of associated species. Thus, the mixture nonideality was attributed exclusively to association. The ideal solution models were incapable of reproducing more complicated phase behavior, including liquid–liquid separation, and are only of historical importance.

Since the work of Dolezalek many theoretical models have been proposed for the interpretation of thermodynamic properties of associated mixtures. In general, assumptions that have to be made to develop an association model concern two aspects (Kehiaian and Treszczanowicz, 1966): (1) thermodynamic functions of association (chemical reaction) and (2) thermodynamic functions of mixing the true molecular species. From the point of view of aspect (1) it is necessary to establish what types of associates can be formed in a mixture. This could be done, in principle, if independent information existed about the structures of possible associates. In fact, spectroscopic and dielectric evidence can provide useful hints at the existence of, e.g., carboxylic acid dimers or alcohol chains. Information from such sources is insufficient, however, to determine the actual numerical values of thermodynamic functions of the formation of associates with satisfactory accuracy. Therefore, it is necessary to resort to empirical association models. The continuous linear association models have been proved to be especially useful for the correlation of spectroscopic and thermodynamic properties of mixtures. Two basic types may be distinguished according to the assumed correlation between the consecutive standard entropies of the reaction:

$$A_i + A_1 = A_{i+1} \tag{3.3-48}$$

For the Mecke–Kempter (MK) model (Kempter and Mecke, 1940) it is assumed that

$$\Delta S^0_{i,i+1} = \Delta S^0 \tag{3.3-49}$$

whereas for the Kretschmer–Wiebe (1954) model (KW)

$$\Delta S^0_{i,i+1} = \Delta S^0 + R \ln\left(\frac{r_{A_{i+1}}}{r_{A_i}}\right) = \Delta S^0 + R \ln\left[\frac{(i+1)}{i}\right] \tag{3.3-50}$$

where r is the volume parameter that is proportional to the number of segments within an associate ($r_{A_i} = i r_{A_1}$). Assuming equal consecutive standard enthalpies of association, one derives equal consecutive association constants for the MK model:

$$K_{i,i+1} = K \tag{3.3-51}$$

while for the KW model:

$$K_{i,i+1} = K\left(\frac{r_{A_{i+1}}}{r_{A_i}}\right) = K\left[\frac{(i+1)}{i}\right] \tag{3.3-52}$$

The assumptions underlying the Mecke–Kempter and Kretschmer–Wiebe models are summarized in Table 3.3-5. Numerous modifications to the above equations have been proposed. Some of them assign a different value to the dimerization constant, thus formulating models with more than one independent association constant (Haskell et al., 1968; Landeck et al., 1977; French and Stokes, 1981). Other authors formulate different correlations for the $K_{i,i+1}$ constants by considering the enthalpy and entropy of association and/or taking into account cyclic associates (Aguirre–Ode, 1986; Gaube et al., 1987; Liu et al., 1989; Hofman, 1990). However, the one-constant MK and KW constants seem to be a good compromise between computational simplicity and physical reality when applied to practically oriented γ–ϕ methods. Therefore, they have been accepted for mixtures containing self-associating substances and extended by accounting for cross-association and complex formation between associating and nonassociating molecules.

After assuming a suitable association model, it is necessary to express the thermodynamic functions of mixing the true species in a tractable way. It is convenient to assume that the activity coefficient can be expressed as a product of two contributions: combinatorial γ^{COM} and residual γ^{RES}:

$$\gamma = \gamma^{\mathrm{COM}}\gamma^{\mathrm{RES}}$$

This is a consequence of separating the combinatorial part of the excess entropy, which represents the effect of molecular size and shape and is independent of temperature (i.e., $R \ln \gamma_i^{\mathrm{COM}} = -s_i^{E,\mathrm{COM}}$). The remaining part of excess entropy and the whole of excess enthalpy produce the residual contribution (i.e., $RT \ln \gamma^{\mathrm{RES}} = h_i^E - Ts_i^{E,\mathrm{RES}}$). This is a general approach used in the excess Gibbs energy models and is independent of association.

In this case the association constant, expressed in terms of activities, becomes

$$K_{i,i+1} = \frac{x_{i+1}\gamma_{i+1}^{\mathrm{COM}}}{x_i x_1 \gamma_i^{\mathrm{COM}}\gamma_1^{\mathrm{COM}}} \frac{\gamma_{i+1}^{\mathrm{RES}}}{\gamma_i^{\mathrm{RES}}\gamma_1^{\mathrm{RES}}} \tag{3.3-53}$$

Table 3.3-5
STANDARD ENTHALPY STANDARD ENTROPY AND EQUILIBRIUM CONSTANT OF ASSOCIATION [REACTION (3.3-48)] FOR THE MECKE–KEMPTER AND KRETSCHMER–WIEBE ASSOCIATION MODELS

Model	$\Delta H^0_{i,i+1}$	$\Delta S^0_{i,i+1}$	$K_{i,i+1}$
Mecke–Kempter	ΔH^0	ΔS^0	K
Kretschmer–Wiebe	ΔH^0	$\Delta S^0 + R\ln(r_{A_{i+1}}/r_{A_i})$	$K(r_{A_{i+1}}/r_{A_i})$

where x_i are the so-called *true* (or *real*) mole fractions of the species A_i defined as

$$x_i = \frac{N_i}{\sum_{\{\text{associates}\}} N_i + \sum_{\{\text{inerts}\}} N_j} \tag{3.3-53a}$$

where N_i is the number of moles of the i-meric species A_i and N_j is the number of moles of the jth (monomeric) inert species B_j.

It is reasonable to adopt a simplifying assumption that the residual chemical potential is not changed during the association reaction, i.e.,

$$\ln \gamma_{i+1}^{\text{RES}} = \ln \gamma_i^{\text{RES}} + \ln \gamma_1^{\text{RES}} \tag{3.3-54}$$

and, subsequently,

$$\frac{\gamma_{i+1}^{\text{RES}}}{\gamma_i^{\text{RES}} \gamma_1^{\text{RES}}} = 1 \tag{3.3-54a}$$

This assumption is in agreement with the solution-of-groups concept (Wilson and Deal, 1962) as the monomeric units within an associate can be treated as groups having the same residual properties. In this case the excess Gibbs energy of an associated mixture can be separated into physical and chemical contributions

$$G^E = G_{\text{ch}}^E + G_{\text{ph}}^E = RT \sum x_i^* \ln \gamma_i^{*,\text{COM}} + RT \sum x_i^* \ln \gamma_i^{*,\text{RES}} \tag{3.3-55}$$

where the asterisk denotes the *apparent* (*nominal*) quantities to differentiate them from the true ones. The apparent mole fraction is defined by

$$x_i^* = \frac{\sum_{\{\text{associates}\}} iN_i}{\sum_{\{\text{associates}\}} iN_i + \sum_{\{\text{inerts}\}} N_j} \tag{3.3-55a}$$

Application of Eq. (3.3-47) to the mixture and pure associating fluid leads to an expression for the apparent activity coefficient in terms of the true ones (Prigogine and Defay, 1954):

$$\gamma_i^* = \frac{x_i \gamma_i}{x_i^0 \gamma_i^0 x_i^*} \tag{3.3-56}$$

In particular, for the combinatorial contribution:

$$\gamma_i^{*,\text{COM}} = \frac{x_i \gamma_i^{\text{COM}}}{x_i^0 \gamma_i^{\text{COM},0} x_i^*} \tag{3.3-56a}$$

where the superscript zero denotes a pure substance.

For a practically oriented model it is important to assume a realistic physical contribution to G^E. Renon and Prausnitz (1967) and Wiehe and Bagley (1967) were the first to derive, on the basis of the Kretschmer–Wiebe model, equations with a physical and a chemical contribution to the excess Gibbs energy and to apply them successfully to calculate vapor–liquid equilibria in binary mixtures. Nitta and Katayama (1973), Nagata (1973, 1977) and Chen and Bagley (1978) extended this method to associated mixtures containing active components, i.e., those incapable of forming self-associates but forming complexes with the associating component. Expressions for the physical contribution to G^E were adopted from either the regular solution and related theories (Renon and Prausnitz, 1967; Treszczanowicz et al., 1973; Treszczanowicz and Treszczanowicz, 1975) or models based on the local composition concept. Among the latter methods, the UNIQUAC associated solution theory developed by Nagata and co-workers (Nagata and Kawamura, 1979; Nagata, 1985, 1990), Nath and Bender (1981, 1983) and Brandani and Evangelista (1984) seems to be most mature and accurate for various classes of mixtures. Therefore, it will be outlined here for some types of mixtures.

The UNIQUAC associated solution theory can be viewed as an extension of the UNIQUAC equation to systems containing associating, active and inert components. Besides the separability of physical and chemical contributions to G^E it is assumed that the structural (volume and surface) parameters of all associated species are additive, e.g.,

$$r_{A_iB} = ir_{A_1} + r_B \tag{3.3-57}$$

$$q_{A_iB} = iq_{A_1} + q_B \tag{3.3-58}$$

The physical contribution is taken as the residual part of the UNIQUAC equation. The chemical contribution is obtained from an athermal associated solution model. Such a model assumes that the excess Gibbs energy (equal to the excess entropy multiplied by T) of the mixture of the true associated species is expressed by the combinatorial part of UNIQUAC. For such a model the activity coefficient of an associating component A in a mixture containing only one associating and any number of inert components is evaluated according to Eq. (3.3-56) and becomes

$$\ln \gamma_A = \ln \frac{\phi_{A_1}}{\phi^0_{A_1} x_A} + r_A\left(\frac{1}{v^0_A} - \frac{1}{v}\right) - \left(\frac{z}{2}\right) q_A \ln \frac{\phi_A}{\theta_A} + 1 - \frac{\phi_A}{\theta_A}$$
$$+ q_A\left(1 - \ln \sum_j \theta_j \tau_{jA}\right) - \sum_j \frac{\theta_j \tau_{Aj}}{\sum_k \theta_k \tau_{kj}} \tag{3.3-59}$$

and that of any inert component B is

$$\ln \gamma_B = \ln \frac{\phi_B}{x_A} + 1 - \frac{r_B}{v_B} - \left(\frac{z}{2}\right) q_B \ln \frac{\phi_B}{\theta_B} + 1 - \frac{\phi_B}{\theta_B}$$
$$+ q_B\left(1 - \ln \sum_j \theta_j \tau_{jB}\right) - \sum_j \frac{\theta_j \tau_{Bj}}{\sum_k \theta_k \tau_{kj}} \tag{3.3-60}$$

θ and ϕ are, as in the UNIQUAC model, the surface and volume fractions, respectively. The true molar volume of the mixture, necessary in the above equations, is

$$\frac{1}{v} = \sum_{i}^{\infty} \frac{\phi_{A_i}}{r_{A_i}} + \sum_{k=1}^{m} \frac{\phi_{B_k}}{r_{B_k}} \tag{3.3-61}$$

where m is the number of inert components B_k. The molar volume of a pure associated substance is

$$\frac{1}{v_A^0} = \sum_{i}^{\infty} \frac{\phi_{A_i}^0}{r_{A_i}} \tag{3.3-62}$$

If the Kretschmer–Wiebe model is assumed, the self-association constant is expressed in terms of volume fractions as

$$K_A = \left(\frac{\phi_{A_{i+1}}}{\phi_{A_i}\phi_{A_1}}\right)\left(\frac{i}{i+1}\right) \tag{3.3-63}$$

For this model the segment fraction of the monomer of the associating substance in the mixture is

$$\phi_{A_1} = \frac{2K_A\phi_A + 1 - (1 + 4K_A\phi_A)^{0.5}}{2K_A^2\phi_A} \tag{3.3-64}$$

and that in the pure substance is

$$\phi_{A_1}^0 = \frac{2K_A + 1 - (1 + 4K_A)^{0.5}}{2K_A^2} \tag{3.3-65}$$

The volumes v and v_A^0 for the KW model are expressed as

$$\frac{1}{v} = \frac{\phi_{A_1}}{r_A(1 - K_A\phi_{A_1})} + \sum_{k=1}^{m} \frac{\phi_{B_k}}{r_{B_k}} \tag{3.3-66}$$

$$\frac{1}{v_A^0} = \frac{\phi_{A_1}^0}{r_A(1 - K_A\phi_{A_1}^0)} \tag{3.3-67}$$

For the Mecke–Kempter model the quantities necessary to calculate the activity coefficients according to Eqs. (3.3-59) and (3.3-60) are as follows (Kehiaian and Treszczanowicz, 1966):

$$K_A = \frac{\phi_{A_{i+1}}}{\phi_{A_i}\phi_{A_1}} \tag{3.3-68}$$

$$\phi_{A1} = \frac{\phi_A}{1 + K_A\phi_A} \tag{3.3-69}$$

$$\phi^0_{A_1} = \frac{\phi^0_A}{1 + K_A \phi^0_A} \tag{3.3-70}$$

$$\frac{1}{v} = -\frac{\ln(1 - K_A \phi_{A_1})}{r_A K_A} + \sum_{k=1}^{m} \frac{\phi_{B_k}}{r_{B_k}} \tag{3.3-71}$$

$$\frac{1}{v^0_A} = -\frac{\ln(1 - K_A \phi_{A1})}{r_A K_A} \tag{3.3-72}$$

This scheme can be generalized by allowing for one or more active components. The simplifications necessary for this extension have been discussed by Brandani and Evangelista (1984). For example, for a ternary mixture of one alcohol and two unassociated active components B and C (e.g., ethers, CCl_4, etc.), the following cross-association (or solvation or complexation) reactions are assumed:

$$A_i + B = A_i B \tag{3.3-73}$$

$$A_i + C = A_i C \tag{3.3-74}$$

and the Kretschmer–Wiebe constants become

$$K_{AB} = \frac{\phi_{A_i B}}{\phi_{A_i} \phi_B} \frac{i}{i r_A + r_B} \tag{3.3-75}$$

$$K_{AC} = \frac{\phi_{A_i C}}{\phi_{A_i} \phi_C} \frac{i}{i r_A + r_C} \tag{3.3-76}$$

The volume is expressed by

$$\frac{1}{v} = \sum_{i=1}^{\infty} \frac{\phi_{A_i}}{r_{A_i}} + \sum_{i=0}^{\infty} \frac{\phi_{A_i B}}{r_{A_i B}} + \sum_{i=0}^{\infty} \frac{\phi_{A_i C}}{r_{A_i C}}$$

$$= \frac{\phi_{A_i}}{r_A (1 - K_A \phi_{A_1})} + \frac{\phi_{B_1}}{r_B}\left(1 + \frac{K_{AB} \phi_{A_1}}{1 - K_A \phi_{A_1}}\right) + \frac{\phi_{C_1}}{r_C}\left(1 + \frac{K_{AC} \phi_{A_1}}{1 - K_A \phi_{A_1}}\right)$$

The monomer segment fractions of all components, ϕ_A, ϕ_B and ϕ_C, are calculated by solving simultaneously the mass balance equations:

$$\phi_A = \sum_{i=1}^{\infty} \phi_{A_i} + \sum_{i=1}^{\infty} \frac{\phi_{A_i B}}{r_{A_i B}} r_{A_i} + \sum_{i=1}^{\infty} \frac{\phi_{A_1 C}}{r_{A_1 C}} r_{A_i}$$

$$= \frac{\phi_{A_1}}{(1 - K_A \phi_{A_1})^2}\left(1 + \frac{K_{AB} \phi_{B_1} r_A}{r_B} + \frac{K_{AC} \phi_{C_1} r_A}{r_C}\right) \tag{3.3-78}$$

$$\phi_B = \sum_{i=0}^{\infty} \frac{\phi_{A_i B}}{r_{A_i B}} r_B = \phi_{B_1}\left(1 + \frac{K_{AB} \phi_{A_1}}{1 - K_A \phi_{A_1}}\right) \tag{3.3-79}$$

$$\phi_C = \sum_{i=0}^{\infty} \frac{\phi_{A_iC}}{r_{A_iC}} r_c = \phi_{C_1}\left(1 + \frac{K_{AC}\phi_{A_1}}{1 - K_A\phi_{A_1}}\right) \tag{3.3-80}$$

Analogous expressions for the MK model can be found in Nagata (1990).

The model can be further generalized by allowing for more than one associating substance. It can be postulated that successive solvation reactions occur leading to the formation of complex multimers A_iB_j, $A_iB_jA_k$, $A_iB_jA_kB_l$ etc., and that the solvation constant of various reactions is independent of the degree of association. In such a case the Kretschmer–Wiebe constants become

$$K_{AB} = \frac{\phi_{A_iB_j}}{\phi_{A_i}\phi_{B_j}} \frac{r_{A_i}r_{B_j}}{r_{A_iB_j}r_A r_B} \quad \text{for} \quad A_i + B_j = A_iB_j \tag{3.3-81}$$

$$= \frac{\phi_{A_iB_jA_k}}{\phi_{A_iB_j}\phi_{A_k}} \frac{r_{A_iB_j}r_{A_k}}{r_{A_iB_jA_k}r_A r_B} \quad \text{for } A_iB_j + A_k = A_iB_jA_k \tag{3.3-82}$$

$$= \frac{\phi_{A_iB_jA_kB_l}}{\phi_{A_iB_jA_k}\phi_{B_l}} \frac{r_{A_iB_jA_k}r_{B_l}}{r_{A_iB_jA_kB_l}r_A r_B} \quad \text{for } A_iB_jA_k + B_l = A_iB_jA_kB_l \tag{3.3-83}$$

The resulting equations for activity coefficients are rather complex and require numerical solution of material balance equations for each component (Hofman and Nagata, 1987; Nagata, 1990).

Overall, the activity coefficients in the UNIQUAC associated solution theory depend on the self-association constants K_A and cross-association constants K_{AB} in addition to the UNIQUAC τ parameters.

ESTIMATION OF ASSOCIATION PARAMETERS FROM PURE COMPONENT DATA

In earlier γ–ϕ methods incorporating association all parameters were determined from binary phase equilibrium data. This led to serious inconsistencies as the self-association constants appeared to depend on the inert solvent. Nath and Bender (1981) were the first to recognize the need for calculating self-association constants from pure-component data. This idea was further developed by Brandani (1983), Książczak and Buchowski (1984), Hofman and Nagata (1986a,b) and Cibulka and Nagata (1987). All these methods consist of comparing selected properties of the associated substance with those of its homomorph, i.e., a compound being incapable of forming associates but having the molecular structure as similar as possible to that of the associating compound. Nath and Bender used vaporization enthalpies for this purpose whereas the remaining authors employed vapor pressures. The difference between the properties of the associating substance and its homomorph can be interrelated to the association constant. The basic relation used to estimate the self-association constant from vapor pressures follows from Eq. (3.3-47) and takes the form (Brandani, 1983)

$$f_A(P_A^{\text{sat}}) = f_H(P_A^{\text{sat}}) x_{A_1} \gamma_{A_1} \tag{3.3-84}$$

where the subscripts A, H and A_1 refer to the associating substance as a whole, its homomorph and its monomer, respectively. The quantities x_{A_1} and γ_{A_1} are related to the association constant for an assumed model according to the formalism outlined before, while Eq. (3.3-84) has to be solved numerically for K_A, Książczak and Buchowski (1984) derived a simplified expression for the association constant for the ideal Mecke–Kempter model:

$$K_A \cong \frac{P_H^{\text{sat}}}{P_A^{\text{sat}} - 1} \tag{3.3-85}$$

Equation (3.3-85), although derived for an ideal Mecke–Kempter model, was further used in conjunction with an athermal model for mixtures.

Unfortunately, the choice of the homomorph for Eqs. (3.3-84) and (3.3-85) remains ambiguous. Most authors use saturated hydrocarbons for this purpose (e.g., the homomorph of an alcohol is a hydrocarbon with a CH_3 group instead of the OH group). Hofman and Nagata (1986a,b) presented evidence that etherlike homomorphs might be a better choice for alcohols.

In general, introduction of the chemical contribution considerably improves the results of correlation. Table 3.3-6 compares the quality of correlation using the UNIQUAC associated solution theory (UNIQUAC-AS) with that obtained from the original UNIQUAC equation. If the self-association constants are determined from pure-component properties, the mixtures containing an associating and an inert component are represented with two binary parameters, i.e., their number is not increased in relation to the original UNIQUAC equation. However, an additional parameter (i.e., the K_{AB} constant) has to be introduced for mixtures containing more than one associating components.

The methods incorporating association offer the possibility of using association parameters generalized for homologous series of compounds. However, this good feature cannot be utilized to establish genuine predictive methods because the physical contribution parameters do not usually show sufficient regularities within homologous series. Therefore, the main advantage of this class of methods lies in the possibility of improving the results of correlating individual vapor–liquid and liquid–liquid equilibrium data. This statement is equally valid for binary as well as ternary and higher systems.

The main drawback of the γ–ϕ methods incorporating association is their specific character. Different equations are used for different classes of mixtures and the estimation of pure-component association constants is, if consistent, not straightforward. Moreover, the computation of activity coefficients for systems containing active or more than one associating components requires the numerical solution of material balance equations. Therefore, the improvement of the accuracy of correlation should be balanced against the increase of the computational effort.

Table 3.3-6
CORRELATION OF BINARY VLE DATA USING THE UNIQUAC ASSOCIATED SOLUTION THEORY (UNIQUAC-AS) AND THE ORIGINAL UNIQUAC MODEL (AFTER BRANDANI AND EVANGELISTA, 1984)

		UNIQUAC		*UNIQUAC-AS*		
System	T (°C)	$\sigma(P)$ torr	$1000 \cdot \sigma(y)$	No. of Parameters	$\sigma(P)$ torr	$1000 \cdot \sigma(y)$
Mixtures of Associating and Inert Components						
Methanol + hexane	35	21.3		2	5.8	
	60	39.7		2	14.8	
	75	52.1		2	30.4	
Ethanol + hexane	25	3.7	15.0	2	0.4	3.1
	50	16.4		2	2.4	
	80	26.7		2	18.5	
1-Propanol + cyclohexane	55	7.0	20.2	2	4.0	19.4
	60	6.1	11.6	2	2.4	8.6
1-Hexanol + hexane	20	2.5		2	0.5	
	60	6.8		2	1.9	
	100	10.0		2	10.4	
Mixtures of Associating and Active Components						
Methanol	20	5.0		3	0.6	
+ tetrachloromethane	50	12.6		3	4.2	
	80	21.6		3	14.2	
Ethanol + benzene	25	0.6	5.0	3	0.2	5.6
	40	2.1	12.2	3	1.1	7.5
	50	2.8	12.1	3	1.4	10.3
	60	3.4	15.5	3	2.0	15.6

3.4 ESTIMATION OF LIQUID PHASE EXCESS FUNCTIONS

3.4.1. Introduction

The excess functions provide a consistent method for the description of thermodynamic properties of multicomponent mixtures. They make it possible to extend experimental binary data to multicomponent mixtures as well as to interpolate and extrapolate them to different concentrations and temperatures.

The equations describing the excess Gibbs energy as a function of mixture composition contain adjustable parameters that are obtained by fitting the models to experimental phase equilibrium data. While vapor–liquid equilibria have been studied experimentally for more than 10,000 systems (Gmehling et al., 1977–; Mączyński et al., 1976–1988; Wichterle et al., 1973–), there is an unlimited number of other systems for which no data are available at all. To make the situation even more complicated, a substantial part of the open literature data is unreliable. Timmermans (1959, Vol. 2, p. V) characterizes the situation in the following way:

> We have collected all the numerical data published about the concentrated binary solutions of organic compounds. In many cases the reader will be shocked by the low degree of precision of the measurements since, when different authors made research on the same subject, the quantitative discrepancies are obvious. This may be due, at least to two kinds of difficulties: different methods on the physical side, and the impurity of the used samples. The possibility of this second cause of error is not considered with sufficient care by most of the authors. Therefore, it is a pity that so much work has been done with too little care in that direction; this diminishes the value of numerous numerical data, whatever the care given to the physical methods used.

When interest is directed at systems for which data are absent or unreliable or, in the best case, incomplete, the safest solution is to take one's own measurements. It is always a challenge to make a good experiment although modern equipment simplifies the necessary procedures. Even a few experimental data are often sufficient to yield a reasonable thermodynamic model of a mixture. There is always a vast difference between a few reliable data and unreliable data or no data at all.

Various types of phase equilibrium data can be used to obtain the parameters of model equations. Vapor–liquid equilibrium data are most reliable when all parameters (composition of vapor and liquid phases in equilibrium, accompanied by pressure and temperature) have been measured. Frequently, so-called incomplete data are used, in most cases consisting of compositions of one phase, temperature and pressure. Such data cannot be checked for thermodynamic consistency. The measuring procedures have to be more accurate in this case and reagents should be treated with special care to prevent contamination, e.g., with moisture or dissolved air leading to systematic errors. The reliability of such data can be checked only during the actual experiment and no independent test is possible.

In principle, experimental data for some thermodynamic property can be used to estimate some other properties. The use of infinite-dilution activity coefficients for this purpose will be described in Section 3.4.3. Another obvious choice is the use of mutual solubility data to estimate activity coefficients, e.g., for vapor–liquid equilibrium computations.

For partially miscible binary mixtures no more than two adjustable parameters can be obtained from mutual-solubility data at one temperature. Any equation representing the Gibbs excess energy as a function of liquid composition is suitable for this purpose. The resulting expressions for activity coefficients should be inserted into the equilibrium conditions for both liquid phases I and II

$$[x_1\gamma_1(x_1, A_1, A_2)]^{\mathrm{I}} = [x_2\gamma_2(x_1, A_1, A_2)]^{\mathrm{II}} \tag{3.4-1}$$

where A_1, A_2 = adjustable parameters.

The obtained parameters A_1 and A_2 are approximate when applied to VLE calculations and always lead to higher values of activity coefficients than those calculated from the measurement of the properties of the liquid phase in the homogeneous region. A useful collection of liquid–liquid equilibrium data was published by Arlt et al. (1979–1988).

Many mixtures exhibit azeotropy, i.e., a situation when the equilibrium composition of the liquid phase is equal to that of the vapor phase. Extensive compilations of azeotropic data are available (Horsley, 1952, 1962, 1973; Ogorodnikov et al., 1971). They can be used to calculate the excess Gibbs energy of a mixture (Malesiński, 1965) and to estimate vapor–liquid equilibria in the whole concentration range using a suitable model equation. For a binary azeotrope $x_1 = y_1$ and $x_2 = y_2$. If the vapor phase nonideality and the Poynting correction are neglected, the activity coefficients can be estimated as

$$\gamma_1 = \frac{P}{P_1^0} \quad \text{and} \quad \gamma_2 = \frac{P}{P_2^0} \tag{3.4-2}$$

where P_1^0 and P_2^0 are pure-component vapor pressures. It is now possible to find two parameters of a correlation equation by solving the equations

$$RT \ln \gamma_1 = f_1(x_1, A, B) \tag{3.4-3}$$

$$RT \ln \gamma_2 = f_2(x_1, A, B) \tag{3.4-4}$$

It should be noted that the accuracy of such calculations depends on the location of the azeotropic point. If the azeotropic composition is in a dilute region of one component (e.g., $x_2 \ll 1$), the activity coefficient of the other component is close to unity ($\gamma_1 \rightarrow 1$ as $x_1 \rightarrow 1$). Subsequently, only one activity coefficient is significant, and it is not possible to calculate two parameters of a correlation equation. In such a case only a one-parameter correlation equation can be applied.

Collections of other excess functions, like excess volume (Handa and Benson, 1979) and heat of mixing (Bielousov and Morachevskii, 1970; Hellwege, 1976; Bielousov et al., 1981; J. Christensen et al., 1982; C. Christensen et al. 1984) are very useful. The excess volume data can be used for the computation of the pressure dependence of the excess Gibbs energy. The heat of mixing data determine the dependence of the excess Gibbs energy on temperature.

Frequently, no experimental data are available and it is necessary to estimate the excess Gibbs energy from some theoretical or empirical correlation.

Theoretical understanding of liquid mixtures is still very poor. There is no useful theory available for mixtures containing larger molecules, especially if they contain various functional groups. This means that the excess Gibbs energy can be reasonably predicted only for systems similar to those used to establish the prediction method. As a matter of fact the practically oriented predictive methods are only empirical. They will be presented in the following sections.

3.4.2. The Regular Solution Model

The concept of regular solutions goes back to the work of Hildebrand and co-workers in 1920s who observed some regularities in solubility curves for mixtures of nonpolar substances. Hildebrand (1929, p. 69) defined a regular solution as "one involving no entropy change when a small amount of of one of its components is transferred to it from an ideal solution of the same composition, the total volume remaining unchanged." Since then, the term *regular solution* has been also attributed some different meanings when used by different authors. The theory and applications of the regular solution model have been reviewed in detail by Hildebrand and Scott (1962), Hildebrand et al. (1970) and, more recently, by Messow and Bittrich (1981) and Barton (1983).

The most straightforward definition of a regular solution states that only the excess enthalpy contributes to the nonideality of a mixture, i.e.,

$$G^E = H^E \neq 0 \tag{3.4-5}$$

$$V^E = S^E = 0 \tag{3.4-6}$$

Van Laar (1910), who derived an expression for G^E from the van der Waals equation of state (EOS), can be viewed as a precursor of the regular solution concept. For a binary mixture the van Laar equation takes the form

$$G^E = \frac{x_1 x_2 v_1 v_2}{x_1 v_1 + x_2 v_2}\left(\frac{a_1^{1/2}}{v_1} - \frac{a_2^{1/2}}{v_2}\right) \tag{3.4-7}$$

Scatchard (1931, 1934) and Hildebrand and Wood (1933) showed that the functional form of Eq. (3.4-7) can be derived in a more general way without using the van der Waals EOS. Scatchard's original assumptions were:

1. The mutual energy of two molecules depends only on the distance between them and their relative orientation and not on the nature of the other molecules between or around them and not on the temperature.
2. The distribution of the molecules is random, i.e., it is independent of temperature and the nature of the other molecules present.
3. The change of volume on mixing at constant pressure is zero.

These assumptions considerably restrict the class of real mixtures that can be treated. For a binary mixture the following expression was obtained:

$$G^E = (x_1 v_1 + x_2 v_2)\phi_1\phi_2[(\delta_1 - \delta_2)^2 + 2l_{12}\delta_1\delta_2] \tag{3.4-8}$$

where ϕ_i is the volume fraction

$$\phi_i = \frac{x_i v_i}{\Sigma\, x_j v_j} \tag{3.4-9}$$

where v_i is the liquid volume and δ_i is the solubility parameter defined as

$$\delta_i = \left(\frac{\Delta U_i}{v_i}\right)^{1/2} \tag{3.4-10}$$

where ΔU_i is the energy required isothermally to evaporate liquid i from the saturated liquid to the ideal gas. At temperatures well below critical

$$\Delta U_i \cong \Delta H_i^v - RT \tag{3.4-11}$$

where ΔH_i^v is the enthalpy of vaporization of pure liquid i at temperature T. The solubility parameters, estimated from Eqs. (3.4-10) and (3.4-11) are temperature dependent although this dependence is relatively weak. Therefore, the initial assumption that $S^E = 0$ is not exactly satisfied. A comprehensive collection of solubility parameters along with methods for their estimation has been compiled by Barton (1983).

Activity coefficients calculated from Eq. (3.4-8) take the form

$$RT \ln \gamma_1 = v_1\phi_2^2[(\delta_1 - \delta_2)^2 + 2l_{12}\delta_1\delta_2] \tag{3.4-12}$$

$$RT \ln \gamma_2 = v_2\phi_1^2[(\delta_1 - \delta_2)^2 + 2l_{12}\delta_1\delta_2] \tag{3.4-13}$$

For a first approximation, the binary parameter l_{12} can be assumed to be equal to zero. In that case, Eqs. (3.4-12) and (3.4-13) contain no binary parameter and activity coefficients can be predicted using only pure component data.

The regular solution model is readily generalized to multicomponent mixtures. For component k

$$RT \ln \gamma_k = v_k \sum_i \sum_j \left(A_{ik} - \frac{A_{ij}}{2}\right)\phi_i\phi_j \tag{3.4-14}$$

where $$A_{ij} = (\delta_i - \delta_j)^2 + 2l_{ij}\delta_i\delta_j \tag{3.4-15}$$

If all binary parameters l_{ij} are assumed equal to zero, Eq. (3.4-14) simplifies to

$$RT \ln \gamma_k = v_k(\delta_k - \bar{\delta})^2 \tag{3.4-16}$$

where

$$\bar{\delta} = \sum \phi_i \delta_i \tag{3.4-17}$$

Equations (3.4-16) and (3.4-17) provide a semiquantitative estimation of activity coefficients provided that the components are nonpolar and not much different in size and shape.

Many authors have tried to improve the predictive capability of the regular solution model. Among them, Orye and Prausnitz (1965) coupled the regular and athermal mixture models to account for differences in molecular size. The Flory–Huggins expression (Flory, 1941, 1942) was added to the Scatchard–Hildebrand formula

$$G^E = (x_1 v_1 + x_2 v_2)\phi_1\phi_2[(\delta_1 - \delta_2)^2] + RT\left[x_1 \ln \frac{\phi_1}{x_1} + x_2 \ln \frac{\phi_2}{x_2}\right] \tag{3.4-18}$$

This modification does not lead to a systematic improvement as the Flory–Huggins term is known to overestimate the athermal entropy of mixing.

The most obvious modification is to adjust the parameter l_{12} to match experimental data. Some authors have attempted to correlate l_{12} with pure component properties to retain the predictive character of the model. Although no general relationship has been found, a satisfactory correlation can be obtained for homologous series of mixtures. Notably, Funk and Prausnitz (1970) found an almost linear variation of l_{12} with the degree of branching of a saturated alkane for several (benzene or toluene) plus alkane mixtures. Schille and Bittrich (1973) extended the Orye–Prausnitz expression and correlated the binary parameter l_{12} in hydrocarbon mixtures with the solubility parameters of pure components:

$$\begin{aligned} G^E = {} & (x_1 v_1 + x_2 v_2)\phi_1\phi_2[(\delta_1 - \delta_2)^2 + (\delta_1 A + B)(\delta_2 A + B)] \\ & + RT\left[x_1 \ln \frac{\phi_1}{x_1} + x_2 \ln \frac{\phi_2}{x_2}\right] \end{aligned} \tag{3.4-19}$$

where $A = 1.103 \pm 0.031$ and $B = -17.784 \pm 0.516\ (\mathrm{J/cm^3})^{1/2}$. This leads to an improvement of prediction, although the results are still only in semiquantitative agreement with the data. The errors of predicting G^E for 82 hydrocarbon systems computed by Messow and Bittrich (1981) are given in Table 3.4-1.

Several authors have tried to extend the regular solution model to polar mixtures. To obtain satisfactory agreement, the classes of components considered have to be restricted. Weimer and Prausnitz (1965) and Helpinstill and

Table 3.4-1
ACCURACY OF PREDICTING THE EXCESS GIBBS ENERGY BY MEANS OF THE REGULAR SOLUTION THEORY AND ITS MODIFICATIONS

	Hildebrand–Scatchard [Eq. (3.4-8)]	Orye–Prausnitz [Eq. (3.4-18)]	Schille–Bittrich [Eq. (3.4-19)]
$\left(\frac{100}{n}\right)\sum\left\lvert\frac{G^E_{\text{exp}} - G^E_{\text{cal}}}{G^E_{\text{exp}}}\right\rvert$	87	103	73

After Messow and Bittrich (1981).

Van Winkle (1968) separated the cohesive energy density (i.e., the square of the solubility parameter) into contributions from nonpolar (dispersive) and polar forces:

$$c_{11} = \delta_1^2 = \tau_1^2 + \lambda_1^2 \tag{3.4-20}$$

$$c_{22} = \delta_2^2 = \tau_2^2 + \lambda_2^2 \tag{3.4-21}$$

$$c_{12} = \delta_1\delta_2(1 - l_{12}) = \lambda_1\lambda_2 + \tau_1\tau_2 + \psi_{12} \tag{3.4-22}$$

where λ_i is the nonpolar solubility parameter (which can be evaluated as the solubility parameter of the homomorph of the polar substance) and τ is the polar solubility parameter. As expected, the binary parameter ψ_{12} is not negligible thus decreasing the predictive capability of the model.

A different extension of the regular solution model has been proposed by Książczak (1986a,b) for mixtures of self-associating compounds with inert solvents. It has been assumed that associates characterized by the mean association number χ are formed in the mixture. The effective mole fraction of the associated substance is then

$$\bar{x} = \frac{x}{\chi - \chi x + x} \tag{3.4-23}$$

and its activity becomes

$$a = \bar{x}\gamma_\chi \tag{3.4-24}$$

where γ_χ, the activity coefficient of the χ-mer, is calculated from a regular solution expression for a mixture of χ-mers and inert molecules:

$$RT \ln \gamma_\chi = v_\chi(1 - \phi_\chi)^2[(\delta_1 - \delta_2)^2 + 2l_{12}\delta_1\delta_2] \tag{3.4-25}$$

where

$$v_\chi = \chi v_1 \tag{3.4-26}$$

$$\phi_\chi = \frac{\bar{x}v_\chi}{(1-\bar{x})v_2 + \bar{x}v_\chi} \tag{3.4-27}$$

and the cohesive solubility parameter is calculated according to (3.4-10) using the cohesive energy ΔU of the homomorph of the associating substance. The method was specifically designed for solid–liquid equilibria for which the mean association number χ can be assigned a constant value along the solubility curve. Książczak and Anderko (1987) found the method to be useful for predicting solid–liquid equilibria in self-associated mixtures from pure-component properties alone.

In general, the regular solution theory and its numerous modifications can be useful for predicting phase equilibria for some restricted classes of mixtures (typically nonpolar). The results are frequently quite accurate for some mixtures, but the overall accuracy of the method is only semiquantitative. The methods based on the regular solution concept can give useful hints about the thermodynamic behavior of mixtures, but their reliability is limited.

3.4.3. Application of Infinite-Dilution Activity Coefficients

The activity coefficient at infinite dilution γ^∞, defined by Eq. (P1.2-47), characterizes the interactions between the solute and solvent molecules without the interference of solute–solute interactions. The γ^∞ coefficient is a characteristic property of the liquid phase of a mixture. As the measurements of γ^∞ are simple and fast, it is a convenient source of information on mixture properties. The γ^∞ values can be easily measured by gas-liquid chromatography (Alessi, 1990) for a more volatile component in a mixture of components, which differ considerably in volatility and by ebulliometry (Rogalski and Malanowski, 1980) for both components in a mixture consisting of components of similar volatility. A collection of activity coefficients at infinite dilution (Gmehling et al., 1986) and a computer data bank (Alessi et al., 1987) are available. Experimental techniques for the determination of infinite-dilution activity coefficients have been reviewed by Alessi (1990).

The γ^∞ coefficients characterize the maximum nonideality of a mixture. This is an immediate consequence of the symmetrical normalization of activity coefficients. Therefore, the values of γ^∞ are particularly useful for calculating parameters needed in expressions for the excess Gibbs energy. In this manner the information contained in the γ^∞ coefficients can be extrapolated to finite concentrations. Calculation of parameters from γ^∞ data is straightforward for two-constant excess Gibbs energy models. For example, the Wilson parameters Λ_{12} and Λ_{21} are found from simultaneous solution of the relations

$$\ln \gamma_1^\infty = -\ln \Lambda_{12} - \Lambda_{21} + 1 \tag{3.4-28}$$

$$\ln \gamma_2^\infty = -\ln \Lambda_{21} - \Lambda_{12} + 1 \tag{3.4-29}$$

Pierotti, Deal and Derr (1959) developed a correlation for predicting γ^∞ for water, hydrocarbons, esters, aldehydes, ketones, nitriles, alcohols etc. in the temperature region from 25 to 100°C. This correlation has been established for numerous homologous series of solutes in some typical solvents such as water, paraffins and alcohols. Different equations have been found, in general, for different homologous series. The pertinent equations and their parameters have been summarized by Treybal (1963) and Reid et al. (1987). The average deviation in γ^{∞} is about 8 percent.

Thomas and Eckert (1984) proposed a method for predicting γ^{∞} from pure component parameters only. The method has been called MOSCED (modified separation of cohesive energy density). The functional form of MOSCED has been inspired by the regular solution theory and its extensions to polar mixtures. The MOSCED method utilizes five parameters associated with different contributions to the cohesive energy density; the dispersion parameter λ, the induction parameter q, the polar parameter τ and the acidity α and basicity β parameters. Additionally, two more parameters ψ and ξ have been introduced to account for the asymmetry of the γ_1^{∞} and γ_2^{∞} coefficients. The ψ and ξ parameters are functions of the remaining ones. In a binary mixture, the infinite-dilution activity coefficient for component 2 is

$$\ln \gamma_2^{\infty} = \frac{v_2}{RT}\left[(\lambda_1 - \lambda_2)^2 + \frac{q_1^2 q_2^2(\tau_1 - \tau_2)^2}{\psi_1} + \frac{(\alpha_1 - \alpha_2)(\beta_1 - \beta_2)}{\xi_1}\right] + \left[\ln\left(\frac{v_2}{v_1}\right)^{aa} + 1 - \left(\frac{v_2}{v_1}\right)^{aa}\right] \tag{3.4-30}$$

where v_2 is the liquid molar volume at 20°C. The γ_1^{∞} coefficient is obtained by interchanging the subscripts. The second term in square brackets on the right-hand side of Eq. (3.4-30) is a modified Flory–Huggins combinatorial term to account for differences in molecular size. This term has been improved by introducing a variable exponent aa. The MOSCED parameters τ, α, β, ψ, ξ and $\alpha\alpha$ are temperature dependent according to the following formulas:

$$\tau = \tau_{293}\left(\frac{293}{T}\right)^{0.4} \tag{3.4-31}$$

$$\alpha = \alpha_{293}\left(\frac{293}{T}\right)^{0.8} \tag{3.4-32}$$

$$\beta = \beta_{293}\left(\frac{293}{T}\right)^{0.8} \tag{3.4-33}$$

$$\psi = \mathrm{POL} + 0.011\alpha\beta \tag{3.4-34}$$

$$\xi = 0.68(\mathrm{POL} - 1) + \{3.4 - 2.4\exp[-0.023(\alpha_{293}\beta_{293})^{1.5}]\}^{(293/T)^2} \tag{3.4-35}$$

$$\mathrm{POL} = q^4[1.15 - 1.15\exp(-0.020\tau^3)] + 1 \tag{3.4-36}$$

$$aa = (0.953 - 0.00968)(\tau_2^2 + \alpha_2\beta_2) \tag{3.4-37}$$

The parameters v, λ, τ, q, α, β, ψ, ξ and aa are listed by Thomas and Eckert (1984) and Reid et al. (1987) for about 145 compounds. For other compounds the parameters can be estimated from approximate relations valid for some classes of compounds. The dispersion parameter λ is related to the refractive index for the sodium D line n_D:

$$\lambda = 10.3\,\frac{n_D^2 - 1}{n_D^2 + 2} + 3.02 \quad \text{(for nonaromatics, except for tertiary amines, nitriles and } CS_2\text{)} \tag{3.4-38}$$

$$\lambda = 19.5\,\frac{n_D^2 - 1}{n_D^2 + 2} + 2.79 \quad \text{(for aromatics)} \tag{3.4-39}$$

The induction parameter q takes the following values:

$$q = 1.0 \quad \text{(for saturated hydrocarbons)} \tag{3.4-40}$$

$$q = 0.9 \quad \text{(for aromatics)} \tag{3.4-41}$$

$$q = 1.0 - 0.5 \quad \text{(number C=C bonds)/(number C atoms)} \quad \text{for unsaturated aliphatics} \tag{3.4-42}$$

The polar, acidity and basicity parameters are estimated from

$$\tau, \alpha, \beta = C_{(\tau,\alpha,\beta)}\left(\frac{4.5}{3.5 + \text{number C atoms}}\right)\left(1 + \frac{\text{number C atoms} - 1}{100}\right) \tag{3.4-43}$$

with the following constants:

	C_τ	C_α	C_β		C_τ	C_α	C_β
Chlorides	2.69	0.42	0.25	Nitroalkanes	5.87	0.35	2.66
Bromides	2.47	0.39	0.38	Alcohols	1.65	7.49	7.49
Iodides	2.02	0.37	0.30	Esters	4.03	0.00	4.64[a]
Nitriles	5.84	0.33	4.00	Ketones	3.93	0.00	4.87[b]

[a]Number C atoms = 1 for methylformate.
[b]Number C atoms = 1 for acetone.

This prediction scheme cannot account for the effects of multifunctionality, secondary or tertiary positioning of the functional groups, chain branching, etc.

Overall, MOSCED provides good predictions of γ^∞. For 3357 γ^∞'s, an average error of 9.1 percent was achieved (with specific parameters for each compound). Moreover, calculations are simple because the model parameters refer only to pure compounds. The parameters can be estimated if they are missing. However, the method is inapplicable to aqueous systems and those for which steric effects are significant (e.g., systems containing triethylamine with an acid). The method works for systems with γ^∞'s below about 100.

If accurate values of γ^∞ are known, vapor–liquid equilibria can be reasonably predicted over the entire range of concentration. For example, Schreiber and Eckert (1971) used the Wilson equation for this purpose and found that the results of prediction are only slightly worse than those obtained by correlating VLE data in the whole concentration range. The average error in vapor composition is about 1.5 to 2 times larger. Schreiber and Eckert also showed that reasonable results are often obtained even when only one infinite-dilution activity coefficient (γ_1^∞ or γ_2^∞) is used.

3.4.4. Group Contribution Methods

The group contribution methods for the calculation of the excess Gibbs energy of a liquid mixture are particularly attractive because they make it possible to predict properties of mixtures for which no experimental data exist. The concept that a chemical molecule is an aggregate of functional groups was first proposed in thermochemistry for the correlation and prediction of pure component heats of formation. Its extension to mixtures is based on the assumption that a mixture of individual chemical compounds can be treated, with reasonable accuracy, as a mixture of functional groups constituting these compounds. Instead of parameters characterizing binary interactions between chemical compounds, a smaller number of parameters characterizing the binary interactions between functional groups is used. The number of functional groups is much smaller than the number of chemical compounds. Thousands of liquid mixtures can be formed from a few dozens of functional groups. The group interaction parameters can be established by regressing the available fluid phase equilibrium data.

A fundamental assumption is the additivity of contributions. A contribution made by one group in the molecule is assumed to be independent of that made by any other group within that molecule. Such assumption is approximate and can be applied only when the contribution of any one group in a molecule can be treated as independent of the nature of other groups within that molecule.

Accuracy of correlation can be improved by distinguishing more and more different groups. The practical number of groups is a result of a compromise. The number of groups must remain small, but should not neglect significant effects of molecular structure on the mixture properties. This problem was carefully evaluated (Fredenslund et al., 1975; Kojima and Tochigi, 1979) for the particular methods.

Two group contribution methods have gained widespread popularity in chemical engineering computations: the analytical solution of groups (ASOG) and UNIFAC (UNIquac FACtor). They are based on the same principle. The difference is in the equation used for representing the Gibbs excess energy of a mixture. For ASOG the Wilson equation is used while for UNIFAC the UNIQUAC equation is employed.

Both methods represent the activity coefficient (γ_i) of a compound i in a mixture as a sum of the combinatorial (γ_i^{COM}) and residual (γ_i^{RES}) contributions. The combinatorial (entropic) part represents the differences in size and shape of the molecules and is computed from pure-component properties while the residual part (group interaction contribution) is due to intermolecular forces and is computed from the properties of mixtures.

The activity coefficient in the ASOG method proposed by Derr and Papadopoulos (1959), Pierotti et al. (1959) and Redlich et al. (1959) is

$$\ln \gamma_i = \ln \gamma_i^{\text{COM}} + \ln \gamma_i^{\text{RES}} \tag{3.4-44}$$

where the combinatorial contribution is given by the Flory–Huggins expression for the athermal entropy of mixing

$$\ln \gamma_i^{\text{COM}} = 1 - \frac{s_i}{\sum_{j=1}^{n} s_j x_j} + \ln \frac{s_i}{\sum_{j=1}^{n} s_j x_j} \tag{3.4-45}$$

where s_j is a measure of the size of a molecule and is equal to the number of nonhydrogen atoms in the molecule, i.e., the number of functional groups.

The entropic contribution to the activity coefficient is usually small but may be significant in the case of molecules of different sizes. The residual part of the activity coefficient is related to group interaction contributions:

$$\ln \gamma_i^{\text{RES}} = \sum_{k=1}^{m} s_{k,i} \ln \frac{\Gamma_k}{\Gamma_{k,i}^0} \tag{3.4-46}$$

where Γ_k is the activity coefficient of group k of component i in a mixture consisting of m group types. Coefficient $\Gamma_{k,i}^0$ is the activity coefficient of group k in a standard state, which is the pure compound i possessing the group. The $\Gamma_{k,i}^0$ coefficient has been introduced to normalize the activity coefficient to unity for pure components.

The group activity coefficient (Γ_k) is, at constant temperature, a function of concentration expressed by the group fractions

$$X_l = \frac{s_l}{\sum_{j=1}^{n} s_j x_j} \tag{3.4-47}$$

The values of group activity coefficients (Γ_k) are computed from the Wilson equation written for the group fractions

$$\ln \Gamma_k = -\ln \sum_{l=1}^{m} X_l A_{l,k} + \left(1 - \sum_{l=1}^{m} \frac{A_{l,k} X_l}{\sum_{l=1}^{m} A_{i,k} x_i} \right) \tag{3.4-48}$$

where x_i is the mole fraction of compound i in the mixture and $A_{i,k}$ are temperature-dependent parameters corresponding to binary interactions between groups i and k ($A_{i,k} \neq A_{k,i}$ and $A_{i,i} = 1$). Kojima and Tochigi (1979) expressed the temperature dependence of the $A_{i,k}$ parameters as

$$\ln A_{i,k} = m_{i,k} + \frac{T}{n_{i,k}} \tag{3.4-49}$$

where $m_{i,k}$ and $n_{i,k}$ are the actual temperature-independent group parameters.

Equation (3.4-48) is also used to calculate the group activity coefficients in the standard state ($\Gamma^0_{k,i}$). For this purpose the pure component i is treated as a mixture of groups and Eq. (3.4-48) is used for the calculation of $\Gamma^0_{k,i}$ When the compound contains only one type of groups, then $\Gamma^0_{k,i} = 1$ (e.g., for cyclohexane).

The values of parameters $A_{i,k}$ depend only on the properties of groups i and k and are independent of the properties of the chemical molecule. The values of $A_{i,k}$ can be regressed from vapor–liquid equilibrium data and used for the prediction of activity coefficients of other mixtures. A collection of ASOG parameters has been published by Kojima and Tochigi (1979). The parameter table has been recently extended by Correa et al. (1989) and Tochigi et al. (1990).

The UNIFAC method proposed by Fredenslund et al. (1975) takes into account the size and shape of molecules in a mixture. It is based on the UNIQUAC equation, described in Section 3.3.3. The UNIFAC model is more elaborate than ASOG. The UNIQUAC combinatorial and residual contributions to the activity coefficient take the following form for UNIFAC:

$$\ln \gamma_i^{\mathrm{COM}} = 1 - \frac{\Phi_i}{x_i} + \ln \frac{\Phi_i}{x_i} - 5q_i\left(1 - \frac{\Phi_i}{\theta_i} + \ln \frac{\Phi_i}{\theta_i}\right) \tag{3.4-50}$$

where the values Φ_i and θ_i correspond to the volume and surface fractions of UNIQUAC, respectively.

$$\Phi_i = \frac{r_i}{\sum_{j=1}^{n} r_j x_j} \quad \text{(summation over all components)} \tag{3.4-51}$$

$$\theta_i = \frac{q_i}{\sum_{j=1}^{n} q_j x_j} \quad \text{(summation over all components)} \tag{3.4-52}$$

$$r_i = \sum_{k=1}^{m} \nu_k^{(i)} R_k \quad \text{(summation over all groups)} \tag{3.4-53}$$

$$q_i = \sum_{k=1}^{m} \nu_k^{(i)} Q_k \quad \text{(summation over all groups)} \tag{3.4-54}$$

The residual contribution is calculated similarly as in the case of ASOG:

$$\ln \gamma_i^{\mathrm{RES}} = \sum_{k=1}^{m} \nu_k^{(i)} (\ln \Gamma_k - \ln \Gamma_k^{(i)}) \quad \text{(summation over all groups)} \tag{3.4-55}$$

where

$$\ln \Gamma_k = Q_k\left(1 - \ln\left(\sum_j \theta_j \Psi_{jk}\right)\right) - \sum_j \left(\frac{\theta_j \Psi_{kj}}{\sum_l \theta_l \Psi_{lj}}\right) \tag{3.4-56}$$

$$\Psi_{lj} = \exp\left(-a_{lj}T\right) \tag{3.4-57}$$

$$\theta_j = \frac{Q_j X_j}{\sum_{l=1}^{m} Q_l X_l} \tag{3.4-58}$$

$$X_j = \frac{\sum_{i=1}^{n} \nu_j^{(i)} x_i}{\sum_{i=1}^{n} \sum_{j=1}^{m} \nu_j^{(i)} x_i} \tag{3.4-59}$$

The number of components in a mixture is denoted by n while m denotes the number of group types. The quantity $\nu_k^{(i)}$ is the number of groups of type k in a molecule of species i. The values of the subgroup parameters R_k and Q_k and the group interaction parameters $a_{j,k}$ have been published by Fredenslund et al. (1977). Since 1977, the UNIFAC parameter matrix has been thoroughly updated by Gmehling et al. (1982), Macedo et al., 1983, Tiegs et al. (1987) and Hansen et al. (1991).

The accuracy of prediction using ASOG and UNIQUAC is similar for the cases when both methods are applicable. UNIQUAC is more universal and the available collection of parameters is more comprehensive.

It has been found that UNIFAC with parameters determined from VLE data (UNIFAC-VLE) does not yield reasonable quantitative prediction of liquid–liquid equilibria. This is due to inherent limitations of G^E models, which are incapable of representing VLE and LLE with sufficient accuracy. Therefore, Magnussen et al. (1981) published the UNIFAC-LLE parameter table with parameters adjusted to match liquid–liquid equilibria at 25°C. The UNIFAC-LLE model is identical to UNIFAC-VLE but contains different parameters.

The last decade witnessed many efforts to improve the predictive ability of UNIFAC. One of the weaknesses of UNIFAC is the deficiency of the combinatorial contribution, which predicts too large negative deviations from ideality when molecules differing in size are considered. This is important for hydrocarbon mixtures for which the residual contribution is zero. The second weakness is the unsatisfactory prediction of heats of mixing and infinite-dilution activity coefficients. To overcome these difficulties, modifications of UNIFAC have been proposed independently by Gmehling and Weidlich (1986) (Mod. UNIFAC–Dortmund) and Larsen et al. (1987) (Mod. UNIFAC–Lyngby).

Gmehling and Weidlich (1986) empirically modified the combinatorial term [Eq. (3.4-50)] by introducing a 3/4 exponent into the volume fraction [Eq. (3.4-51)]:

$$\Phi_i = \frac{r_i^{3/4}}{\sum_{j=1}^{n} r_j^{3/4} x_j} \tag{3.4-60}$$

Larsen et al. (1987) replaced Eq. (3.4-50) by a Flory–Huggins-type expression with a noninteger exponent in the volume fraction:

$$\ln \gamma_i^{\mathrm{COM}} = 1 - \frac{\Phi_i}{x_i} + \frac{\ln \Phi_i}{x_i} \tag{3.4-61}$$

$$\Phi_i = \frac{r_i^{2/3}}{\sum_{j=1}^{n} r_j^{2/3} x_j} \tag{3.4-62}$$

The above expressions improved the results of predicting VLE and infinite-dilution activity coefficients for mixtures containing hydrocarbons of moderate chain length. More recently, Elbro et al. (1988) further improved the γ_i^{COM} term by including both combinatorial and free-volume effects:

$$\ln \gamma_i^{\mathrm{COM}} = 1 - \frac{\Phi_i^{fv}}{x_i} + \frac{\ln \Phi_i^{fv}}{x_i} \tag{3.4-63}$$

where

$$\Phi_i^{fv} = \frac{x_i(v_i - v_i^*)}{\sum_j x_j(v_j - v_j^*)} \tag{3.4-64}$$

where v^* is the hard-core volume and v is the molar volume. This model has been shown to better reproduce activity coefficients in hydrocarbon mixtures including those containing polymers.

The residual contribution has been modified by introducing temperature-dependent group interaction parameters:

$$a_{jk} = a_{jk,1} + a_{jk,2}(T - T_0) + a_{jk,3}\left[T \ln\left(\frac{T_0}{T}\right) + T - T_0\right] \tag{3.4-65}$$

where the reference temperature T_0 is 298.15 K. This method, called SUPERFAC (Larsen et al., 1987), made it possible to improve the prediction of excess enthalpies. A disadvantage of this method is that a special parameter table has to be used. This table is less comprehensive than the original UNIFAC parameter table. Table 3.4-2 compares the predictive ability of the original UNIFAC, modified UNIFAC (Dortmund), modified UNIFAC (Lyngby) and ASOG.

It should be noted that the infinite-dilution activity coefficients have been used for determining the parameters of only the modified UNIFAC–Dortmund. Table 3.4-2 shows that the modified versions of UNIFAC do not offer much improvement for VLE predictions. The predicted excess enthalpies are generally better although this improvement is not always significant. Inclusion of the γ^∞ data into the data base of the Dortmund modification resulted in an improvement in predicting this quantity without affecting the prediction of VLE at finite concentrations. A drawback is once more the necessity to use a special parameter table.

Table 3.4-2
COMPARISON OF UNIFAC, MODIFIED UNIFAC (DORTMUND), MODIFIED UNIFAC (LYNGBY) AND ASOG WITH RESPECT TO THE PREDICTION OF VAPOR–LIQUID EQUILIBRIA, HEATS OF MIXING AND INFINITE-DILUTION ACTIVITY COEFFICIENTS FOR MIXTURES OF ALKANES WITH ALKANES, AROMATICS, ALCOHOLS, KETONES AND ETHERS (AFTER GMEHLING ET AL., 1990).

Mixtures (alkane + ···)	Data Sets	UNIFAC δQ	Mod. UNIFAC Dortmund δQ	Mod. UNIFAC Lyngby δQ	ASOG δQ
		Vapor–Liquid Equilibria			
Alkanes	49	1.52	1.36	1.65	1.36
Aromatics	95	1.94	1.44	1.34	1.28
Alcohols	106	5.54	3.42	5.32	5.59
Ketones	53	4.10	2.99	2.61	2.75
Ethers	24	6.01	2.00	5.97	8.19
		Heats of Mixing			
Alkanes	382	76.7	83.1	76.7	90.2
Aromatics	150	50.9	6.5	44.0	16.2
Alcohols	331	17.9	11.6	39.9	21.1
Ketones	126	31.5	9.0	37.3	44.4
Ethers	88	28.3	14.5	25.8	107.1
		Infinite-Dilution Activity Coefficients			
Alkanes	1077	21.5	6.1	7.4	24.9
Aromatics	159	15.6	5.5	27.4	16.1
Alcohols	621	37.7	18.3	31.3	37.6
Ketones	313	19.4	12.8	15.2	20.2
Ethers	75	17.2	3.5	16.4	34.3

After Gmehling et al. (1990).

$$\delta Q = \left(\frac{100}{n}\right) \sum_{i=1}^{n} \left| \frac{Q_i^{\text{calc}} - Q_i^{\text{exp}}}{Q_i^{\text{exp}}} \right| \quad \text{where } Q = y \text{ or } H^E \text{ or } \gamma^{\infty}.$$

The parameters of UNIFAC and its modifications are being constantly updated. Therefore, we do not give them here. The latest versions can be obtained from the authors.

To illustrate the predictive capability of UNIFAC, the accuracy of predicting vapor–liquid equilibria in ternary systems using the UNIFAC method is compared in Table 3.4-3 with the accuracy of predicting the equilibria using the Wilson, NRTL and UNIQUAC equations with parameters fitted to corresponding *binary* data. To show the maximum accuracy that can be obtained, the results of correlating individual *ternary* data using the local composition equations are also listed.

The systems shown in Table 3.4-3 were not used to establish the UNIFAC parameter table. The errors of the correlation of ternary data and prediction from binary data are almost identical for the local composition

Table 3.4-3
PREDICTION OF TERNARY VAPOR–LIQUID EQUILIBRIA USING THE UNIFAC METHOD AND LOCAL COMPOSITION MODELS WITH PARAMETERS FITTED TO CORRESPONDING BINARY DATA[a,b]

System	T (K) or P (MPa)	*UNIFAC (mixture of groups)*		*Wilson*		*NRTL (binary mixtures)*		*UNIQUAC*	
		$\delta(P)$	$\delta(y)$	$\delta(P)$	$\delta(u)$	$\delta(P)$	$\delta(y)$	$\delta(P)$	$\delta(y)$
Prediction									
Ethanol + chloroform + hexane	328	5.43	2.40	1.22	0.87	1.36	1.01	1.74	0.63
Acetone + chloroform + hexane	308	4.02	2.19	1.14	0.77	1.44	1.00	2.34	0.80
Acetone + chloroform + 4-methyl-2-pentanone	0.1	—	1.04	—	0.53	—	0.77	—	—
Correlation (direct fit to ternary data)									
Ethanol + chloroform + hexane	328	—	—	0.66	0.58	0.57	0.63	0.61	0.66
Acetone + chloroform + hexane	308	—	—	1.70	0.86	1.69	0.78	1.64	0.76
Acetone + chloroform	0.1	—	—	—	0.79	—	0.80	—	0.78

[a] Direct correlation using the local composition models is shown for comparison.

[b] $\delta(P) = \left(\frac{100}{n}\right) \sum_{i=1}^{n} \left| \frac{P_i^{calc} - P_i^{exp}}{P_i^{exp}} \right| \qquad \delta(y) = \left(\frac{100}{n}\right) \sum_{i=1}^{n} |y_i^{calc} - y_i^{exp}|.$

equations. The errors of the UNIFAC method are twice as large as those of the direct correlation, which is still a very good result.

A group contribution model for gas solubilities at pressures below about 15 bar has been developed by Sander et al. (1983) on the basis of UNIFAC. The method makes use of the unsymmetric convention for calculating VLE:

$$y_i \phi_i^G P = x_i \gamma_i^* H_{i,s} \tag{3.4-66}$$

where $H_{i,s}$ is the Henry constant of gas i in solvent s and γ_i^* is the activity coefficient in the unsymmetric convention. This relation can be rewritten using the symmetric activity coefficient γ_i:

$$y_i \phi_i^G P = \frac{x_i H_{i,r} \gamma_i}{\gamma_{i,r}^{\infty}} \tag{3.4-67}$$

where $H_{i,r}$ is the Henry constant in a reference solvent and $\gamma_{i,r}^{\infty}$ is the infinite-dilution activity coefficient of gas i in the reference solvent r. Sander et al. applied a group contribution method to calculate γ_i and $\gamma_{i,r}^{\infty}$.

Extensions of UNIFAC have been developed for mixtures containing electrolytes (Sander et al., 1986) and polymers (Oishi and Prausnitz, 1978). UNIFAC and its modifications have been also used in methods for predicting surface tension (Suarez et al., 1989) and in flash point calculations (Wu et al., 1988). The various methods based on UNIFAC have been reviewed by Fredenslund and Rasmussen (1988), Fredenslund (1989) and Gmehling et al. (1990).

A group contribution method of a different type has been proposed by Kehiaian (1977, 1985). The model, called DISQUAC (DISpersive QUASI-Chemical), is an extended quasi-chemical pseudolattice group contribution model in terms of group surface interactions. DISQUAC resembles ASOG and UNIFAC in that it separates the combinatorial and residual (here called *interactional*) contributions to the excess Gibbs energy. The interactional terms in G^E, as well as in H^E, are given as the sum of the dispersive contribution, $G_{\text{int}}^{E,\text{dis}}$ or $H^{E,\text{dis}}$, and the quasi-chemical contribution $G_{\text{int}}^{E,\text{quac}}$ and $H^{E,\text{quac}}$:

$$G^E = G_{\text{comb}}^E + G_{\text{int}}^{E,\text{dis}} + G_{\text{int}}^{E,\text{quac}} \tag{3.4-68}$$

$$H^E = H^{E,\text{dis}} + H^{E,\text{quac}} \tag{3.4-69}$$

where G_{comb}^E is expressed by a Flory–Huggins combinatorial term and the dispersive contributions are given for a binary mixture by

$$G_{\text{int}}^{E,\text{dis}} = (q_1 x_1 + q_2 x_2)\xi_1 \xi_2 g_{12}^{\text{dis}} \tag{3.4-70}$$

$$H^{E,\text{dis}} = (q_1 x_1 + q_2 x_2)\xi_1 \xi_2 h_{12}^{\text{dis}} \tag{3.4-71}$$

where

$$g_{12}^{\text{dis}} = -\frac{1}{2}\sum_s \sum_t (\alpha_{s1} - \alpha_{s2})(\alpha_{t1} - \alpha_{t2}) g_{st}^{\text{dis}} \tag{3.4-72}$$

$$h_{12}^{\text{dis}} = -\frac{1}{2}\sum_s \sum_t (\alpha_{s1} - \alpha_{s2})(\alpha_{t1} - \alpha_{t2}) h_{st}^{\text{dis}} \tag{3.4-73}$$

g_{st}^{dis} and h_{st}^{dis} are the dispersive interchange parameters of the (s, t) contact, α_{si} is the molecular surface fraction of surface of type s on a molecule of type i, q_i is the total relative molecular area of a molecule of type i and ξ_i is the surface fraction of component i in a mixture. Equations (3.4-72) and (3.4-73) bear resemblance to the regular solution equations, rewritten in terms of group surface fractions instead of volume fractions. The quasi-chemical contributions (cf. Guggenheim, 1952) are

$$G_{\text{int}}^{E,\text{quac}} = x_1 z q_1 \sum_s \alpha_{s1} \ln\left(\frac{X_s \alpha_{s1}}{X_{s1}\alpha_s}\right) + x_2 z q_2 \sum_s \alpha_{s2} \ln\left(\frac{X_s \alpha_{s2}}{X_{s2}\alpha_s}\right) \tag{3.4-74}$$

$$H^{E,\text{quac}} = \frac{1}{2}(q_1 x_1 + q_2 x_2)\sum_s \sum_t [X_s X_t - (\xi_1 X_{s1} X_{t1} + \xi_2 X_{s2} X_{t2})\eta_{st} h_{st}^{\text{quac}} \tag{3.4-75}$$

where

$$\eta_{st} = \exp\left(\frac{-g_{st}^{\text{quac}}}{zRT}\right) \tag{3.4-76}$$

g_{st}^{quac} and h_{st}^{quac} are the quasi-chemical interchange parameters of the (s, t) contact and z is the lattice coordination number. The quantities X_s and X_t are obtained by solving the system of λ equations (λ is the number of contact surfaces)

$$X_s\left(X_s + \sum_t X_t \eta_{st}\right) = \alpha_s \tag{3.4-77}$$

X_{si} and X_{ti} are the solutions of the system of equations (3.4-77) for a pure component i. The dispersive contributions [Eqs. (3.4-72) and (3.4-73) are obtained from the quasi-chemical equations-chemical equations (3.4-74) and (3.4-75) when $z \to \infty$. The temperature dependence of the dispersive or quasi-chemical parameters g_{st} has been expressed as

$$\frac{g_{st}(T)}{RT} = C_{st,1} + C_{st,2}\left[\left(\frac{T^0}{T}\right) - 1\right] \tag{3.4-78}$$

$$\frac{h_{st}}{RT} = C_{st,2}\left(\frac{T^0}{T}\right) \tag{3.4-79}$$

where $T^0 = 298.15$ K. The coefficients $C_{st,1}$ and $C_{st,2}$ are correlated within homologous series. If both groups, s and t, are nonpolar, the contact (s, t) is characterized by dispersive coefficients only. If one group is polar, both dispersive and quasi-chemical coefficients are used. The shapes of G^E and H^E curves depend on the relative amounts of dispersive and quasichemical terms.

Contrary to ASOG and UNIFAC, the purpose of DISQUAC is not to arrive at an engineering-oriented model of wide applicability. This is not a universal predictive method with a large number of groups. The aim of DISQUAC is to analyze, interpret and critically evaluate VLE and H^E data for mixtures, mainly in homologous series. The subjects of analysis include the effects of molecular environment on the functional group and its characteristic parameters. The group contribution formalism is applied to study the effects of the vicinity of branched alkyl groups, the inclusion of a functional group in a ring, proximity of two functional groups etc.

Other group contribution methods have been proposed by Vera and Vidal (1984) (the SIGMA method), Abusleme and Vera (1985, 1989a), Koukios et al. (1984) and Eckart et al. (1986). They are essentially based on modifications of the quasi-chemical theory. However, their range of applicability is limited in comparison with ASOG and UNIFAC. In general, the latter two methods can be recommended for routine engineering calculations.

PROBLEM OF SELECTING THE FUNCTIONAL GROUPS

All the group contribution methods described above have been established by dividing molecules into a number of functional groups. This division has been guided by some general knowledge of the structure of molecules. However, Wu and Sandler (1991) showed that it is possible to develop a relatively simple theoretical prescription for defining functional groups on the basis of quantum mechanical ab initio molecular orbital calculations for single molecules. The authors postulated that:

1. The geometry of the functional group should be the same independent of the molecules in which the group occurs.
2. Each atom in a functional group should have approximately the same charge in all molecules in which the group occurs, and the group should be approximately electroneutral. This condition results from the fact that the intermolecular interaction energy is affected by the distribution of charges. Therefore, if a group is always to make the same contribution to a molecule in a mixture, its net charge must be unchanging.
3. More obviously, each group should be the smallest entity so that a molecule can be divided into a collection of electroneutral groups.

To satisfy the above conditions, Wu and Sandler (1991) selected groups that are in some cases different from those used in UNIFAC. The new groups were shown to lead to better UNIFAC predictions, especially for mixtures in which proximity effects are important (i.e., two functional groups are likely to interact with each other within one molecule). However, a number of cases still remain in which the group contribution methods are likely to be inaccurate due to the strong influence of intramolecular proximity effects on intermolecular interactions. Wu and Sandler (1991) showed that this is the case, for example, for mixtures of water with cyclic ethers. The hydrogen bond energy between a water molecule and an ether group is different for mixtures of H_2O with tetrahydrofuran (with one ether group) and 1,3-dioxolane (with two neighboring groups). The energy for H_2O + 1,4-dioxane is intermediate because the two ether groups in 1,4-dioxane are separated from each other. Therefore, the coupling of intermolecular hydrogen bonding and intramolecular proximity effects may result in a change in the intermolecular interactions that cannot be accounted for by redefining the groups. This inherent limitation of group contribution methods should be taken into account when applying the methods described above to fluids with more complex interactions.

REFERENCES

Abbott, M. M. and H. C. Van Ness, 1975. *AIChE J.* 21: 62.

Abbott, M. M., J. K. Floess, G. E. Walsh and H. C. Van Ness, 1975. *AIChE J.* 21: 72.

Abrams, D. C. and J. M. Prausnitz, 1975. *AIChE J.* 21: 116.

Abusleme, J. A. and J. H. Vera, 1985. *Fluid Phase Equilibria* 22: 123.

Abusleme, J. A. and J. H. Vera, 1989a. *Fluid Phase Equilibria* 44: 273.

Abusleme, J. A. and J. H. Vera, 1989b. *AIChE J.* 21: 116.

Aguirre-Ode, F., 1986. *Fluid Phase Equilibria* 30: 315.

Alessi, P., 1990. In S. Malanowski and A. Anderko (Eds.), *Thermodynamics of Fluids: Measurement and Correlation*, World Scientific, Singapore.

Alessi, P., M. Fermeglia and I. Kikic, 1987. Activity Coefficients at Infinite Dilution Data Bank, University of Trieste, Trieste, Italy.

Arlt, W., M. A. E. Macedo, P. Rasmussen and J. M. Sorensen, 1979–1988. *Liquid-Liquid Equilibrium Data Collection*. Vols. 1–4. Chemistry Data Series, Dechema, Frankfurt/Main.

Barker, J. A., 1953. *Austral. J. Chem.* 6: 207.

Barton, A. F. M., 1983. *Handbook of Solubility Parameters and Other Cohesion Parameters*, CRC Press, Boca Raton, FL.

Beattie, J. A., 1949. *Chem. Rev.* 44: 141.

Bielousov, V. P. and A. G. Morachevski, 1970. *Teploty smesenija zidkostej* (in Russian), Nauka, Leningrad.

Bielousov, V. P., A. G. Morachevskii and M. Y. Panov, 1981. *Teplotnye svoistva* (in Russian), Nauka, Leningrad.

Bondi, A., 1968. *Physical Properties of Molecular Crystals, Liquids, and Glasses*, Wiley, New York.

Brandani, V., 1983. *Fluid Phase Equilibria*, 12: 87.

Brandani, V. and F. Evangelista, 1984. *Fluid Phase Equilibria* 17: 281.

Brown, I., 1952. *Aust. J. Sci. Res. Ser. A* 5: 530.

Bruin, S. and J. M. Prausnitz, 1971. *Ind. Eng. Chem. Proc. Des. Dev.*, 10: 562.

Chen, S. A. and E. B. Bagley, 1978. *Chem. Eng. Sci.*, 33: 153, 161.

Choi, P. B. and E. McLaughlin, 1983. *AIChE J.*, 29: 150.
Choliński, J., A. Szafrański and D. Wyrzykowska-Stankiewicz, 1986. Computer-Aided Second Virial Coefficient Data for Organic Individual Compounds and Binary Systems, PWN–Polish Scientific Publishers, Warszawa.
Christensen, C., J. Gmehling, P. Rasmussen and U. Weidlich, 1984. *Heats of Mixing Data Collection*, Vols. 1–2, Chemistry Data Series, Dechema, Frankfurt/Main.
Christensen, J., R. W. Hanks and R. M. Izatt, 1982. *Handbook of Heats of Mixing*, Wiley, New York.
Cibulka, I. and I. Nagata, 1987. *Fluid Phase Equilibria* 35: 19.
Correa, A., J. Tojo, J. M. Correa and A. Blanco, 1989. *Ind. Eng. Chem. Res.* 28: 609.
Derr, E.L. and M. N. Papadopoulos, 1959. *J. Am. Chem. Soc.* 81: 2285.
Dolezalek, F., 1908. *Z. Phys. Chem.* 64: 727.
Dymond, J. H. and E. B. Smith, 1980. *The Virial Coefficients of Gases and Gaseous Mixtures*, Clarendon Press, Oxford.
Eckart, D. E., D. W. Arnold, R. A. Greenkorn and K. C. Chao, 1986. *AIChE J.* 32: 307.
Elbro, H. S., A. Fredenslund and P. Rasmussen, 1988. Report SEP 8819, Istituttet for Kemiteknik, DTH. Lyngby, Denmark.
Flory, P. J., 1941. *J. Chem. Phys.* 9: 660.
Flory, P. J., 1942. *J. Chem. Phys.* 10: 51.
Fredenslund, A., 1989. *Fluid Phase Equilibria* 52: 135.
Fredenslund, A. and P. Rasmussen, 1988. Report SEP 8807, Instituttet for Kemiteknik, DTH, Lyngby, Denmark.
Fredenslund, A., R. L. Jones and J. M. Prausnitz, 1975. *AIChE J.* 21: 1086.
Fredenslund, A., J. Gmehling and P. Rasmussen, 1977. *Vapor-Liquid Equilibria Using UNIFAC: a Group-Contribution Method*, Elsevier, Amsterdam.
French, H. T. and R. H. Stokes, 1981. *J. Phys. Chem.* 85: 3347.
Funk, E. W. and J. M. Prausnitz, 1970. *Ind. Eng. Chem.* 62: 8.
Gaube, J., L. Karrer and P. Spellucci, 1987. *Fluid Phase Equilibria* 33: 223.
Glasstone, S., 1946. *Textbook of Physical Chemistry*, 2nd ed., Van Nostrand, Princeton, NJ.
Gmehling, J. and U. Weidlich, 1986. *Fluid Phase Equilibria* 27: 171.
Gmehling, J., Onken, U., Arlt, W., Rarey-Nies, J. and Tiegs, D., Vapour-Liquid Equilibrium Data Collection, Chemistry Data Series; Dechema, Frankfurt/Main.
Gmehling, J. (Ed.), 1977. *Dortmund Data Bank*, Lehrstuhl Technische Chemie B, Universität Dortmund.
Gmehling, J., P. Rasmussen and A. Fredenslund, 1982. *Ind. Eng. Chem. Proc. Des. Dev.* 21: 118.
Gmehling, J., D. Tiegs, A. Medina, M. Soares, J. Bastos, P. Alessi and I. Kikic, 1986. Activity Coefficients at Infinite Dilution, Chemistry Data Series, Dechema, Frankfurt/Main.
Gmehling, J., D. Tiegs and U. Knipp, 1990. *Fluid Phase Equilibria* 54: 147.
Gmehling, J. and U. Weidlich, 1986. *Fluid Phase Equilibria* 24: 171.

Guggenheim, E. A., 1952. *Mixtures*, Clarendon Press, Oxford.

Hala, E., J. Pick, V. Fried and O. Vilim, 1967. *Vapor-Liquid Equilibrium*, Pergamon Press, Oxford.

Hala, E., I. Wichterle, J. Polak and T. Boublik, 1968. *Vapour-Liquid Equilibrium Data at Normal Pressure*, Pergamon Press, London.

Handa, Y. P. and G. C. Benson, 1979. *Fluid Phase Equilibria* 3: 185.

Hansen, H. K., P. Rasmussen, A. Fredenslund, M. Schiller and J. Gmehling, 1991. *Ind. Eng. Chem. Res.*, 30: 2241.

Haskell, R. W., H. B. Holliger and H. C. Van Ness, 1968. *J. Phys. Chem.* 13: 4534.

Hellwege, K. H., 1976. Mischungs- und Lösungswärmen. In, *Landolt-Börnstein: Zahlenwerte und Funktionen aus Naturwissenschaften und Technik*, 4th ed., New Series, Group IV, Vol. II; Springer-Verlag, Heidelberg.

Helpinstill, J. G. and M. Van Winkle, 1968. *Ind. Eng. Chem. Process Des. Dev.* 2: 213.

Hildebrand, J. H., 1929. *J. Am. Chem. Soc.* 51: 66.

Hildebrand, J. H. and R. L. Scott, 1962. *Regular Solutions*, Prentice-Hall, Englewood Cliffs, NJ.

Hildebrand, J. H. and S. E. Wood, 1933. *J. Chem. Phys.* 1: 817.

Hildebrand, J. H., J. M. Prausnitz and R. L. Scott, 1970. *Regular and Related Solutions*, Van Nostrand Reinhold, New York.

Hirata, M., S. Ohe and K. Nagahama, 1975. *Computer Aided Data Book of Vapour-Liquid Equilibria*, Kodansha-Elsevier, Tokyo.

Hofman, T., 1990. *Fluid Phase Equilibria* 55: 39.

Hofman, T. and I. Nagata, 1986a. *Fluid Phase Equilibria* 25: 113.

Hofman, T. and I. Nagata, 1986b. *Fluid Phase Equilibria* 28: 233.

Hofman, T. and I. Nagata, 1987. *Fluid Phase Equilibria* 33: 29.

Horsley, L. H., 1952, 1962, 1973. *Azeotropic Data, Advances in Chemistry Series*, Vols. 6, 35, 116, ACS, Washington, D.C.

Huggins, M. L., 1941. *J. Chem. Phys.* 9: 440.

Huggins, M. L., 1942. *Ann. N.Y. Acad. Sci.* 1: 431.

IUPAC, 1979. *Manual of Symbols and Terminology for Physicochemical Quantities and Units, Pure & Appl. Chem.*, 51: 1.

Kehiaian, H. V., 1977. *Ber. Bunsenges. Phys. Chem.* 81: 908.

Kehiaian, H. V., 1985. *Pure & Appl. Chem.* 57: 15.

Kehiaian, H. and A. Treszczanowicz, 1966. *Bull. Acad. Pol. Sci., Ser. Sci. Chim.*, 14: 891.

Kempter, H. and R. Mecke, 1940. *Z. Phys. Chem.* B46: 229.

Kogan, V. B., V. M. Fridman and V. V. Kafarov, 1966. *Ravnovesie mezdu zidkostju i parom* (in Russian), Nauka, Leningrad.

Kojima, K. and K. Tochigi, 1979. *Prediction of Vapor-Liquid Equilibrium by the ASOG Method*, Kodansha Ltd., Elsevier, Amsterdam.

Koukios, E. J., C. H. Chien, R. A. Greenkorn and K. C. Chao, 1984. *AIChE J.*, 30: 662.

Kretschmer, C. B. and R. Wiebe, 1954. *J. Chem. Phys.* 22: 1697.

Książczak, A., 1986a. *Fluid Phase Equilibria* 28: 39.

Książczak, A., 1986b. *Fluid Phase Equilibria* 28: 57.

Książczak, A. and A. Anderko, 1987. *Fluid Phase Equilibria* 35: 127.
Książczak, A. and H. Buchowski, 1984. *Fluid Phase Equilibria* 16: 353, 361.
Ladurelli, A. J., C. H. Eon and G. Guiochon, 1975. *Ind. Eng. Chem. Fundam.* 14: 191.
Landeck, H., H. Wolff and R. Götz, 1977. *J. Phys. Chem.* 81: 718.
Landolt-Börnstein, *Zahlenwerte und Funktionen aus Physik, Chemie, Astronomie, Geophysik und Technik*, 4th ed., Vol. II, Part 2a; Vol. IV, Part 4, "New Series," Group IV, Vol. III, Springer, Berlin.
Larsen, B. L., P. Rasmussen and A. Fredenslund, 1987. *Ind. Eng. Chem. Res.* 26: 2274.
Liu, A., F. Kohler, L. Karrer, J. Gaube and P. Spellucci, 1989. *Pure & Appl. Chem.* 61: 1441.
Macedo, E. A., U. Weidlich, J. Gmehling and P. Rasmussen, 1983. *Ind. Eng. Chem. Proc. Des. Dev.* 22: 676.
Mączyński, A., A. Biliński, A. Skrzecz, Z. Mączyńska, T. Treszczanowicz, P. Oracz and K. Dunajska, 1976–1988. *Verified Vapour-Liquid Equilibrium Data*, Vols. 1–10, PWN–Polish Scientific Publishers, Warszawa.
Mączyński, A. (Ed.) 1988–. *Floppy-Book, Vapor-Liquid Equilibrium*, Instytut Chemii Fizycznej P.A.N., Warszawa.
Magnussen, T. P., P. Rasmussen and A. Fredenslund, 1981. *Ind. Eng. Chem. Proc. Des. Dev.* 20: 331.
Malanowski, S., 1974. *Równowaga Ciecz-Para* (in Polish), PWN–Polish Scientific Publishers, Warszawa.
Malesiński, W., 1965. *Azeotropy and other Theoretical Problems of Vapour–Liquid Equilibrium*, Wiley, London.
Margules, M., M., 1895. Sitzber. *Akad. Wiss. Wien, Math. Naturw.* (*2A*) 104: 1234.
Messow, U. and H.-J. Bittrich, 1981. In, H.-J. Bittrich (Ed.), *Modellierung von Phasengleichgewichten als Grundlage von Stofftrennprozessen*, Akademie Verlag, Berlin.
Morris, J. W., P. J. Mulvey, M. M. Abbott and H. C. Van Ness, 1975. *J. Chem. Eng. Data* 20: 403.
Nagata, I., 1973. *Z. Phys. Chem.* (*Leipzig*) 254: 273.
Nagata, I., 1977. *Fluid Phase Equilibria* 1: 93.
Nagata, I., 1985. *Fluid Phase Equilibria* 19: 153.
Nagata, I., 1990. In, S. Malanowski and A. Anderko (Eds.), *Thermodynamics of Fluids: Measurement and Correlation*, World Scientific Publishing, Singapore.
Nagata, I. and Y. Kawamura, 1979. *Chem. Eng. Sci.* 34: 601.
Nath, A. and E. Bender, 1981. *Fluid Phase Equilibria* 7: 275, 289.
Nath, A. and E. Bender, 1983. *Fluid Phase Equilibria* 10: 43.
Nitta, I. and T. Katayama, 1973. *J. Chem. Eng. Japan* 6: 1.
Ogorodnikov, S. K., T. M. Lesteva and W. B. Kogan, 1971. *Azeotropnye smesi, spravotchnik* (in Russian), Khimia, Leningrad.
Oishi, T. and J. M. Prausnitz, 1978. *Ind. Eng. Chem. Proc. Des. Dev.* 17: 333.
Orye, R. V. and J. M. Prausnitz, 1965. *Ind. Eng. Chem.* 57: 18.
Pierotti, G. J., C. H. Deal and E. L. Derr, 1959. *Ind. Eng. Chem.* 51: 95.

Prausnitz, J. M., R. N. Lichtenthaler and E. G. de Azevedo, 1986. *Molecular Thermodynamics of Fluid Phase Equilibria*, 2nd ed., Prentice-Hall, Englewood Cliffs, NJ.
Preston, G. T., E. W. Funk and J. M. Prausnitz, 1971. *J. Phys. Chem.* 75: 2345.
Prigogine, I. and R. Defay, 1954. *Chemical Thermodynamics*, Longmans, London.
Redlich, O. and T. A. Kister, 1948. Ind. Eng. Chem. 40: 345.
Redlich, O., E. L. Derr and G. J. Pierotti, 1959. *J. Am. Chem. Soc.* 81: 2283.
Reid, R. C., J. M. Prausnitz and B. E. Poling, 1987. *The Properties of Gases and Liquids*, 4th ed., McGraw-Hill, New York.
Renon, H., L. Asselineau, G. Cohen and C. Raimbault, 1971. *Calcul sur ordinateur des equilibres liquide-vapeur et liquide-liquide*, Technip, Paris.
Renon, H. and J. M. Prausnitz, 1967. *Chem. Eng. Sci.* 22: 299, 1891.
Renon, H. and J. M. Prausnitz, 1968. *AIChE J.* 14: 135.
Rogalski, M. and S. Malanowski, 1977. *Fluid Phase Equilibria* 1: 137.
Rogalski, M. and S. Malanowski, 1980. *Fluid Phase Equilibria* 5: 97.
Sander, B., A. Fredenslund and P. Rasmussen, 1986. *Chem. Eng. Sci.* 41: 1171.
Sander, B., S. Skjold-Joergensen and P. Rasmussen, 1983. *Fluid Phase Equilibria* 11: 105.
Scatchard, G., 1931. *Chem. Rev.* 8: 321.
Scatchard, G., 1934. *J. Am. Chem. Soc.* 56: 995.
Scatchard, G. and W. J. Hamer, 1935. *J. Am. Chem. Soc.*, 57: 1805.
Schille, W. and H.-J. Bittrich, 1973. *Chem. Techn.* 25: 292.
Schreiber, L. B. and C. A. Eckert, 1971. *Ind. Eng. Chem. Proc. Des. Dev.* 10: 572.
Scott, R. L., 1956. *J. Chem. Phys.* 25: 193.
Suarez, J. T., C. Torres-Marchal and P. Rasmussen, 1989. *Chem. Eng. Sci.* 44: 782.
Thomas, E. R. and C. A. Eckert, 1984. *Ind. Eng. Chem. Proc. Des. Dev.* 23: 194.
Tiegs, D., P. Rasmussen, J. Gmehling and A. Fredenslund, 1987. *Ind. Eng. Chem. Res.* 26: 159.
Timmermans, J., 1959. *The Physico-chemical Constants of Binary Systems in Concentrated Solutions*, Vols. 1–4, Interscience, New York.
Tochigi, K., D. Tiegs, J. Gmehling and K. Kojima, 1990. *J. Chem. Eng. Japan.* 23: 453.
Treszczanowicz, A. and T. Treszczanowicz, 1975. *Bull. Acad. Pol. Sci., Ser. Sci. Chim.* 23: 169.
Treszczanowicz, A., T. Treszczanowicz and M. Rogalski, 1973. *Proc. Third. Int. Conf. Chem. Thermodyn.*, Baden, Austria, Vol. III, 11, 18.
Treybal, R. E., 1963. *Liquid Extraction*, 2nd ed., McGraw-Hill, New York.
Tsonopoulos, C., 1974. *AIChE J.* 20: 263.
Tsonopoulos, C., 1975. *AIChE J.* 21: 827.
Van Laar, J. J., 1910. *Z. Phys. Chem.*, 72: 723.

Van Ness, H. C. and M. M. Abbott, 1982. *Classical Thermodynamics of Nonelectrolyte Solutions with Applications to Phase Equilibria*, McGraw-Hill, New York.

Vera, J. H. and J. Vidal, 1984. *Chem. Eng. Sci.* 39: 651.

Warowny, W. and J. Stecki, 1979. *The Second Cross Virial Coefficients of Gaseous Mixtures*, PWN–Polish Scientific Publishers, Warszawa.

Weimer, R. F. and J. M. Prausnitz, 1965. *Hydrocarb. Process.* 44: 237.

Wichterle, I., J. Linek and E. Hala, 1973–. *Vapour-Liquid Equilibrium Data Bibliography & Supl.*, 1–4, Elsevier, Amsterdam.

Wiehe, I. A. and E. B. Bagley, 1967. *Ind. Eng. Chem. Fundam.* 6: 209.

Wilson, G. M., 1964. *J. Am. Chem. Soc.* 86: 127.

Wilson, G. M. and C. H. Deal, 1962. *Ind. Eng. Chem. Fundam.* 1: 20.

Wong, K. F. and C. A. Eckert, 1971. *Ind. Eng. Chem. Fundam.* 10: 20.

Wu, H. S. and S. I. Sandler, 1991. *Ind. Eng. Chem. Res.* 30: 881, 889.

Wu, D. T., L. Longinger and J. C. Klein, 1988. Presented at XIX FATIPEC Congress, Aachen.

4

Equation-of-State Methods

4.1 INTRODUCTION

A great variety of thermodynamic properties can be computed for pure substances and mixtures if their P-v-T relations, i.e., the equations of state (EOS), are known. In this chapter we shall review the equation-of-state methods for the modelling of phase equilibria. In this section the formal thermodynamic tools, which are useful for the calculation of phase equilibria from EOS models, will be summarized.

For this purpose it is convenient to use the residual thermodynamic functions defined as

$$\mathscr{X}^r = \mathscr{X} - \mathscr{X}^0 \tag{4.1-1}$$

where $\mathscr{X}$ is any thermodynamic function (e.g., $\mathscr{X} = G, A, H, S, U, V, \mu$), and $\mathscr{X}^0$ is its value in the ideal-gas state.

Two methods exist for the computation of residual functions from P-v-T properties. The first one, based on the Helmholtz energy, is convenient if the P-v-T relations are represented by a pressure-explicit equation. The second one, based on the Gibbs energy, is convenient if a volume-explicit equation is used (e.g., the volume-explicit virial EOS, discussed in Section 2.3, and some correlations based on the corresponding-states principle). The overwhelming majority of equations of state of practical importance are pressure explicit. In this section, the method based on the Helmholtz energy will be presented.

For constant temperature and composition, the differential of the Helmholtz energy (1.2-58) reduces to

$$dA = -P\,dV \tag{4.1-2}$$

Integrating at constant temperature and composition from the reference total ideal-gas volume V^0 to the actual volume of the system V we get

$$A - A^0 = -\int_{V^0}^{V} P\,dV = -\int_{\infty}^{V} P\,dV - \int_{V^0}^{\infty} P\,dV \tag{4.1-3}$$

The first integral on the right-hand side of Eq. (4.1-3) depends on the real fluid properties, whereas the second depends only on the properties of the reference state, i.e., the ideal-gas state. Adding and subtracting $\int_{\infty}^{V} (NRT/V)\,dV$ to the right-hand side of Eq. (4.1-3) we obtain

$$A - A^0 = -\int_{\infty}^{V} \left(P - \frac{NRT}{V}\right) dV - NRT \ln\left(\frac{V}{V^0}\right) \tag{4.1-4}$$

Other residual functions are obtained from the residual Helmholtz energy according to (1.2-57) and (1.2-61)

$$S - S^0 = -\frac{\partial}{\partial T}(A - A^0)_V = \int_{\infty}^{V} \left[\left(\frac{\partial P}{\partial T}\right)_V - \frac{NR}{V}\right] dV + NR \ln\left(\frac{V}{V^0}\right) \tag{4.1-5}$$

$$H - H^0 = (A - A^0) + T(S - S^0) + NRT(z - 1) \tag{4.1-6}$$

$$U - U^0 = (A - A^0) + T(S - S^0) \tag{4.1-7}$$

$$G - G^0 = (A - A^0) + NRT(z - 1) \tag{4.1-8}$$

The residual functions $H - H^0$ and $U - U^0$ are independent of the reference pressure P^0 (or, equivalently, $V^0 = RT/P^0$). On the other hand, the functions $A - A^0$, $S - S^0$ and $G - G^0$ depend on P^0. To calculate these functions, it is convenient to assign a unit pressure to P^0 (e.g., 1 atm) or to assume that the reference pressure is equal to the system pressure ($P = P^0$).

The chemical potential of component i is related to the Helmholtz energy by (1.2-59):

$$\mu_i \equiv \left(\frac{\partial A}{\partial N_i}\right)_{T,V,N_{j\neq i}} \tag{4.1-9}$$

where N_i is the number of moles of component i. Rewriting Eq. (4.1-9):

$$\mu_i - \mu_i^0 = \frac{\partial}{\partial N_i}(A - A^0)_{T,V,N_{j\neq i}} \tag{4.1-10}$$

and using the definition of fugacity (1.2-199)

$$\mu_i - \mu_i^0 = RT \ln\left(\frac{f_i}{f_i^0}\right) \tag{4.1-11}$$

we obtain

$$\begin{aligned} RT \ln\left(\frac{f_i}{f_i^0}\right) &= \frac{\partial}{\partial N_i}(A - A^0)_{T,V,N_{j\neq i}} \\ &= \frac{\partial}{\partial N_i}\left[-\int_{\infty}^{V}\left(P - \frac{NRT}{V}\right) dV - NRT \ln\left(\frac{V}{V^0}\right)\right]_{T,V,N_{j\neq i}} \end{aligned} \tag{4.1-12}$$

Taking into account that the ideal-gas state has been chosen as the reference state, the reference fugacity is $f_i^0 = P^0 y_i$, and an expression for the fugacity coefficient ϕ_i of a component of a mixture is obtained from

$$RT \ln \phi_i = RT \ln \frac{f_i}{Py_i} = -\int_{\infty}^{V} \left[\left(\frac{\partial P}{\partial N_i} \right)_{T,V,N_{j \neq i}} - \frac{RT}{V} \right] dV - RT \ln z = \mu_i^r - RT \ln z \tag{4.1-13}$$

where μ_i^r is the residual chemical potential.

If the mixture as well as its pure components are in the same phase, the excess thermodynamic functions can be calculated from the residual ones:

$$Z^E = Z - Z^0 - \sum x_i (Z_i - Z_i^0) \tag{4.1-14}$$

In particular, the excess Gibbs energy of a liquid mixture can be expressed using fugacity coefficients (Smith and Van Ness, 1975):

$$G^E = RT \left(\ln \phi_{\text{mix}} - \sum x_i \ln \phi_i^0 \right) \tag{4.1-15}$$

According to Eq. (4.1-11), fugacity is directly related to the chemical potential of a substance. Thus, fugacity is an intensive property and satisfies the general condition of phase equilibrium [Eq. (1.2-22)].

The above equations show that the P-v-T relationships make it possible to calculate fugacities (and, subsequently, phase equilibria) as well as other thermodynamic functions such as enthalpy, entropy and molar volume of a phase. The problem of computing phase equilibria consists in solving the condition of equal fugacities in coexisting phases α and β:

$$f_i^\alpha = f_i^\beta \qquad i = 1, \ldots, n \tag{4.1-16}$$

where n is the number of components and α and β denote separate phases coexisting in the state of equilibrium. There is no conceptual difference between vapor–liquid and liquid–liquid equilibrium calculations. In particular, three or more phase equilibria can be computed using the same isofugacity criterion [Eq. (4.1-16)].

Equations of state are equally applicable to the calculation of solid–liquid equilibria provided that the difference between the chemical potential of a pure subcooled liquid solute i and that of the solute in saturated solution ($\mu_i^{0L} - \mu_i^L$) in the general equation (P1.2-62) is expressed by corresponding fugacities. According to the definition of fugacity [Eq. (1.2-199)], we obtain

$$\mu_i^{0L} - \mu_i^L = RT \ln \frac{f_i^{0L}}{f_i^L} \tag{4.1-17}$$

where f_i^{0L} and f_i^L are the fugacities of pure subcooled liquid i and the solute i

in the solution, respectively. The solubility equation then takes the form:

$$\ln \frac{f_i^{0L}}{f_i^L} = \frac{\Delta h_f}{RT_m}\left(\frac{T_m}{T} - 1\right) - \frac{\Delta C_p}{R}\left(\frac{T_m}{T} - 1\right) + \frac{\Delta C_p}{R} \ln \frac{T_m}{T} \tag{4.1-18}$$

where Δh_f is the enthalpy of fusion at the melting temperature T_m and ΔC_p is the heat capacity of fusion. Both quantities f_i^{0L} and f_i^L can be computed from an equation of state.

From the thermodynamic point of view the equation-of-state method is more general and comprehensive than the γ–ϕ method. Equation (4.1-15) shows that the γ–ϕ method can be regarded as a special case of the EOS method. On the other hand, calculations by means of equations of state are usually more complicated and time consuming than those using the γ–ϕ methods. Another, more fundamental, problem associated with the use of equations of state is the still unsatisfactory knowledge of the functional form of the equations and the methods used for their extension to binary and multicomponent mixtures.

Despite their long history, equations of state continue to be an important subject of research in applied thermodynamics. The traditional, over 100 years old, quest for equations representing quantitatively the fluid state has been revived by industrial needs for accurate process design calculations and, on the other hand, by the great advances in computational techniques.

First, equations of state were used mainly for pure components. When applied to mixtures, equations of state were used up to early 1970s virtually only for mixtures of nonpolar and slightly polar compounds. As EOS were not applicable to strongly polar and hydrogen-bonding fluids, only the asymmetric method utilizing activity coefficients and standard state fugacities was used for this class of compounds. Things have changed over the past decade and a multitude of papers appeared on the construction of equation-of-state models applicable to systems containing strongly polar components. Nevertheless, Han, Lin and Chao justifiably stated in 1988 (p. 2327) that "the usefulness of EOS to polar substances is still limited, and the asymmetric convention employing activity coefficients and vapor pressures remains the standard procedure. .... Rapid progress ... is being made to apply EOS to these substances. Their general description at low pressures as well as high pressures is at a state of flux."

The primary objective of this chapter is to review equation-of-state methods including those applicable to mixtures containing strongly polar and associating components. This chapter will deal primarily with applications of the equations to vapor–liquid and liquid–liquid equilibria although the majority of methods presented can be also applied to supercritical fluid equilibria (cf. a review by Brennecke and Eckert, 1989). The subject is extremely broad and any review will be unavoidably biased by personal prejudices of the authors. Nevertheless, an attempt will be made to outline the main trends of research.

4.2 PURE-COMPONENT EQUATIONS OF STATE

4.2.1. Van der Waals-Type Equations of State

The first equation of state that reasonably represented both the gas and liquid phases was proposed by van der Waals in 1873. The original formulation of the van der Waals equation was

$$\left(P + \frac{a}{v^2}\right)(v - b) = (1 - a)(1 - b)(1 + \alpha t) \tag{4.2-1}$$

Van der Waals based his derivation on a paper by Clausius (1857) who argued that deviations from Boyle's (1660) law were caused by intermolecular attractions (represented by the constant a) and repulsions (constant b). Initially, the purpose of his work was to analyze the real gas behavior in the vicinity of the critical point rather than to reproduce the *PVT* behavior. Van der Waals made an analysis of the equation.

$$(P + r)(v - \psi) = RT \tag{4.2-2}$$

formulated for the first time by Hirn (1863) and found out that the constant ψ (denoted by van der Waals as b) was probably four times larger than the volume of molecules and the intermolecular distance r decreased rapidly with an increase in fluid volume.

The van der Waals equation was derived in an intuitive manner. The molecules were assumed to have a finite diameter, thus making a part of the volume not available to molecular motion. This increases the number of collisions with the vessel walls and, subsequently, increases the pressure. Therefore, the actual volume available to molecular motion is $v - b$ where b is a characteristic constant for each fluid. Another factor is the intermolecular attraction, which decreases the pressure. The pressure decrease is proportional to the number of molecules in a volume unit and is inversely proportional to volume. Therefore, the effect of mutual attractions (i.e., cohesion) is inversely proportional to square volume. The real pressure is expressed as the ideal-gas pressure corrected by the term a/v^2:

$$P_{id} = P - \frac{a}{v^2} \tag{4.2-3}$$

Therefore, the product of the term $(v - b)$ and Eq. (4.2-3) corresponds to the ideal-gas law. This gives the van der Waals equation in its present form as a sum of two terms:

$$p = \frac{RT}{v - b} - \frac{a}{v^2} \tag{4.2-4}$$

The two terms on the right-hand side of Eq. (4.2-4) are called, due to the intermolecular forces they reflect, the repulsive and attractive terms, respectively. The van der Waals equation reduces to the ideal-gas law in the infinite volume limit. The most important features of the van der Waals EOS are:

1. The constants a and b are valid for both gas and liquid phases.
2. The equation predicts the vapor–liquid critical point.

Obviously, the van der Waals equation can be treated only as a crude approximation. The *PVT* properties of the liquid phase are reproduced much less satisfactorily than those of the gas phase. Nevertheless, the van der Waals equation played a very important role in the past in the development of the theory of fluids and is still used for their semiquantitative representation. According to Smoluchowski (1915), the intuitive method of van der Waals was more correct than purely mathematical methods and, what is very important, never gave absurd results, which are likely to occur in the case of mathematical methods without a physical background. The work of van der Waals was severely criticized by some physicists (e.g., Maxwell, 1874) while numerous chemists and physical chemists such as Ostwald, van't Hoff, Nernst and Boltzmann appreciated it (Partington, 1949). Clausius (1881) was the first to recognize the importance of van der Waals's work and develop an important modification of it. The van der Waals EOS initiated more research on the fluid phase than any other fluid state theory. Among the earliest researchers, who modified the van der Waals EOS, Berthelot (1899) introduced an equation with a temperature dependent a constant ($a_T = a/T$). A review of earlier modifications can be found in the book by Partington (1949).

The van der Waals EOS can be rewritten in the form

$$v^3 - \left(\frac{RT}{P} + b\right)v^2 + \frac{av}{P} - \frac{ab}{P} = 0 \tag{4.2-5}$$

It is evident that the equation is cubic in volume. Therefore, it can be solved analytically for volume, thus facilitating computations. A large family of equations inspired by the van der Waals EOS are also cubic in volume.

The currently used van der Waals-type equations of state resemble their common predecessor in that all of them contain a repulsion term z_{rep} and an attractive term z_{attr}. They can be expressed as

$$z = z_{\text{rep}} + z_{\text{attr}} \tag{4.2-6}$$

CONTEMPORARY CUBIC EQUATIONS OF STATE

Among the van der Waals-type of equations of state, cubic equations proved to be especially useful due to their simplicity, short computation time and reliability. A cubic equation is the simplest polynomial form capable of yielding the ideal-gas limit at $v \to \infty$ and of representing both liquid and gas phases. The cubic equation in their general form have been analyzed by Martin (1967,

1979), Abbott (1973, 1979) and Vera et al. (1984). The most general form of a cubic EOS contains five parameters and takes the form (Abbott, 1979)

$$p = \frac{RT}{v - b} - \frac{a(v - \eta)}{(v - b)(v^2 + \delta v + \epsilon)} \tag{4.2-7}$$

where the adjustable parameters b, a, η, δ and ϵ can be, in general, functions of temperature. The efforts to formulate a workable cubic EOS concentrated on establishing its most appropriate functional form or/and adjusting its parameters and their temperature dependence to match the thermophysical properties of interest. The majority of cubic EOS are constrained to satisfy the critical points conditions

$$\frac{\partial P}{\partial v} = \frac{\partial^2 P}{\partial v^2} = 0 \tag{4.2-8}$$

Application of Eq. (4.2-8) makes it possible to calculate the values of parameters at the critical point. If a cubic EOS contains only two parameters, a and b, their values at the critical point are completely determined by T_c and P_c. For example, solution to Eq. (4.2-8) for the van der Waals (vdW) EOS [Eq. (4.2-4)] yields

$$a_{c(\mathrm{vdW})} = \frac{\frac{27}{64} R^2 T_c^2}{P_c^2} \tag{4.2-9}$$

and

$$b_{c(\mathrm{vdW})} = \frac{\frac{1}{8} R T_c}{P_c} \tag{4.2-10}$$

For a two-parameter cubic EOS the value of the compressibility factor at the critical point (called the critical compressibility factor) assumes a constant value because v_c is determined by substituting the a_c and b_c values into the equation of state. For the original vdW EOS, the critical compressibility factor is $z_c = \frac{3}{8}$. If a three- or more-parameter cubic EOS is used, the critical compressibility factor is no longer fixed. However, it is not advantageous to adjust it to match the experimental z_c value because this entails a deterioration of results for the rest of the critical isotherm.

The critical compressibility factor predicted by cubic EOS is usually different from its experimental value and is frequently used as an adjustable parameter. Abbott (1979) described the compromise that has to be made between the reproduction of the critical compressibility factor and the critical isotherm in the low- and high-density region. The influence of parameters of Eq. (4.2-7) on the representation of the critical isotherm was also studied. The author discussed the proper choice of independent variables for analyzing cubic EOS and suggested also a complimentary relationship between the dimensionless second virial coefficient and dimensionless co-volume of a cubic EOS.

Martin (1979) established that a compromise was needed between the reproduction of the critical compressibility factor z_c and the reduced second virial coefficient at the critical isotherm B_c, i.e., that the difference $\Sigma = z_c - B_c$ should be minimized. Vera et al. (1984) tried to establish the best set of coefficients δ, ϵ and η with respect to errors in the low- and high-pressure region. They concluded that minimization of the quantity Σ to obtain the best critical isotherm does not ensure the existence of a region with equal liquid and vapor phase fugacities. The equations that correctly represent the phase behavior use either $\eta = b$ in Eq. (4.2-7) or, when η is relaxed, they use $\epsilon = 0$ to obtain a two-phase region characterized by equal fugacities of the liquid and vapor at low reduced temperatures.

Although it is evident that cubic equations cannot represent simultaneously all thermodynamic functions with desirable accuracy, their simplicity and practical success in calculating phase equilibria continue to stimulate further refinements. The multitude of variations on the cubic form precludes an exhaustive review. Some features and applications of the more recent cubic EOS have been reviewed by Vidal (1983, 1984), Tsonopoulos and Heidman (1985, 1986), Kolasińska (1986) and Saito and Arai (1986). Therefore, only some developments relevant to phase equilibrium calculations will be reviewed here as an illustration of the main trends of research.

Redlich and Kwong (1949) proposed the first cubic EOS (RK) that was widely accepted as a tool for routine engineering calculations of fugacity. The equation was proposed to satisfy the boundary conditions in the low- and high-density limit. In the low-density limit it was postulated that the equation give a reasonable second virial coefficient:

$$B = b - \frac{\text{constant}}{T^{3/2}} \tag{4.2-11}$$

At high densities it was noted that as p tends to infinity, the reduced volume is approximately 0.26. This suggested a simple relation for the excluded volume: $b = 0.26V_c$. To meet also the critical point conditions the equation has the form

$$P = \frac{RT}{v - b} - \frac{a}{T^{1/2}v(v + b)} \tag{4.2-12}$$

Thus, a simple physical picture and reasonable boundary conditions led to the formulation of the first quantitatively successful van der Waals-type equation of state. The success of the Redlich–Kwong EOS stimulated numerous investigators to propose various methods for improving the equation. Horvath (1974) reviewed the modifications up to early 1970s. The improvements of the RK EOS fall within two main trends of research: They either adopt a better temperature dependence of the parameters or change the functional form of the $P(v)$ dependence.

The temperature dependence of the a parameter is essential for the reproduction of vapor pressures. The early methods for adjusting the parameters to enable the EOS to match experimental vapor pressures were reviewed by Tsonopoulos and Prausnitz (1969), Wichterle (1978) and Abbott (1979). Among them, Wilson (1964) first introduced a general form of the temperature dependence of the a parameter for both phases

$$a = a_c \alpha \tag{4.2-13}$$

$$\alpha = T_R[1 + (1.57 + 1.62\omega)(T_R^{-1})] \tag{4.2-14}$$

where a_c is the value of a at the critical point, $T_R = T/T_c$ is the reduced temperature and ω is the acentric factor [Eq. (2.1-32)]. This correlation was established by forcing the EOS to give reasonable values of the terminal slope of the vapor pressure curve. The first method for expressing the temperature dependence that gained widespread popularity due to its accuracy and simplicity was proposed by Soave (1972):

$$\alpha = [1 + mT_R^{1/2}]^2 \tag{4.2-15}$$

where m was expressed as a quadratic function of the acentric factor. Soave obtained his relation more directly than Wilson, by forcing the equation to reproduce vapor pressures for nonpolar substances at $T_R = 0.7$. Soave's function was then used by many investigators [among others by Peng and Robinson (1976), Schmidt and Wenzel (1980), Patel and Teja (1982), Adachi et al. (1983a,b), Watson et al. (1986)] who changed only the $m(\omega)$ function to accommodate it into their own equations of state. Corrections to Eq. (4.2-15) were introduced by, among others, Harmens and Knapp (1980) who added one more term to improve the performance of the equation at low temperatures.

$$\alpha = \left[1 + A(\omega)(1 - T_R^{1/2}) - B(\omega)\left(1 - \frac{1}{T_R}\right)\right]^2 \tag{4.2-16}$$

Soave's function was found to be incorrect at high reduced temperature as it does not always decrease monotonically. Equation (4.2-15) produces a zero value at high reduced temperatures and then increases. The approach of real-gas behavior to that of the ideal gas at high temperatures requires that $a \to 0$ as $T_R \to \infty$. Therefore, Graboski and Daubert (1978, 1979) found that Eq. (4.2-15) failed to give accurate results for hydrogen, which is normally at high reduced temperatures, and recommended an exponential form

$$\alpha = C_1 \exp(-C_2 T_R) \tag{4.2-17}$$

Heyen (1980) recommended a similar function

$$\alpha = \exp[C(1 - T_R^n)] \tag{4.2-18}$$

Patel and Teja (1982) found, however, that Eqs. (4.2-17) and (4.2-18) do not offer any advantages over the Soave form when applied to vapor pressures in typical reduced temperature ranges. Another exponential form of the temperature dependence was proposed by Yu and Lu (1987)

$$\alpha = 10^{M(\omega)(A_0 + A_1 T_R + A_2 T_R^2)(1 - T_R)} \tag{4.2-19}$$

where A_i are adjustable parameters and $M(\omega)$ is a function of the acentric factor.

In the case of nonpolar fluids the parameters of the above equations can be generalized in terms of the acentric factor. For polar substances, however, the equations are also applicable provided that the parameters are determined from the actual vapor pressures of the polar compound and not correlated with ω (e.g., Georgeton et al., 1986). Although no general characteristic parameter exists for polar compounds that could replace ω, correlation with the reduced dipole moment can give satisfactory results (Guo et al., 1985). In fact, vapor pressures of both nonpolar and polar compounds can be very accurately represented by cubic EOS if the temperature dependence of the parameter a is sufficiently flexible. Numerous equations were proposed for this purpose. They usually contain one to three empirical constants to be derived from vapor pressures of individual compounds. Among them, Mathias and Copeman (1983) proposed a three-parameter form:

$$\alpha = [1 + c_1(1 - T_R^{1/2}) + c_2(1 - T_R^{1/2})^2 + c_3(1 - T_R^{1/2})^3]^2 \tag{4.2-20}$$

Soave (1984) proposed a two-parameter equation:

$$\alpha = 1 + m(1 + T_R) + n(T_R^{-1} - 1) \tag{4.2-21}$$

and Stryjek and Vera (1986a) proposed a one-parameter form:

$$\alpha = \{1 + [k_0(\omega) + k_1(1 + T_R^{1/2})(0.7 - T_R)](1 - T_R^{1/2})\}^2 \tag{4.2-22}$$

Androulakis et al. (1989) analyzed several two- and three-parameter forms of the temperature dependence of a with respect to their capability of correlating experimental data and extrapolating to low temperatures. They recommended a three-parameter form:

$$\alpha = 1 + d_1(1 - T_R^{2/3}) + d_2(1 - T_R^{2/3})^2 + d_3(1 - T_R^{2/3})^3 \tag{4.2-23}$$

and its simplified version with $d_2 = 0$. For supercritical temperatures they recommended an equation providing a smooth transition between the saturated and supercritical regions. Malhem et al. (1989) also discussed several forms of the temperature dependence and recommended a two-parameter equation that correlates the data with similar accuracy as that of Soave (1984) but extrapolates better to supercritical temperatures.

$$\alpha = \exp[m(1 - T_R) + n(1 - T_R^{-1})]^2 \tag{4.2-24}$$

The authors presented an extensive list of the m and n parameters. If a selected vapor pressure point is to be reproduced with a cubic EOS, a method of Sugie et al. (1989), who developed an analytical expression for α / T_R, can be used.

While an appropriate temperature dependence of the a parameter is sufficient for representing accurately the vapor pressure, modifications of the $P(v)$ functional dependence are necessary to improve the prediction of volumetric properties. The simplest approach is to change the form of the attractive term of the RK EOS without adding any additional parameters. Along this line, Peng and Robison (1976) (PR) recognized that the critical compressibility factor of the RK EOS ($z_c = 0.333$) is overestimated thus impairing the liquid volume calculations. They postulated an equation reducing z_c to 0.307.

$$P = \frac{RT}{v - b} - \frac{a(T)}{v(v + b) + b(v - b)} \tag{4.2-25}$$

This form improved the representation of liquid density in relation to the Soave (1972) modification (SRK) of the RK EOS. Harmens (1977) modified the attractive term in the van der Waals EOS lowering z_c to 0.286.

The two-parameter cubic EOS that satisfy exactly the critical condition [Eq. (4.2-8)] predict a constant compressibility factor for all components. A third parameter was therefore used by several authors to introduce a component-dependent critical compressibility factor and thus to enhance the equation flexibility. Although a three-parameter equation can be forced to predict the correct critical compressibility factor, the isotherms at low and high pressures are then distorted much more than can be tolerated for the purposes of modelling PVT relations. Better overall results are usually obtained when the apparent (calculated) critical compressibility factor ζ is greater than the real one z_c. Among the three-parameter cubics, the equation of Fuller (1976) contains a third parameter $c(T)$:

$$P = \frac{RT}{v - b(T)} - \frac{a(T)}{v[v + c(T)b(T)]} \tag{4.2-26}$$

The co-volume in Fuller's EOS varies with temperature, which is also the case for the Heyen (1980) EOS:

$$p = \frac{RT}{v - b(T)} - \frac{a(T)}{v^2 + [b(T) + c]v - b(T)c} \tag{4.2-27}$$

Heyen considered three options for ζ: experimental z_c, adjustable or a function of ω. Schmidt and Wenzel (1980) recognized that the SRK EOS satisfactorily predicts liquid volumes for compounds characterized by low values of the acentric factor ω whereas the PR EOS is at its best for intermediate values of ω. They analyzed a more general form,

$$P = \frac{RT}{v - b} - \frac{a(T)}{v^2 + ubv + wb^2} \tag{4.2-28}$$

and found the parameters ($w = -3\omega$ and $u = 1 - w$) that interpolate between the SRK and PR EOS. Harmens and Knapp (1980) found also a three-parameter form

$$P = \frac{RT}{v - b} - \frac{a(T)}{v^2 + vcb - (c - 1)b^2} \tag{4.2-29}$$

designed to be a generating function for successful two-parameter equations. Correlations for the apparent critical compressibility factor in terms of the acentric factor was proposed for both Eqs. (4.2-28) and (4.2-29).

Patel and Teja (1982) used a form proposed previously by Heyen (1980)

$$P = \frac{RT}{v - b} - \frac{a(T)}{v(v + b) + c(v - b)} \tag{4.2-30}$$

and treated ζ as an adjustable parameter (temperature dependent for $0.9 < T_R < 1.0$) to obtain an optimum reproduction of densities. Yu et al. (1986) and Yu and Lu (1987) analyzed deviation contours for volumetric properties on the $u - w$ diagram and arrived at an optimum relation different from that of Schmidt and Wenzel, i.e., $u - w = 3$. The proposed equation takes the form

$$P = \frac{RT}{v - b} - \frac{a(T)}{v(v + c) + b(3v + c)} \tag{4.2-31}$$

where $c = wb$ and $a(T)$ is given by Eq. (4.2-19).

In the development of the above three-parameter forms use was made of the fact that acceptable prediction of both low- and high-density regions requires that ζ be treated as an empirical parameter. The simplest, yet powerful, method to achieve this is the so-called volume translation, which can be used to separate the application of an EOS to vapor–liquid equilibrium calculations from the application to density calculations. The concept, proposed by Martin (1979) and elaborated further by Péneloux et al. (1982) consists in a translation along the volume axis

$$v_i^{\text{translated}} = v_i^{\text{original}} - c_i \tag{4.2-32}$$

It the translation parameter c for a mixture is related to composition by a linear mixing rule, the partial molar volumes are translated analogously and the fugacities are multiplied by a composition-independent coefficient so that calculations of phase equilibria are not affected by the volume translation. Péneloux et al. showed that the method improves markedly the prediction of liquid volume except in the vicinity of the critical point. The original van der Walls equation assumes, after the volume translation, a form essentially identical to the Clausius EOS (1881)

$$p = \frac{RT}{V - b} - \frac{a}{(V + c)^2} \tag{4.2-33}$$

The interest in the Clausius EOS was revived by Martin (1979) who found it to be the best cubic equation with respect to gas phase properties. Joffe (1981) and Kubic (1982) used the Clausius equation in modified forms. The apparent critical compressibility factor ζ was treated as a function of the acentric factor. While Kubic's volume translation parameter is temperature dependent, Soave (1984) recommended the use of a constant value of c obtained from liquid densities at low temperatures. Salerno et al. (1986), Watson et al. (1986), Czerwienski et al. (1988) and Androulakis et al. (1989) presented a translated van der Waals EOS, called vdW-711, containing a temperature-dependent volume translation. The translating function was fitted to experimental liquid volumes, including the critical volume, and was generalized in terms of ω. Mathias et al. (1989) proposed a different form of the volume translation that also reproduces the critical volume and applied it to the PR EOS.

While considerable progress was achieved using the three-parameter equations, some authors tried to improve the flexibility of cubics by using four or five parameters. Notably, Adachi et al. (1983b) used the four parameters of their equation

$$P = \frac{RT}{v - b_1} - \frac{a(T)}{(v - b_2)(v + b_3)} \tag{4.2-34}$$

as well as the apparent critical compressibility factor ζ to optimize the representation of the critical isotherm. Similarly as in the case of three-parameter cubics [Eqs. (4.2-26)–(4.2-33)], ζ appeared to be larger than z_c. To improve further the accuracy of saturated liquid densities, Sugie et al. (1989) proposed to fit exactly two liquid density data of an individual compound. Trebble and Bishnoi (1987a,b) also selected a four-parameter form

$$p = \frac{RT}{v - b} - \frac{a(T)}{v^2 + (b + c)v - (bc + d^2)} \tag{4.2-35}$$

to be able to optimize not only ζ but also the critical co-volume b_c, which determines the "hardness" of the EOS [i.e., the slope of $(\partial P/\partial v)_T$] at elevated pressures. The authors established the functional form of the EOS by calculating the critical isotherm with ζ set equal to z_c and optimized b_c.

The largest number of parameters in a cubic EOS is five. The five-parameter cubics may be advantageous when a large body of pure-component *PVT* data is studied. Adachi et al. (1986) analyzed the shapes of the attractive term versus density curves and concluded that the flexibility of a five-parameter form

$$p = \frac{RT}{v - b} - \frac{a(T)[v - c(T)]}{(v - b)[v - d(T)][v + e(T)]} \tag{4.2-36}$$

is necessary for an accurate representation of the isotherms. Kumar and Starling (1982) developed a five-parameter equation

$$z = \frac{1 + d_1(T)\rho + d_2(T)\rho^2}{1 + d_3(T)\rho + d_4(T)\rho^2 + d_5(T)\rho^3} \tag{4.2-37}$$

without isolating the repulsive and attractive terms. Kumar and Starling's equation, unlike all the above equations, does not satisfy exactly the critical point conditions. The equation gives accurate results that are on par with the generalized 11-parameter Benedict–Webb–Rubin EOS. However, the neglect of the critical constraints leads to the overestimation of the critical temperature and pressure, e.g., by 5 K and 6 bars for CO_2 as reported by Vidal (1984).

Most investigators made only the parameter a temperature dependent in the above equations as it is sufficient to reproduce vapor pressures. Two or more temperature-dependent parameters may be useful for optimizing the reproduction of some properties other than vapor pressures. Among the earlier methods, Joffe et al. (1970) and Chang and Lu (1970) suggested that the temperature dependence of the a and b parameters of the Redlich–Kwong EOS be determined by simultaneously matching liquid density and vapor pressures. This procedure resulted, however, in a discontinuity of parameters at the critical temperature. In the case of more recent modifications that use temperature-dependent parameters to fit vapor pressure and liquid density, i.e., those by Fuller (1976) and Heyen (1980), the parameters are well-behaved, continuous functions of temperature. Another approach was taken by Kubic (1982) who fit vapor pressures and second virial coefficients by making his a and c parameters temperature dependent. Trebble and Bishnoi (1986) examined the consequences of using more than one temperature-dependent parameter and found that the Heyen and Fuller equations are inconsistent in that they predict negative heat capacity in the one-phase region. The authors showed that the second derivative of b with respect to temperature should remain negative at all temperatures and constructed their own EOS to satisfy this condition. Yu et al. (1986) also warned against using more than one temperature-dependent parameter claiming that the results are then generally inferior due to a transportation of errors to properties other than those used as input in the fitting procedure.

In general, cubic equations can be treated as convenient engineering tools due to their simplicity and reliability. The equations with generalized parameters provide reasonable predictions of thermophysical properties as illustrated in Table 4.2-1. Considerable progress was achieved in the last period with respect to the representation of saturated region properties. It should be noted, however, that the results quoted in Table 4.2-1 do not provide an unequivocal comparison of the equations as the results are sensitive to the data base used.

Obviously, the equations with parameters fitted separately to individual compounds' properties rather than generalized provide a more accurate reproduction of the data. It is often said that although the repulsive and attractive terms of cubic EOS are incorrect, the equations work as a whole because of a fortunate cancellation of errors. A question may then arise: What is the effect of changing both terms on the reproduction of the properties of interest? An unequivocal answer is difficult, but some hints have been given by Adachi et al.

Table 4.2-1
REPRESENTATION OF SATURATED VAPOR PRESSURES AND LIQUID AND VAPOR VOLUMES USING SELECTED GENERALIZED CUBIC EOS.[a,b]

Model	Equations	$\delta(P_{sat})$	$\delta(V_{liq})$	$\delta(V_{vap})$
Soave (1972)	[(4.2-12), (4.2-15)]	1.5	17.2	3.1
Peng–Robinson (1976)	(4.2-25)	1.3	8.2	2.7
Fuller (1976)	(4.2-26)	1.3	2.0	2.8
Schmidt–Wenzel (1980)	(4.2-28)	1.0	7.9	2.6
Harmens–Knapp (1980)	[(4.2-29), (4.2-16)]	1.5	6.6	3.0
Heyen (1980)	[(4.2-27), (4.2-18)]	5.0	1.9	7.2
Patel–Teja (1982)	(4.2-30)	1.3	7.5	2.6
Kubic (1982)	(4.2-33)	3.5	7.4	15.9
Adachi et al. (1983b)	(4.2-34)	1.1	7.4	2.5
CCOR (Kim et al., 1986)	(4.2-54)	1.8	4.3	8.2
Trebble–Bishnoi (1987a)	(4.2-35)	2.0	3.0	3.1
Ishikawa et al. (1980)	(4.2-53)	2.5	4.3	3.4
Vol.-translated PR (Yu et al., 1986)	[(4.2-25), (4.2-32)]	1.2	3.8	1.9
Yu–Lu (1987)	[(4.2-31), (4.2-19)]	1.3	3.3	2.2

[a]The results for the first 12 equations were taken from Trebble and Bishnoi (1986).
[b]$\delta(Q) = (100/N)\Sigma\,|Q_i^{\text{cal}} - Q_i^{\text{exp}}|/Q_i^{\text{exp}}$.

(1983a). The authors evaluated 16 three-parameter EOS for the representation of saturation properties and the high liquid density region. The equations were obtained by combining the repulsive term due to van der Waals and those proposed by Scott (1971):

$$z_{\text{rep}} = \frac{v+b}{v-b} \tag{4.2-38}$$

Guggenheim (1965):

$$z_{\text{rep}} = \frac{1}{(v-b)^4} \tag{4.2-39}$$

and Carahan–Starling (1969):

$$z_{\text{rep}} = \frac{v^3 + bv^2 + b^2v - b^3}{(v-b)^3} \tag{4.2-40}$$

with the attractive terms of Redlich–Kwong, Clausius, Peng–Robinson and Harmens–Knapp. Only some of the combinations yield cubic equations of state. It was found that the combination of the simple van der Waals term with the Redlich–Kwong term is the best choice. This is by no means a definite result, but it shows the good features of cubic equations of state with respect to the representation of saturation properties and the high liquid density region.

The results of several authors indicate that cubic EOS require at least three parameters to accurately represent vapor and liquid volumetric properties and at least one of them should be temperature dependent. Temperature dependence of other parameters, especially the co-volume, may lead to spurious results. From the practical point of view, an important feature of cubics is their predictive capability and their possibility of being "tailor-made" to describe selected properties of interest. However, extrapolation of the equations beyond the range of properties and conditions for which they were designed may be unreliable.

AUGMENTED VAN DER WAALS-TYPE EQUATIONS OF STATE

In view of the practical success of simple van der Waals-type models it seems attractive to construct semiempirical equations of state on the basis of more theoretically justified expressions for the contributions of molecular repulsions and attractions to thermodynamic properties of fluids. This has been the subject of the generalized van der Waals theory. The separation of the repulsive and attractive contributions to pressure [Eq. (4.2-6)] was theoretically established for some model intermolecular potentials (Zwanzig, 1954; Hemmar et al., 1964; Barker and Henderson, 1972). Subsequently, more practically oriented generalized van der Waals theory has been developed to rationalize equation-of-state models (Vera and Prausnitz, 1972; Anderson and Prausnitz, 1980; Sandler, 1985; Lee et al., 1985; Lee and Sandler, 1987; Abbott and Prausnitz, 1987). It has been long recognized that the repulsive term of the van der Waals EOS is an oversimplification. Therefore, better expressions for z_{rep} have been proposed that simulate the behavior of hard bodies. As such expressions are used as reference terms in equations of state, they should represent properties of idealized hard-cord systems that are as similar as possible to systems of interest. A widely accepted expression for the compressibility factor of hard spheres was developed by Carnahan and Starling (1969) [Eq. (4.2-40)]. For nonspherical hard convex bodies, Boublík and Nezbeda (1977) derived a more general form

$$z_{rep} = \frac{1 + (3\alpha - 2)\xi + (3\alpha^2 - 3\alpha + 1)\xi^2 - \alpha^2\xi^3}{(1 - \xi)^3} \qquad (4.2\text{-}41)$$

where ξ is the reduced density $\xi = 0.74048v^0/v$, v^0 is the close-packed volume, α is a nonsphericity parameter of a hard convex body and N is the number of particles. The nonsphericity parameter is defined as $\alpha = R_0S_0/V_0$ where R_0, S_0 and V_0 are the mean curvature, mean surface and volume of the hard convex body. For $\alpha = 1$ Eq. (4.2-41) reduces to the Carnahan–Starling (1969) expression for hard spheres. Expressions describing the effect of hard-core geometry on molecular repulsions in a different way have been also formulated (e.g., Naumann et al., 1981). For dipolar hard-sphere fluids Bryan and Prausnitz (1987) proposed a reference system with the same density dependence as that given by the Carnahan–Starling equation but with coefficients expressed as functions of temperature through the reduced dipole moment.

In the early attempts to utilize the hard-core repulsive terms to construct equations for real fluids, the simple van der Waals or Redlich–Kwong attractive terms were employed as perturbations (Carnahan and Starling, 1972). Henderson (1979) strongly promoted this type of equations as they better reproduce real fluid isotherms on the basis of critical data than the cubic equations. Other investigators noted, however, that although the difference between the Carnahan–Starling and van der Waals repulsive terms is very large, it has little influence on practical calculations (e.g., Tsonopoulos and Heidman, 1985). It should be noted that an EOS with the Carnahan–Starling repulsive term is not cubic and thus requires more computation time. The equations containing a hard-core repulsive term and simple attractive terms were further studied by Anderson and Prausnitz (1980) and Dimitrelis and Prausnitz (1986). Wong and Prausnitz (1985) proposed a simple attractive term interpolating between the van der Waals and Redlich–Kwong forms. The Carnahan–Starling hard-sphere EOS was also used with attractive terms expressed by a truncated virial expansion (Bienkowski et al., 1973; Bienkowski and Chao, 1975; Orbey and Vera, 1986, 1989). To obtain a more accurate equation of state, several authors adopted expressions for the attractive contribution obtained by fitting an appropriate power series to computer simulation data. For example, Alder et al. (1972) obtained the following expression for a square-well fluid:

$$z_{\mathrm{attr}} = \sum_{n,m} mD_{nm}\left(\frac{u}{kT}\right)^n\left(\frac{v^0}{v}\right)^m \tag{4.2-42}$$

where u is the characteristic energy. Chen and Kreglewski (1977; Kreglewski and Chen, 1978) combined Eqs. (4.2-41) and (4.2-42) into the BACK (Boublik–Alder–Chen–Kreglewski) equation of state. The authors determined the parameters D_{nm} [Eq. (4.2-42)] from the residual energy and volumetric data of argon and treated them as universal constants. Both hard-core volume and characteristic energy u were assumed to be temperature dependent:

$$v^0 = v^{00}\left[1 - C\exp\left(-\frac{3u^0}{kT}\right)\right]^3 \tag{4.2-43}$$

$$\frac{u}{k} = \left(\frac{u^0}{k}\right)\left(1 + \frac{\eta}{kT}\right) \tag{4.2-44}$$

The equation contains five constants: v^{00}, α, u^0/k, η/k and C. Three of them must be determined from experimental data (v^{00}, α, u^0/k) and the remaining two are assigned fixed values. The BACK equation was further studied by Simnick et al. (1979) and Machát and Boublík (1985) who found correlations of the parameters v^{00}, α, u^0/k with critical properties and the acentric factor.

Beret and Prausnitz (1975) and Donohue and Prausnitz (1978) utilized the Carnahan–Starling and Alder et al. results to construct an equation for molecules treated as chains of segments (perturbed hard chain theory, or PHCT). The authors factorized the canonical partition function as follows:

$$Q = \left(\frac{1}{N!}\right)\left(\frac{V}{\Lambda^3}\right)^N (q_{\text{rep}} q_{\text{attr}} q_{r,v})^N \tag{4.2-45}$$

where $q_{r,v}$ is the contribution of rotational and vibrational motion of a molecule, N is the number of molecules and Λ is the length of the thermal de Broglie wave. As the rotational and vibrational movement of chain molecules is hindered by the presence of other molecules, some of rotational and vibrational degrees of freedom depend not only on temperature but also on density. This can be written as

$$q_{r,v} = (q_{r,v})_{\text{ext}} (q_{r,v})_{\text{int}} \tag{4.2-46}$$

where $(q_{r,v})_{\text{ext}}$ and $(q_{r,v})_{\text{int}}$ are the external (density-dependent) and internal (density-independent) contributions to the partition function, respectively. The crucial assumption in PHCT is that the external contributions to the partition function from rotational and vibrational motions can be calculated as contributions from equivalent translational motions. This assumption goes back to the works of Prigogine et al. (1957) on the cell model of chainlike substances. In the PHCT model each translational degree of freedom contributes $(q_{\text{rep}} q_{\text{attr}})$ to the partition function. Assuming that the total number of external degrees of freedom is $3c$, the partition function (4.2-45) becomes

$$Q = \left(\frac{1}{N!}\right)\left(\frac{V}{\Lambda^3}\right)^N (q_{\text{rep}} q_{\text{attr}})^{Nc} (q_{r,v})_{\text{int}} \tag{4.2-47}$$

where the last term depends only on temperature and does not contribute to the equation of state. The form of Eq. (4.2-47) has been postulated so that the ideal-gas partition function be recovered in the low-density limit. For the q_{rep} contribution the Carnahan–Starling expression is used and for q_{attr} Eq. (4.2-44) is employed with constants fitted to experimental data on methane. The equation contains three parameters for each substance: the characteristic temperature T^* related to the depth of the potential well, the characteristic volume v^* related to the size of segments and one-third of the external degrees of freedom c. The parameters were found to correlate smoothly within homologous series of hydrocarbons.

Morris et al. (1987) modified the PHCT EOS by replacing the original perturbation for the square-well potential by a perturbation for the Lennard–Jones potential. The resulting equation, called PSCT (perturbed soft chain theory), contains the same molecular parameters as PHCT. Since these parameters correlate smoothly, Jin et al. (1986) recast the working equations of PSCT in terms of group contributions. Such an equation should be useful especially for polymers and intermediate molecular weight compounds for which little experimental data exists. The next refinement of the perturbed chain theory (Vimalchand and Donohue, 1985; Vimalchand et al., 1986) incorporates the contributions of anisotropic forces z^{ani} to yield the PACT (perturbed anisotropic chain theory) EOS:

$$z = 1 + c(z^{\text{rep}} + z^{\text{iso}} + z^{\text{ani}}) \tag{4.2-48}$$

where z^{iso} is a contribution of isotropic forces calculated as perturbation for the Lennard–Jones potential and z^{ani} is calculated from a perturbation expansion of Gubbins and Twu (1978) for anisotropic multipolar interactions. Kim et al. (1986) developed a simplified version of PHCT (simplified perturbed hard chain theory, of SPHCT) by replacing the complex attractive part of PHCT by a simpler expression combining ingredients of the radial distribution function of square-well molecules with those of lattice statistics of chain molecules:

$$z_{\text{att}} = -Z_M \frac{cv^* Y}{v + v^* Y} \quad \text{with} \quad Y = \exp\left(\frac{\epsilon q}{2ckT}\right) - 1 \tag{4.2-49}$$

where Z_M is the maximum coordination number, ϵ is the characteristic energy per unit surface area and q is the normalized surface area. As the parameters of the SPHCT EOS correlate within homologous series of nonpolar compounds, Georgeton and Teja (1988) developed a group contribution equation of state based on SPHCT. To improve the accuracy of PHCT at low densities, Cotterman et al. (1986) separated the attractive Helmholtz energy into low- and high-density contributions:

$$a_{\text{att}} = a^{\text{sv}}(1 - F) + a^{\text{df}} F \tag{4.2-50}$$

where the a^{sv} contribution is calculated from a virial expansion and a^{df} is a dense fluid contribution calculated from PHCT extended to polar fluids using the multipolar expansion of Gubbins and Twu (1978). For intermediate densities, a continuous function F is used to interpolate between the two density limits.

Chien et al. (1983) developed their chain of rotators (COR) equation of state by adopting a different approximation for the chainlike structure of molecules than that used in the perturbed chain theories (PCT). After Prigogine et al. (1957) they suggested treating the first segment of a chainlike species like a free molecule. The subsequent segments are assumed to rotate about their neighbors. Therefore, the configurational canonical partition function is written as

$$Q_{\text{conf}} = Q_t q_r^{Nc} Q_{\text{attr}} \tag{4.2-51}$$

where Q_t is the translational partition function calculated as in the case of a spherical molecule, q_r is the partition function of an elementary rotator and c is the number of rotational degrees of freedom of the molecular chain. Function

Q_t was calculated from the Carnahan–Starling (4.2-40) expression whereas q_r was evaluated by forcing Eq. (4.2-51) to represent the partition function of a hard dumbbell Q_{db} as given by the Boublík–Nezbeda theory, i.e., $Q_{\mathrm{db}} = Q_t q_r^{2N}$. The obtained expression has been assumed to be valid also for chains containing more than two segments. This assumption has been shown later by Fischer and Lustig (1985) to be a good approximation. The Q_{attr} contribution is calculated from the Alder et al. expansion (4.2-42) with the constants D_{nm} determined by fitting the equation to the *PVT*, $U - U_0$ and vapor pressure data on methane. The final expression for the COR equation is

$$z = 1 + \frac{4(\bar{v}/\tau)^2 - 2(\bar{v}/\tau)}{(\bar{v}/\tau - 1)^3} + \frac{c}{2}(\alpha - 1)\frac{3(\bar{v}/\tau)^2 + 3\alpha(\bar{v}/\tau) - (\alpha + 1)}{(\bar{v}/\tau - 1)^3}$$
$$+ \left[1 + \left(\frac{c}{2}\right)\left(B_0 + \frac{B_1}{\bar{T}} + B_2\bar{T}\right)\right] \sum_{n,m} \frac{mA_{nm}}{\bar{T}^n \bar{v}^m} \tag{4.2-52}$$

where $\alpha = 1.073$, $\tau = \pi \times 2^{1/2}/6$, B_0, B_1, B_2 and A_{nm} are universal constants, $\bar{v} = v/v^0$ and $\bar{T} = T/T^*$.

The COR EOS was also extended to polar fluids (Masuoka and Chao, 1984) by adding a polar pressure term derived from Monte Carlo simulations of Stockmayer molecules (Yao et al., 1982).

Deiters (1981a,b) developed a different augmented van der Waals EOS by deriving a semiempirical approximation for the radial distribution function of square-well fluids, involving corrections for nonspherical molecular shape, "soft" repulsive potential and three-body effects. The equation was shown to be capable of representing the *PVT* behavior in a very large pressure range using only three parameters determined from T_c, P_c and v_c.

The virtue of the "augmented" van der Waals equations of state is their ability to correlate accurately the saturated properties of fluids and *PVT* data outside the saturation region using only a few (typically three) equation constants. This is due primarily to the introduction of more realistic repulsive and attractive terms. Moreover, most equations of this kind contain large sets of universal parameters fitted to the properties of selected reference fluids (e.g., argon or methane) that ensure the correct shape of the equation also for other similar fluids. This feature makes the equations attractive in comparison with the multiparameter virial-type equations. Unlike the cubic equations of state, their parameters are not correlated with the critical properties and acentric factor although some regularities are observed for several classes of compounds. From the practical point of view, the main drawback of the augmented van der Waals equations is their complexity, which makes them cumbersome to use in routine process simulation calculations. This is especially important when only some selected properties are dealt with for which a tailor-made cubic EOS can be used. Moreover, the equations from the PHCT family are not constrained to reproduce the experimental critical point. Usually, T_c and P_c are overestimated by several Kelvins and bars, respectively. Only the Deiters equation reproduces correctly T_c, P_c and v_c.

SIMPLIFICATIONS OF THE AUGMENTED VAN DER WAALS-TYPE EQUATIONS OF STATE

Some investigators proposed equations combining the simplicity of cubic EOS with some of the theoretical background of the augmented van der Waals treatment. Ishikawa et al. (1980) developed a cubic equation of state by coupling the Scott (1971) expression for the repulsive contribution with the Redlich–Kwong attractive term.

$$P = \frac{RT(2v + b)}{v(2v - b)} - \frac{a}{T^{1/2}v(v + b)} \tag{4.2-53}$$

The Scott expression is compatible with the cubic form. Guo et al. (1985a,b) and Kim et al. (1986) presented a simplified version of the COR EOS, called the cubic chain of rotators (CCOR) EOS. The authors derived analytical approximations for the hard-core and rotational pressure contributions.

$$p = \frac{RT(1 + 0.77b/v)}{v + 0.42b} + \frac{c^R(0.055RTb/v)}{v - 0.42b} - \frac{a(T)}{v[v + c(T)]} - \frac{bd}{v[v + c(T)](v - 0.42b)} \tag{4.2-54}$$

The equation, with ζ set equal to z_c and two temperature-dependent parameters, was fitted to vapor pressures and liquid densities at subcritical temperatures and $P - v$ isotherms at supercritical temperatures. These forms, although put on a firmer theoretical basis, are subject to the same limitations as purely empirical cubic EOS and give comparable results (Table 4.2-1). Another attempt to develop a simple equation with some theoretical background was made by Kubic (1986) who developed a quartic equation of state by simplifying the Beret–Prausnitz generalized van der Waals partition function. For the repulsive part he used the approximate hard-core equation as developed for the CCOR EOS.

$$p = \frac{RT}{v} + \frac{1.19cbRT}{v(v - 0.42b)} - \frac{a}{(v + d)^2} \tag{4.2-55}$$

The results were comparable with those obtained from the more recent cubic equations of state.

To summarize the review of the van der Waals-type equations of state, the typical accuracy of calculating vapor pressure, orthobaric liquid density, second virial coefficient, isothermal compressibility coefficient $\beta_T = (-1/v)(\partial v/\partial P)_T$ and compressed liquid volume $V(P)$ (up to ca. 50 bars) is shown in Table 4.2-2. The table contains results obtained by Gregorowicz (1990) who applied the Peng–Robinson (4.2-25), Yu–Lu (4.2-31) and COR (4.2-52) equations to calculate the properties of seven hydrocarbons. In the case of the Peng–Robinson and Yu–Lu equations the parameters generalized in terms of T_c, P_c and ω were used. As the parameters of the COR equation are not generalized, Gregorowicz developed a semipredictive technique. This technique consists in

Table 4.2-2
REPRESENTATION OF VAPOR PRESSURE, ORTHOBARIC LIQUID DENSITY, SECOND VIRIAL COEFFICIENT, ISOTHERMAL COMPRESSIBILITY $\beta_T = (-1/v)(\partial v/\partial P)_T$ AND COMPRESSED LIQUID VOLUME BY MEANS OF THE PENG–ROBINSON (4.2-25), YU–LU (4.2-31) AND COR (4.2-52) EQUATIONS OF STATE FOR HYDROCARBONS[a]

	$(100/N)(\Sigma[(Q_i^{cal} - Q_i^{exp})/Q_i^{exp}]^2)^{1/2}$		
Property	**Peng–Robinson (generalized parameters)**	**Yu–Lu (generalized parameters)**	**COR (v_*^0, c generalized T from vapor pressure)**
Vapor pressure (up to 1 bar)	3.5	2.6	0.3
Orthobaric liquid volume (up to 333 K)	1.1	1.2	0.6
Second virial coefficients	25.0	25.0	31.0
Isothermal compressibility (up to 333 K)	12.0	14.0	31.0
Compressed liquid volume (up to 333 K and 50 bars)	1.0	0.8	0.8

After Gregorowicz (1990).

[a]The hydrocarbons studied were toluene, ethylbenzene, propylbenzene, butylbenzene, 1,2,4-trimethylbenzene and methylcyclohexane.

calculating the hard-core volume v^0 from a generalized correlation with Bondi's (1968) values of the van der Waals volume, the c parameter from a correlation with the acentric factor and the energetic parameter T^* from a single vapor pressure data point.

Tables 4.2-1 and 4.2-2 give a feel of the accuracy that can be obtained using the van der Waals-type equations of state. When no experimental data are available, the cubic equations of state can be recommended without doubt to estimate the thermodynamic properties using only T_c, P_c and ω as input. The augmented van der Waals equations lack this predictive capability. The van der Waals-type equations, cubic or noncubic, can reproduce vapor pressures quite accurately if their parameters are specific (i.e., adjusted to some experimental data) rather than generalized. Therefore, the cubic equations of state can be recommended for practical calculations if the calculations are restricted to the saturation region (as is the case in most phase equilibrium modelling tasks). It is not important which equation is used for vapor pressure calculations; it is the temperature dependence of parameters that matters. On the other hand, the results obtained for volumetric properties are more sensitive to the choice of an equation as discussed before. The augmented van der Waals-type equations can be advantageous when the PVT properties in far-from-saturation regions need to be reproduced simultaneously with saturation properties.

Moreover, the augmented van der Waals-type equations show regular (frequently linear) dependencies of their parameters on the molar mass or carbon number in homologous series of compounds (e.g., alkanes, substituted aromatics, polynuclear aromatics, etc.). This makes it possible to extrapolate the parameters to polymeric and ill-defined fluids (heavy oils, asphaltenes). For such compounds the parameters T_c, P_c and ω are not known and, very often, experimentally inaccessible. Therefore, the use of corresponding-states based equations is difficult and the augmented van der Waals-type equations provide a safe method of extrapolation.

EQUATIONS OF STATE BASED ON THE LATTICE AND HOLE MODELS

Besides the augmented van der Waals-type equations described above, we shall briefly mention a related group of equations based on the lattice, cell or hole theories. The advanced lattice theories were developed in the 1950s (Barker, 1952) for the representation of excess properties of nonelectrolyte mixtures (cf. Section 1.3). Barker's and related models based on the idea of a rigid lattice without vacancies could not reproduce the volumetric properties of mixtures and their pressure and temperature dependence. Cell lattice theories developed by Prigogine (1957) and Orwoll and Flory (1967) made it possible to arrive at an equation of state for the liquid phase. The hole theories (Sanchez and Lacombe, 1976; Panayiotou and Vera, 1981, 1982; Nies et al., 1983; Smirnova and Victorov, 1987) allow for the presence of empty sites on the lattice. These models are valid for both liquid and gas phases.

Sanchez and Lacombe (1976) developed a model for pure fluids and mixtures with arbitrary molecular size on the assumption of complete randomness of distribution of molecules and holes on the lattice. Panayiotou and Vera (1982) and Nies et al. (1983) included the effects of ordering through the quasi-chemical approximation. Jain and Simha (1979) applied the hole models to liquids composed of chain molecules. Smirnova and Victorov (1987) developed a hole model taking into account molecular size and shape along with orientational effects.

VAN DER WAALS-TYPE EQUATIONS INCORPORATING ASSOCIATION

Representation of the thermodynamic properties of pure associating and polar substances can be improved by explicitly allowing for association. A useful approach is to treat a pure substance as a mixture of monomers and multimers formed in the consecutive association reaction

$$A_i + A_1 = A_{i+1} \qquad (4.2\text{-}56)$$

where i is the number of monomers in a multimer. The multimers in equilibrium are regarded as components of a mixture that can be treated by equation-of-state method for nonpolar mixtures. Therefore, a thorough analysis of methods for incorporating association into the formalism of van der Waals-type equations has to be deferred until the methods for mixtures are explained.

Here, we should only note that the compressibility factor can be split into two parts: the physical $z^{(ph)}$ and the chemical $z^{(ch)}$ one. In this notation (Anderko, 1989a, 1990a) the compressibility factor is

$$z = z^{(ph)} + z^{(ch)} - 1 \tag{4.2-57}$$

where $z^{(ph)}$ is the equation of state of a nonreacting monomer of the associating substance and is expressed by an equation of state for nonpolar compounds. The $z^{(ch)}$ contribution can be expressed analytically for some association models. The combined equation has been called AEOS (association + equation of state).

If the multimers are assumed to be formed according to the Mecke–Kempter continuous linear association model (cf. Section 3.3.5), the $z^{(ch)}$ contribution becomes (Ikonomou and Donohue, 1986, 1987; Anderko, 1989a, 1990a)

$$z^{(ch)} = \frac{2}{1 + (1 + 4RTK/v)^{1/2}} \tag{4.2-58}$$

where K is the association constant. If only monomers and dimers are assumed to exist in the mixture (a reasonable assumption for carboxylic acids), we obtain

$$z^{(ch)} = \frac{2 - 2RTK/v}{1 - 4RTK/v + (1 + 8RTK/v)^{1/2}} \tag{4.2-59}$$

For the special case of water, which is known to form three-dimensional molecular aggregates, Anderko (1991a) deduced the form

$$z^{(ch)} = \frac{1}{1 + RTK/v + \alpha(RTK/v)^2} \tag{4.2-60}$$

where α is an empirical parameter.

The temperature dependence of the association constant K is expressed assuming that the standard heat capacity of the reaction of association is independent of temperature:

$$\ln K = \frac{-\Delta H^0(T_0) + \Delta C_p^0 T_0}{RT} + \frac{1}{R}[\Delta S^0(T_0) - \Delta C_p^0 - \Delta C_p^0 \ln T_0] + \frac{\Delta C_p^0}{R} \ln T \tag{4.2-61}$$

where $\Delta H^0(T_0)$ and $\Delta S^0(T_0)$ are the standard enthalpy and entropy of association at the reference temperature T_0, respectively. Alternatively, a simplified temperature dependence with $\Delta C_p^0 = 0$ may be employed. The equation of state with association built in usually has five or six parameters for each associating compound. In addition of $\Delta H^0(T_0)$, $\Delta S^0(T_0)$ and ΔC_p^0 they are the characteristic parameters of the physical contribution to the compressibility factor. If a cubic equation of state is used for the physical contribution, the critical temperature T_c', critical pressure P_c' and acentric factor ω' have to be found for a model compound with physical interactions identical with those in the association substance but not having the capability of forming associates. The parameters of such a compound can be approximated by those of the homomorph of the associated substance (cf. Section 3.3.5). Since the properties of a homomorph provide only a rough estimate of the properties of the nonassociating monomer of the associating substance, the critical properties and acentric factors of existing homomorphs (hydrocarbons or ethers) cannot be used directly but provide convenient initial estimates of the parameters T_c', P_c' and ω'. The pure-component parameters of the combined equation of state can be calculated by fitting the equation to experimental vapor pressure and liquid density data because such data are usually most widely accessible and precise. The parameters are fitted to the properties of individual compounds and no attempt has been made to generalize them.

The combination of a van der Waals-type equation with an association model makes it possible to obtain a good reproduction of experimental data in the saturation region. Table 4.2-3 shows the deviations obtained by calculating vapor pressures, liquid densities, second virial coefficients and vaporization enthalpies by means of the AEOS (Anderko, 1989a) for seven representative compounds. As the physical contribution, the equation of Yu and Lu (4.2-31) was used.

The association parameters, i.e., the association constants and enthalpies are physically reasonable. This has been proved by predicting the pressure dependence of gas phase thermal conductivity, a property not related to phase equilibrium, using the parameters determined from vapor pressure and liquid density (Anderko, 1989a). The pressure dependence of thermal conductivity is

Table 4.2-3
REPRESENTATION OF VAPOR PRESSURES, ORTHOBARIC LIQUID DENSITIES, SECOND VIRIAL COEFFICIENTS, VAPORIZATION ENTHALPIES AND PRESSURE DEPENDENCE OF GAS PHASE THERMAL CONDUCTIVITY BY MEANS OF THE AEOS EQUATION OF STATE[a]

Property	$(100/N)\Sigma\,\lvert(Q_i^{\mathrm{cal}} - Q_i^{\mathrm{exp}})/Q_i^{\mathrm{exp}}\rvert$
Vapor pressure	0.4
Liquid density	2.2
Second virial coefficient	6.9
Vaporization enthalpy	0.6
Pressure dependence of thermal conductivity	1.0

After Anderko (1989a).
[a]The compounds studied were ethanol, 1-propanol, 2-propanol, 2-propanone, pyridine and ethanoic acid.

sensitive to association and makes it possible to validate the association model in an independent way.

While the results of calculating simultaneously several pure-component properties are good, the actual merit of equations of state incorporating association is their greatly improved performance for mixtures. This will be discussed in detail in Section 4.3.7. While association improves the results for pure components, the improvement is achieved at the cost of making the computational procedure more difficult. Notably, an equation constructed from a cubic EOS and an association model is no longer cubic, and the optimization procedure to obtain the parameters requires a judicious choice of initial estimates. If a user is interested exclusively in pure-component properties, he can achieve the same goal by using simpler tools as described in this Section (e.g., temperature-dependent parameters, volume translation etc.). The benefits of using association manifest themselves mostly for mixtures.

EQUATIONS OF STATE FOR ILL-DEFINED COMPOUNDS

The parameters of all the equations discussed in this chapter are clearly identified for specific compounds. However, many mixtures of technical interest contain, for example, heavier hydrocarbons and hydrocarbon fractions having no strictly defined stoichiometry. Then, the practical calculations should also involve a subdivision into a number of pseudocomponents. When equations of state are used for calculating phase equilibria in such mixtures, characterization of the heavy hydrocarbons has to be performed. If cubic equations are used for this purpose (this is the case for most practical applications), characterization of heavy hydrocarbons implies estimation of critical temperature and pressure and acentric factor for C_{7+} fractions due to the unavailability of relevant experimental data. Experimental data for heavy hydrocarbon mixtures can be obtained from freezing point depression giving the molecular weight, specific gravity and chromatographic analysis and distillation giving the boiling points of fractions. Frequently, the available data are limited to molecular weight and specific gravity of various fractions.

Various methods have been proposed to estimate critical temperature and pressure and acentric factor of heavy hydrocarbon mixtures. A detailed review of these methods is beyond the scope of this work. The interested reader is referred to the books by Chorn and Mansoori (1989) and Pedersen et al. (1989).

4.2.2. Virial-Type Equations of State

As discussed in detail in Section 2.3 the virial equation of state is an expansion in terms of density (the Leiden form):

$$z = 1 + B\rho + C\rho^2 + \cdots \qquad (2.2\text{-}5)$$

or pressure (the Berlin form)

$$z = 1 + B'P^2 + C'P^2 + \cdots \qquad (2.2\text{-}6)$$

In practical calculations Eqs. (2.2-5) and (2.2-6) are used in a truncated form for systems containing one fluid phase or for the representation of the gas phase in two-phase systems (e.g., in the γ–ϕ approach). As the virial equation is an expansion about the compressibility factor of an ideal gas, it is formally valid only in the vicinity of $\rho = 0$ or $P = 0$. Higher coefficients of the expansion are required in the high-density region. However, the third and higher virial coefficients cannot be estimated due to our insufficient knowledge of intermolecular forces, nor can they be determined empirically in the regions of slow convergence of the series. Therefore, the virial expansions can merely serve as an inspiration for the development of empirical equations of state valid over wide ranges of density.

The EOS of Beattie and Bridgeman (1927)

$$p = RT\rho + \left(B_0RT - A_0 - \frac{CR}{T^2}\right)\rho^2 + \left(aA_0 - bB_0RT - \frac{cB_0R}{T^2}\right)\rho^3 + \left(\frac{bB_0cR}{T^2}\right)\rho^4 \qquad (4.2\text{-}62)$$

was a step in this direction, although the equation was accurate only in the gaseous region. The five constants A_0, B_0, a, b and c were determined from gas phase PVT data.

The celebrated Benedict–Webb–Rubin (BWR) equation (1940, 1942)

$$p = RT\rho + \left(B_0RT - A_0 - \frac{C_0}{T^2}\right)\rho^2 + (bRT - a)\rho^3 + \alpha a\rho^6 + \left(\frac{c\rho^3}{T^2}\right)(1 + \gamma\rho^2)\exp(-\gamma\rho^2) \qquad (4.2\text{-}63)$$

was the first virial-type EOS capable of representing the PVT properties up to the high-density region and vapor–liquid equilibria in industrially important mixtures. The eight constants of the BWR EOS were initially determined for light hydrocarbons, primarily from gas phase PVT and vapor pressure data. To extend the predictive capability of the BWR equation, the parameters were generalized by Joffe (1949) in terms of the critical temperature and pressure. Opfell et al. (1956), Cooper and Goldfrank (1967) and Edmister et al. (1968) improved the generalization for normal fluids by introducing the acentric factor as the third parameter.

The BWR EOS was found, however, to give inaccurate results in the high-density region, in the low-temperature region where vapor pressure is low and around the critical point. These drawbacks are due, in part, to the shape of the isotherms predicted by BWR and similar EOS. While in the case of a van der Waals-type isotherm the derivative $dp/d\rho$ has at most one minimum and no maximum in the density region corresponding to vapor–liquid equilibrium (VLE) calculations (Topliss et al., 1988), the BWR-type equations may display more complicated behavior. Further developments of the BWR EOS either improved the equation for a specific substance or modified its generalized form for larger groups of compounds. The two paths of refinement have been described in detail by Saito and Arai (1986). The first group of equations contains the 16-constant equation of Strobridge (1962), developed initially for nitrogen, and its modifications. Among them the 20-constant equation of Bender (1975) is particularly successful for cryogenic applications. Vennix and Kobayashi (1969) developed an accurate 25-constant equation for methane. Goodwin (1979) developed an equation of state for methane that is nonanalytic at the critical point due to the introduction of a function designed to give a large increase in the constant-volume heat capacity upon close approach to the critical point, which is consistent with experiment. Gallagher and Haar (1978) developed an equation of this kind for ammonia and Keenan et al. (1969) and Grigull et al. (1984) for water. In general, such accurate multiparameter equations of state can be utilized for mixture calculations as reference fluid equations within the framework of the corresponding-states principle. This type of equations is also used in an extended form for the representation of extensive data for compounds (*IUPAC Thermodynamic Tables of the Fluid State*).

An extended generalized BWR EOS, called HCBKS (Han–Cox–Bono–Kwok–Starling) was developed by Cox et al. (1971) and Starling and Han (1972) who added three constants to improve the temperature dependence of parameters in the low-temperature region:

$$p = \rho RT + \left(B_0 RT - A_0 - \frac{C_0}{T^2} + \frac{D_0}{T^3} - \frac{E_0}{T^4}\right)\rho^2 + \left(bRT - a - \frac{d}{T}\right)\rho^3 + \alpha\left(a + \frac{d}{T}\right)\rho^6 + \left(\frac{c\rho^3}{T^2}\right)(1 + \gamma\rho^2)\exp(-\gamma\rho^2) \tag{4.2-64}$$

where ρ is the density and the parameters B_0, A_0, C_0, D_0, E_0, b, a, d, c, α, γ are generalized for normal fluids in terms of the critical temperature, critical pressure and acentric factor. Nishiumi and Saito (1975) modified the above equation to improve its accuracy for heavier hydrocarbons and in the low-temperature region. Four more constants e, f, g and h were added.

$$p = \rho RT + \left(B_0 RT - A_0 - \frac{C_0}{T^2} + \frac{D_0}{T^3} - \frac{E_0}{T^4}\right)\rho^2 + \left(bRT - a - \frac{d}{T} - \frac{e}{T^4} - \frac{\rho}{T^{23}}\right)\rho^3 + \alpha\left(a + \frac{d}{T} + \frac{e}{T^4} + \frac{f}{T^{23}}\right)\rho^6 + \left(\frac{c}{T^2} + \frac{g}{T^8} + \frac{h}{T^{17}}\right)\rho^3(1 + \gamma\rho^2)\exp(-\gamma\rho^2) \tag{4.2-65}$$

Nishiumi (1980) extended the equation to polar substances. The authors recommended the equation for the gas and liquid region except for the intervals $1 \le T_R \le 1.3$ and $1 \le \rho_R \le 3$.

Lee and Kesler (1975) used a modified 12-constant BWR EOS as a reference fluid equation with their corresponding-states method:

$$z = 1 + \left(b_1 - \frac{b_2}{T_R} - b_3 T_R^2 - b_4 T_R^3\right)\rho_R + \left(c_1 - \frac{c_2}{T_R} + c_3 T_R^3\right)\rho_R^2 + \left(d_1 + \frac{d_2}{T_R}\right)\rho_R^5 + \left(\frac{C_4}{T_R^3}\right)\rho_R^2(\beta + \gamma\rho_R^2)\exp(-\gamma\rho_R^2) \tag{4.2-66}$$

where subscript R denotes reduced variables. The parameters were determined from the properties of argon and methane, which were used as reference fluids.

The more recent developments retain the virial-type form as it is convenient for differentiation and integration of the equation of state. Notably, Vetere (1982, 1983) coupled the concepts pertinent to van der Waals-type and virial-type models. The author assumed a van der Waals-type attractive contribution to the internal energy $E = ae^{-kT}/V$ and utilized well-known thermodynamic relations $p = T(\partial S/\partial V)_T$ and $(\partial p/\partial T)_V = (\partial S/\partial V)_T$. The equation of state

$$p = \frac{RT}{v} - a\frac{e^{-kT}}{v^2} + a_1\frac{e^{-k_1 T}}{v^m} + b\frac{e^{-k_2 T}}{v^n} \tag{4.2-67}$$

contains eight constants, four of which are directly calculated from pure-component properties. As the temperature dependence of the EOS is not sufficient to represent both the liquid and vapor phases with the same set of parameters, accurate representation of the saturation region was obtained by forcing the b constant to reproduce the liquid volume versus temperature data as calculated from the Rackett equation. Vetere (1982) found the equation to compare favorably with the BWR EOS.

Saito and Arai (1986) evaluated the above BWR-type equations of state with respect to the reproduction of vapor pressures and saturated liquid and vapor densities. In general, the multiparameter BWR-type equations offer the possibility of a very accurate representation of pure-compound properties. If their parameters are generalized in terms of T_c, p_c and ω, their accuracy becomes, however, comparable to that of the van der Walls-type EOS whose parameters were fitted also to density data. The equations with parameters generalized on the basis of the three-parameter corresponding-states principle cannot be used for polar compounds.

4.2.3. Equations of State in a Corresponding-States Format

The principle of corresponding states (PCS) is one of the most useful approaches to the prediction of fluid properties. The principle is utilized in all the methods described before that express the equation-of-state parameters as functions of critical properties and acentric factor. Some important equations of state are formulated, however, in an explicit corresponding-states format. Two different approaches have been taken along this line. The first one introduces shape factors $\theta_{\alpha\alpha,0}$ and $\phi_{\alpha\alpha,0}$ in the scaling parameters to account for the deviations from the simple two-parameter PCS (Leach et al., 1968)

$$z_\alpha = z_0\left(\frac{T}{f_{\alpha\alpha,0}}, \frac{V}{h_{\alpha\alpha,0}}\right) \tag{4.2-68}$$

or

$$z_\alpha = z_0\left(\frac{T}{f_{\alpha\alpha,0}}, \frac{Ph_{\alpha\alpha,0}}{f_{\alpha\alpha,0}}\right) \tag{4.2-69}$$

where the scale factors $f_{\alpha\alpha,0}$ and $h_{\alpha\alpha,0}$ are given by

$$f_{\alpha\alpha,0} = \left(\frac{T^c_\alpha}{T^c_0}\right)\theta_{\alpha\alpha,0} \tag{4.2-70}$$

$$h_{\alpha\alpha,0} = \left(\frac{V^c_\alpha}{V^c_0}\right)\phi_{\alpha\alpha,0} \tag{4.2-71}$$

The volume- and temperature-dependent shape factors define the state of the reference fluid (denoted by 0) that corresponds to the state of the fluid of interest (α). Analytical formulas for the shape factors were determined from data on normal hydrocarbons from C_2 to C_{15} by Leach et al. and expressed as functions of critical properties and acentric factor. Methane was chosen as a reference substance because of the high accuracy and wide range of data.

Mentzer et al. (1980) reviewed the theoretical basis and applications of the shape-factor-based PCS. In particular, the properties of light hydrocarbons in the saturation region were calculated by Mollerup and Rowlinson (1974) and Mollerup (1977) using Goodwin's (1974) equation as a reference fluid EOS. Mentzer et al. (1980) extended the calculations to heavier hydrocarbons (C_{16}). For pure fluids the method can be used in a purely predictive manner although various schemes were proposed to improve the accuracy by expressing the acentric factor as a function of temperature (e.g., Gunning and Rowlinson, 1973; Singh and Teja, 1976). In general, calculations using this method are best for low molecular weight hydrocarbons that are similar to the reference substance, methane. The accuracy decreases when the size, shape and polarity of the compound of interest is significantly different from those of the reference substance. Such method cannot be recommended for strongly polar or associating compounds (Mentzer et al., 1980).

Mollerup (1980) noted that although the method offers high accuracy it is time consuming due to the nonanalytic reference equation and the volumetric dependence of Leach's shape factor correlation. Therefore, Mollerup applied the corresponding-states approach to simple two-constant van der Waals-type equations of state and separated them into a shape factor correlation and an equation for the reference fluid. Such shape factor correlation can be used with any reference fluid.

The second approach consists in expressing the compressibility factor as a perturbation about its value for a reference fluid. Pitzer et al. (1955) applied a first-order perturbation about a spherical reference fluid

$$z_\alpha = z_0\left(\frac{T}{f_{\alpha\alpha,0}}, \frac{P}{g_{\alpha\alpha,0}}\right) + \omega\left(\frac{\partial z_0}{\partial \omega}\right) \tag{4.2-72}$$

where z_0 is the compressibility factor of a spherical reference fluid such as argon ($\omega = 0$) and the second term represents deviations from the two-parameter corresponding-states principle. The scaling factors are defined as ratios of critical parameters: $f_{\alpha\alpha,0} = T^c_\alpha / T^c_0$ and $g_{\alpha\alpha,0} = p^c_\alpha / p^c_0$, i.e., analogously to the simple two-parameter corresponding-states principle. In Pitzer's original formulation, the functions z_0 and $(\partial z_0/\partial \omega)$ were given in a graphical form. Lee and Kesler (1975) replaced the derivative $(\partial z_0/\partial \omega)$ by its finite-difference approximation and obtained

$$z_\alpha = z_0\left(\frac{T}{f_{\alpha\alpha,0}}, \frac{P}{g_{\alpha\alpha,0}}\right) + \frac{\omega}{\omega_r}\left[z_r\left(\frac{T}{f_{\alpha\alpha,r}}, \frac{P}{g_{\alpha\alpha,r}}\right) - z_0\left(\frac{T}{f_{\alpha\alpha,0}}, \frac{P}{g_{\alpha\alpha,0}}\right)\right] \tag{4.2-73}$$

where the subscript r refers to the second, nonspherical reference fluid, octane. The compressibility factors z_0 and z_r are expressed by a 12-constant BWR-type equation [Eq. (4.2-59)]. The choice of the reference fluids (i.e., argon and octane) enabled the equation to be accurate for nonpolar fluids.

The more recent developments aimed at extending the applicability of the corresponding-states formulation to a wider spectrum of fluids either by changing the reference fluids or by introducing new polarity parameters, complementing or superceding the acentric factor. As a better accuracy of the expansion can be expected if the reference fluids are similar to the fluid under study (i.e., the perturbation is small), Teja (1980) and Teja et al. (1981) proposed a generalized corresponding-states method with variable reference fluids

$$z_\alpha = z_{r1}\left(\frac{T}{f_{\alpha\alpha,r_1}}, \frac{P}{g_{\alpha\alpha,r_1}}\right) + \frac{\omega - \omega_{r_1}}{\omega_{r_1} - \omega_{r_2}}\left[z_{r_2}\left(\frac{T}{f_{\alpha\alpha,r_2}}, \frac{P}{g_{\alpha\alpha,r_2}}\right)\right] \tag{4.2-74}$$

where subscripts r_1 and r_2 represent two reference fluids that are chosen such that they are similar to the fluid of interest. Equation (4.2-74) is applicable to a wide variety of fluids including polar and associated ones if r_1 and r_2 are appropriately selected. The acentric factor is retained as the third parameter in PCS. This approach, initially proposed for liquid densities, was used by Wong et al. (1983, 1984) and Rosenthal et al. (1987) for VLE.

Another method for extending PCS to polar fluids consists in applying specific factors describing polarity effects besides the acentric factor. Wu and Stiel (1985) employed a polarity factor Y in the expansion for z:

$$z = z_0(T_r, p_r) + \omega z_1(T_r, p_r) + Yz_2(T_r, p_r) \tag{4.2-75}$$

where Y is calculated from PVT data for the fluid. The functions z_0 and z_1 were obtained from the Lee–Kesler EOS [i.e., $z_1 = (z_r - z_0)/\omega_r$] and z_2 was calculated from Keenan's (1969) multiparameter EOS for water

$$z_2 = z_w - (z_0 + \omega_w z_1) \tag{4.2-76}$$

where w denotes water. Thus, Wu and Stiel's corresponding-states scheme is based on three fixed reference fluids rather than two variable ones in Eq. (4.2-74). A similar approach was taken by Wilding and Rowley (1986) and Wilding et al. (1987) who developed another extension of the Lee–Kesler method for polar fluids. They wrote z in terms of size/shape, α, and polar, β, parameters

$$z = z_0(T_r, p_r) + \alpha z_1(T_r, p_r) + \beta z_2(T_r, p_r) \tag{4.2-77}$$

The z_1 and z_2 contributions were obtained from

$$z_1 = \frac{z_{r_1} - z_{r_0}}{\alpha_{r_1}} \tag{4.2-78}$$

and

$$z_2 = \frac{z_{r_2} - [z_{r_0} + \alpha_{r_2}(z_{r_1} - z_{r_0})/\alpha_{r_1}]}{\beta_{r_2}} \tag{4.2-79}$$

where subscripts r_0, r_1 and r_2 refer to the reference fluids argon, octane and water, respectively. The size parameter α was correlated as a function of the radius of gyration, and the polarity parameter β was obtained from a density data point using Eqs. (4.2-77)–(4.2-79).

Lee et al. (1979) developed a perturbation expansion in terms of an orientation factor γ that can be approximated by ω for nonpolar fluids. The expansion is similar to that of Pitzer [Eq. (4.2-72)]:

$$z(T^*, \rho^*, \gamma) = z_0(T^*, \rho^*) + \gamma z_1(T^*, \rho^*) \tag{4.2-80}$$

where $T^* = kT/\epsilon$ and $\rho^* = \rho\sigma^3$ are functions of molecular parameters ϵ and σ simply correlated with critical properties T_c and V_c. For both contributions the HCBKS EOS (4.2-58) was used with constants separated into two parts—one isotropic and one anisotropic part:

$$B_i = a_i + \gamma b_i \tag{4.2-81}$$

where B_i is a parameter of the Han–Starling EOS. Twu et al. (1980) applied Eq. (4.2-80) to nonpolar fluids (i.e., by setting $\gamma = \omega$) and expressed the reference compressibility factor by a Strobridge-type equation:

$$\begin{aligned} P = {} & RT\rho + \left(RC_1T + C_2 + \frac{C_3}{T} + \frac{C_4}{T^2} + C_5T^4\right)\rho^2 + (RC_6T + C_7)\rho^3 + C_8T\rho^4 \\ & + C_{15}\rho^6 + \rho^3\left(\frac{C_9}{T^2} + \frac{C_{10}}{T^3} + \frac{C_{11}}{T^4}\right)\exp(-\gamma\rho^2) \\ & + \rho^5\left(\frac{C_{12}}{T^2} + \frac{C_{13}}{T^3} + \frac{C_{14}}{T^4}\right)\exp(-\gamma\rho^2) \end{aligned} \tag{4.2-82}$$

Twu (1983) used the normal boiling point as an easily accessible third parameter for corresponding states. Another corresponding-states method applicable also to ill-defined mixtures has been proposed by Brulé et al. (1982) and Brulé and Starling (1984).

Chung et al. (1984) refined Eq. (4.2-80) by changing the parameterization technique for polar fluids. They expressed the compressibility factor as

$$z = 1 + \lambda z_{\text{conf}}^{(0)} + (\lambda - 1) z_{\text{conf}}^{(p)} \tag{4.2-83}$$

where λ is a structural parameter and $z_{\text{conf}}^{(0)}$ and $z_{\text{conf}}^{(\text{p})}$ are the isotropic reference and the perturbation contribution, respectively. Both contributions have the same Strobridge-type form with parameters expressed as

$$A_i = B_i^{(0)} + (\lambda - 1) B_i^{(p)} \tag{4.2-84}$$

where B_i are universal constants, $\rho^* = \rho V^*$, $T^* = kT/\epsilon$ and V^* are characteristic volume and energy parameters, respectively. The energy parameter for polar compounds is temperature dependent:

$$\frac{\epsilon}{k} = \frac{\epsilon^0}{k} + \frac{D}{T} \tag{4.2-85}$$

where D is a polar parameter. Thus, similar to the case of augmented van der Waals EOS, a pure substance is characterized by three parameters ϵ, V^* and λ and, additionally, D for polar compounds.

Platzer and Maurer (1989) developed a generalization of the Bender EOS to polar fluids in terms of T_c, V_c, ω and the polar factor χ of Halm and Stiel (1967). They wrote the parameters of Bender's EOS as a nonlinear expansion

$$e_i = g_{4,i} + g_{1,i}\omega + g_{2,i}\chi + g^* g_{3,i}\omega\chi + g_{5,i}\chi^2 \tag{4.2-86}$$

The authors also proposed an extension of Eq. (4.2-74) incorporating the reference fluids:

$$z = z_0 + \lambda(z_1 - z_0) + \mu(z_2 - z_0) \tag{4.2-87}$$

$$\lambda = \frac{1}{(\omega_1 - \omega_0)[(\omega - \omega_0) - \mu(\omega_2 - \omega_0)]} \tag{4.2-88}$$

$$\mu = \frac{(\omega - \omega_0)(\chi_1 - \chi_0) - (\omega_1 - \omega_0)(\chi - \chi_0)}{(\omega_2 - \omega_0)(\chi_1 - \chi_0) - (\omega_1 - \omega_0)(\chi_2 - \chi_0)} \tag{4.2-89}$$

Equation (4.2-87) uses, in contrast to Eq. (4.2-74), two descriptors of a substance, i.e., ω and χ. The authors concluded that the accuracy achieved for polar fluids is comparable to that for nonpolar fluids obtained from the corresponding-states methods.

The main advantage of using equations of state in the corresponding-states format consists in the possibility of employing very accurate reference fluid equations. The properties of the fluid of interest can be then accurately reproduced provided that the substance is not very different from the reference fluids.

4.2.4 Nonclassical Corrections in the Near-Critical Regions

It has been long recognized that an analytic equation of state does not yield the correct limiting properties at critical points of any type (e.g., Rowlinson and Swinton, 1982; Sengers and Levelt-Sengers, 1986). For example, analytic EOS are unable to reproduce the shape of the critical isotherm, the weakly divergent heat capacity C_v and the phase coexistence curve near the critical point. The impossibility of describing the critical behavior with analytic equations of state has led to compromises. The equation either reproduces correctly the slope of the orthobaric density curve and overshoots T_c and P_c or predicts too steep slopes when it is constrained to reproduce T_c and P_c. On the other hand, the critical region whose size is of the order of only 10^{-3} for the reduced temperature difference $(T_c - T)/T_c$ can be described by the scaling laws. But the equations valid for near-critical behavior are not useful outside the critical region; they do not reduce to the ideal-gas law at low density, for example. This behavior prompted some investigators to combine nonanalytic and analytic EOS to represent the thermodynamic properties in the critical as well as noncritical region. Chapela and Rowlinson (1974) suggested a solution to this problem by proposing a mixing function that blends the pressure from the scaling law and that from a classical EOS. Apart from different equations for the near-critical and far-from-critical regions, the method also required an appropriate switching function. Levelt-Sengers et al. (1983a,b) and Albright et al. (1987) forced their scaled equation of state to match the applicable limits of an analytic equation of state. Fox (1983) transformed an analytic equation of state into a nonanalytic equation by using parametric equations depending on the displacement from the critical point. The transformation satisfies the scaling laws at the critical point and ensures a smooth transition into the classical region. Erickson and Leland (1986) and Erickson et al. (1987) modified the method of Fox to obtain a smooth transition of derivative properties. The above methods, although accurate for pure components, are very complex and require a large number of adjustable constants. Their extension to mixtures would be difficult.

An alternative method of improving analytical EOS in the near-critical region consists in expressing the Helmholtz energy as a sum of classical and nonclassical contributions, the latter being represented by empirical functions designed to be negligible except for small values of $T - T_c$ and $(\rho - \rho_c)$ (Pitzer and Schreiber, 1988). While the critical exponents predicted by such an equation are incorrect, the EOS yields the near-critical behavior to the level of practical interest. Similarly, Chou and Prausnitz (1989) introduced a near-critical contribution to the Helmholtz energy to a volume-translated cubic EOS. The method was also extended to mixtures.

In general, the nonclassical corrections are necessary to improve the performance of multiparameter EOS for pure fluids near their critical points. The results of Chou and Prausnitz indicate, however, that the near-critical contribution does not have an appreciable effect on calculated phase equilibria of mixtures because the binary parameters have a dominant effect on calculated results, and the near-critical effects are masked by the choice of the binary parameters.

4.3 MIXTURES

4.3.1. Classical Quadratic Mixing Rules and Other Mixing Rules for Nonpolar Mixtures

Equations of state are now widely used for modelling the phase equilibria and *PVT* properties of mixtures of nonpolar and slightly polar substances. In the past they were used only for high-pressure calculations whereas the γ–ϕ methods were retained for the representation of phase equilibria at low pressures. This was due primarily to inaccurate predictions of pure-component vapor pressures, which obscured the capability of equations of state to correlate mixture data. The modern equations of state with temperature-dependent parameters fitted to vapor pressure are applicable to low pressure as well as high pressure.

The most widely used method for extending equations of state to nonpolar mixtures is to use the classical "one-fluid" approach first proposed by van der Waals in 1890. It is assumed that the properties of a fluid mixture are the same as those of a hypothetical pure fluid at the same temperature and pressure but having the characteristic constants appropriately averaged over the composition. Two- or more-fluid concepts (cf. Section 1.3) have received less attention with respect to the formulation of equation-of-state composition dependence and are used mostly for equations in the explicit corresponding-states format. The functions used for averaging the pure-component parameters for mixtures (mixing rules) are quadratic in mole fraction

$$\text{const} = \sum \sum x_i x_j \, \text{const}_{ij} \tag{4.3-1}$$

where (const_{ii}) is a constant of the equation for pure component i and const_{ij} ($i \neq j$) is determined by an appropriate combining rule with or without binary parameters.

The quadratic mixing rules (4.3-1) are strongly supported by the results of Leland et al. (1968a,b, 1969) who derived them from the theory of radial distribution functions. For a fluid mixture with a potential energy of interaction between the i and j molecules in the form

$$u_{ij}(r) = \epsilon_{ij} f\left(\frac{r}{\sigma_{ij}}\right) \tag{4.3-2}$$

where ϵ_{ij} is the interaction energy and σ_{ij} is the interaction distance, the following mixture parameters are obtained:

$$\sigma^3 = \sum \sum x_i x_j \sigma_{ij}^3 \tag{4.3-3}$$

$$\epsilon\sigma^3 = \sum \sum x_i x_j \epsilon_{ij} \sigma_{ij}^3 \tag{4.3-4}$$

Equation (4.3-3) and (4.3-4) suggest a quadratic composition dependence of the co-volume and energetic interaction parameters, respectively. In the case of three- or more-parameter empirical equations, the mixing rules for the remaining parameters can be guided merely by the requirement that the second virial coefficient be a quadratic function of composition, the third virial coefficient be a cubic function etc. Therefore, the mixing rules for the third, fourth etc. parameters of cubic EOS are at most quadratic or, more frequently, linear [by setting $\text{const}_{ij} = (\text{const}_i + \text{const}_j)/2$ in Eq. (4.3-1)] functions of composition.

$$\text{const} = \sum x_i \, \text{const}_i \tag{4.3-5}$$

where const $= b$, c. Most frequently, the co-volume is also represented as a linear average. The standard method for introducing a binary parameter into the mixing rule is to assume a corrected geometric mean rule for the a parameter in cubic equations of state.

$$a_{ij} = (a_i a_j)^{1/2}(1 - k_{ij}) \tag{4.3-6}$$

Mixing rules for the virial-type EOS are postulated empirically to reproduce the theoretically justified composition dependence of the second and third virial coefficients. Therefore, they are linear or quadratic or cubic functions of the mole fraction. For example, the mixing rules for the HCBKS EOS (4.2-58) take the form

$$\begin{aligned}
B_0 &= \sum x_i \, B_{0i} \\
A_0 &= \sum \sum x_i x_j A_{0i}^{1/2} A_{0j}^{1/2} (1 - k_{ij}) \\
C_0 &= \sum \sum x_i x_j C_{0i}^{1/2} C_{0j}^{1/2} (1 - k_{ij})^3 \\
D_0 &= \sum \sum x_i x_j D_{0i}^{1/2} D_{0j}^{1/2} (1 - k_{ij})^4 \\
E_0 &= \sum \sum x_i x_j E_{0i}^{1/2} E_{0j}^{1/2} (1 - k_{ij})^5 \\
\gamma &= \left(\sum x_i \gamma_i^{1/2} \right)^2
\end{aligned} \tag{4.3-7}$$

and for the remaining constants

$$\text{const} = \left(\sum x_i \, \text{const}_i^{1/3} \right)^3$$

where const $= b$, a, α, c, d.

The performance of the classical mixing rules has been thoroughly tested for several equations of state by Han et al. (1988). Among the seven equations of state studied by the authors, five are cubic, namely the equations of Soave–Redlich–Kwong [(4.2-12) with (4.2-13) and (4.2-15)], Peng–Robinson (4.2-25), Kubic (4.2-33), Heyen (4.2-27) and CCOR (4.2-54). The pure component parameters of the above equations are generalized in terms of the critical temperature, critical pressure and acentric factor. Han et al. also evaluated the HCBKS equation (4.2-64) as an example of a generalized virial-type equation of state and the COR equation (4.2-52), which is an equation based on the augmented van der Waals theory. All the equations were used with one adjustable binary parameter except for the CCOR EOS, which requires two parameters. In the case of Kubic's EOS the combining rule was defined in a different way, namely a_{ij} was calculated from the same equation as a_{ii} using a pseudocritical temperature $T_{cij} = (1 - k_{ij})(T_{ci}T_{cj})^{1/2}$ where k_{ij} is an adjustable parameter. Table 4.3-1, compiled from the results of Han et al., compares the accuracy of calculating binary vapor–liquid equilibria by means of the seven aforementioned equations. The results are shown for symmetric mixtures (with two subcritical components) and those containing methane, hydrogen, nitrogen and carbon dioxide. As a measure of the quality of correlation, the average relative deviations in K factors are listed.

$$\delta K = \left(\frac{1}{N}\right) \frac{\sum (K_i^{\text{cal}} - K_i^{\text{exp}})}{K_i^{\text{exp}}} \tag{4.3-8}$$

Not surprisingly, the results obtained from various equations are not very different because the mixing rules are similar. The above equations of state with classical mixing rules are, with the exception of Heyen's equation, capable of representing vapor–liquid equilibria quite accurately with only one adjustable binary parameter. They are at their best for symmetric mixtures, but

Table 4.3-1
DEVIATIONS OF CALCULATED K FACTORS $\delta K \times 100$ FROM EXPERIMENTAL DATA CALCULATED BY HAN ET AL. (1988) FOR REPRESENTATIVE BINARY SYSTEMS[a]

Type of Systems	RKS	PR	K	H	CCOR	HCBKS	COR
Symmetric mixtures	1.96	2.45	3.94	13.6	2.93	2.97	2.30
Hydrogen mixtures	7.31	5.93	6.33	12.6	6.47		
Methane mixtures	4.06	3.65	3.94	8.17	4.11	7.93	3.92
Carbon dioxide mixtures	4.13	4.31	3.85	12.5	5.36	5.24	4.60
Nitrogen mixtures	6.71	4.83	8.17	12.1	5.52		6.26

[a]By means of the Redlich–Kwong–Soave (RKS) [Eq. (4.2-12) with (4.2-13) and (4.2-15)], Peng–Robinson (PR [Eq. (4.2-25)], Kubic (K), [Eq. (4.2-33)], Heyen (H) [Eq. (4.2-27)], cubic chain of rotators (CCOR) [Eq. (4.2-54)], Han–Cox–Bono–Kwok–Starling (HCBKS) [Eq. (4.2-64)] and chain of rotators (COR) [Eq. (4.2-52)] equations of state.

highly asymmetric mixtures, e.g., those containing hydrogen can be also satisfactorily represented. In the case of hydrogen-containing systems cubic equations of state are also able to reproduce a less common type of phase behavior, viz. the retrograde behavior of the second kind (Tsonopoulos and Heidman, 1986). Moreover, it is also possible to predict vapor–liquid equilibria for two similar components (e.g., aliphatic hydrocarbons) without a binary parameter, i.e., by setting it equal to zero. In some cases it is necessary to include a second binary parameter in the mixing rule for the co-volume. Gray et al. (1983) showed that the use of the second parameter improved the temperature dependence of VLE predictions in asymmetric binaries containing hydrogen. However, the additional parameter did not improve the temperature dependence of VLE predictions for symmetric binaries. Moreover, the additional parameter did not improve the fit in the critical region.

When applied to mixtures, the more complex virial-type equations do not offer any advantages over cubic EOS. The HCBKS equation is even markedly inferior to the Soave and Peng–Robinson equations when applied to hydrogen-containing systems. In general, the multiparameter virial-type EOS are at their best for representing pure fluid properties when abundant experimental data are available. They are also less attractive for mixtures because of computational difficulties.

In the case of the augmented van der Waals equations the mixing rules are either used on the basis of the one-fluid approach or derived from more sophisticated theories. Expressions for the equation of state of hard-core mixtures were derived from the scaled-particle and Percus–Yevick theories (Boublík, 1970; Mansoori et al., 1971). For a hard-sphere mixture the compressibility factor takes the form

$$z = \frac{1 + (3DE/F - 2)\xi + (3E^3/F^2 - 3DE/F + 1)\xi^2 - (E^3/F^2)\xi^3}{(1 - \xi)^3} \qquad (4.3\text{-}9)$$

with $F = \Sigma\, x_i \sigma_i^3$, $E = \Sigma\, x_i \sigma_i^2$ and $D = \Sigma\, x_i \sigma_i$. Dimitrelis and Prausnitz (1986) found Eq. (4.3-9) superior to the one-fluid approximation when applied to the correlation of phase equilibria in binary mixtures of nonpolar molecules differing significantly in size. Among the more elaborate augmented van der Waals equations, one-fluid mixing rules have been used for the BACK [Eqs. (4.2-41)–(4.2-44)] and COR EOS [Eq. (4.2-52)]. The mixing rules for the BACK equation parameters u and v^0 are equivalent to Eq. (4.3-1) whereas the nonsphericity parameter α is represented either by a simple rule (Simnick et al., 1979)

$$\alpha = \sum x_i \alpha_i \qquad (4.3\text{-}10)$$

or by an expression proposed by Machát and Boublík (1985)

$$\alpha = \sum x_i (\alpha_i v_i^0)^{1/3} \frac{\sum x_i (\alpha_i v_i^0)^{2/3}}{\sum x_i v_i^0} \qquad (4.3\text{-}11)$$

Although the BACK EOS satisfactorily represents phase equilibria in systems with low relative volatility, it is less accurate for more asymmetric mixtures (Machát and Boublík, 1985). The mixing rules for the COR equation of state

$$
\begin{aligned}
uv^0 &= \sum \sum x_i x_j u_{ij} v^0_{ij} \\
u_{ij} &= (1 - k_{ij})(u_i u_j)^{1/2} \\
v^0 &= \sum \sum x_i x_j v^0_{ij} \\
v^0_{ij} &= \frac{v^0_i + v^0_j}{2} \\
c &= \sum x_i c_i
\end{aligned}
\tag{4.3-12}
$$

are closely related to the mixing rules of Leland et al. [Eqs. (4.3-3) and (4.3-4)] as u and V^0 are proportional to ϵ and σ^3, respectively. Han et al. (1988) found that the COR equation (4.2-52) with these mixing rules does not offer any advantages over the cubic equations of state with respect to the correlation of VLE. In the case of the PHCT EOS [(4.2-45)–(4.2-47)] and its subsequent modifications the one-fluid approach is not used. The mixing rules are based on a combination of results of the perturbation theory for a mixture of square-well molecules (Henderson, 1974) with mixing rules derived from the lattice theory. The latter can be written as

$$
\epsilon q r \sigma^3 = \sum x_i x_j \epsilon_{ij} q_i r_j \sigma^3_{ji} \tag{4.3-13}
$$

where r is the number of segments and q is the normalized surface area. Mixing rules based on the perturbation theory were also used for the description of dipole–dipole and dipole–quadrupole interactions in the PACT EOS (4.2-48). Vimalchand et al. (1986) found the PACT model to yield reasonable predictions of VLE in systems containing dipolar and quadrupolar fluids. Peters et al. (1988) applied the SPHCT EOS (4.2-49) to the prediction of the phase behavior of mixtures of short and long n-alkanes and showed its superiority over the Soave cubic EOS. Deiters (1982) developed mixing rules based on Guggenheim's (1952) quasi-chemical model and found them to reproduce phase equilibria at very high pressures (over 1000 bars) with reasonable accuracy.

For typical mixtures containing nonpolar or weakly polar components, the results of correlating VLE using the augmented hard-core EOS are, on average, similar to those obtained using simple cubic EOS. This has been illustrated by Han et al. (1988) for the COR EOS (4.2-52). Chao and Lin (1985) presented an extensive comparison of several EOS with the VLE data of asymmetric mixtures and concluded that complex equations such as BACK [(4.2-41)–(4.2-44)] were no better than the cubic equations. No such comparison is available for the equations from the PCT family, but Han et al. (1988) pointed out some examples of inadequacy of the PHCT EOS [(4.2-45)–

(4.2-47)]. These results indicate that a more complex theoretical background of an equation of state does not necessarily lead to an improvement of the results of VLE calculations over purely empirical equations. On the other hand, the PCT equations may be advantageous for some highly asymmetric, nonpolar mixtures (Peters et al., 1988).

In general, equations of state with the mixing rules described above are reliable for the computation of phase equilibria in mixtures of nonpolar and weakly polar components. Many more difficulties are encountered in the case of systems containing strongly polar or associated components. The simplest modification is to use two binary parameters. Adachi and Sugie (1985) and Iwai et al. (1988) showed that the classical mixing rules with two parameters are adequate for mixtures that contain polar components but do not deviate strongly from ideality (e.g., alcohol–alcohol, water–lower alcohols). In such cases the quadratic mixing rules may yield results similar to those obtained from methods designed specifically for polar mixtures. Therefore, phase equilibria can be predicted for such mixtures using two binary parameters calculated from infinite-dilution activity coefficients obtained from the UNIFAC group contribution model (Schwartzentruber et al., 1986). This method cannot work, however, for more strongly nonideal mixtures that cannot be represented by the quadratic mixing rules.

A classical example of the failure of classical quadratic mixing rules for strongly nonideal mixtures is provided by water–hydrocarbon systems. Fairly accurate results can be obtained for such mixtures only when different values of binary parameters are used for the water-rich and hydrocarbon-rich phases (Peng and Robinson, 1980; Robinson et al., 1985). This leads, however, to a thermodynamic inconsistency arising from the fact that the liquid phase fugacity is calculated by an integral from zero density to liquid density. In effect, two different values of a_{ij} are used for the vapor phase at low pressures. Another example has been provided by Trebble (1988) who analyzed the correlation of VLE data for binary mixtures of 1-alkanols and hexane with the Peng–Robinson [Eq. (4.2-25)] and the Trebble–Bishnoi [Eq. (4.2-35)] equations of state: The mixing rules used for the Trebble–Bishnoi EOS contain at most four binary parameters:

$$\begin{aligned}
a &= \sum \sum x_i x_j (a_i a_j)^{1/2}(1 - k_{aij}) \\
b &= \sum \sum x_i x_j \left(\frac{b_i + b_j}{2}\right)(1 - k_{bij}) \\
c &= \sum \sum x_i x_j \left(\frac{c_i + c_j}{2}\right)(1 - k_{cij}) \\
d &= \sum \sum x_i x_j \left(\frac{d_i + d_j}{2}\right)(1 - k_{dij})
\end{aligned} \tag{4.3-14}$$

Trebble found that the use of more than two binary parameters yielded small benefits in fitting binary VLE for alcohol–n-hexane systems. Consequently, two binary parameters k_{aij} and k_{bij} were used for the Trebble–Bishnoi equation

Table 4.3-2
REGRESSION OF BINARY VAPOR–LIQUID EQUILIBRIA FOR ALCOHOL–ALKANE SYSTEMS BY MEANS OF CUBIC EQUATIONS OF STATE WITH CLASSICAL QUADRATIC MIXING RULES[a]

System	Case	$(100/N)\,\Sigma\,\lvert P^{cal}-P^{exp}\rvert/P^{exp}$	$(100/N)\,\Sigma\,\lvert y^{cal}-y^{exp}\rvert$
Methanol–hexane	TB	17.42	10.70
	PR	22.97	15.12
Ethanol–hexane	TB	8.95	5.05
	PR	11.63	7.92
1-Propanol–hexane	TB	2.48	1.21
	PR	10.39	4.81
1-Pentanol–hexane	TB	5.18	0.91
	PR	10.56	2.74

After Trebble (1988).
[a]TB = Trebble–Bishnoi EOS with a two-parameter mixing rule; PR = Peng–Robinson EOS with a one-parameter mixing rule.

(i.e., k_{cij} and k_{dij} were set equal to zero) whereas the Peng–Robinson equation was used with the classical one-parametric mixing rules [Eq. (4.3-6)]. Trebble's results are shown in Table 4.3-2.

In both cases unsatisfactory results were obtained especially for systems forming azeotropes. Simple, unconstrained regression of VLE data resulted in the prediction of nonexisting phase splitting for all equations of state. It was possible to correlate the data so that the incorrect miscibility gap was avoided, but this resulted in very large systematic deviations from experimental data. According to Trebble significant improvement in the fit was not possible even using four binary parameters in the Trebble–Bishnoi EOS (4.2-35). Somewhat better results were obtained for the wide-boiling alcohol–hydrocarbon systems, which do not form azeotropes. Another example of the failure of classical mixing rules for associated systems was provided by Georgeton et al. (1986) who employed the Patel–Teja (4.2-30) EOS as well as the Peng–Robinson EOS to describe liquid–liquid equilibria. Both equations were found to be very poor in their ability to reproduce liquid–liquid equilibrium (LLE) with one-parameter quadratic mixing rules. For example, for the methanol–heptane system they predicted one liquid phase formed by essentially pure methanol.

Various methods for overcoming these difficulties will be presented in Sections 4.3.3–4.3.7.

4.3.2. Mixing Rules for EOS in the Corresponding-States Format

Mixing rules used in conjunction with the shape factor approach are different from those used in the expansions about a reference fluid. In the case of the shape factor method the scaling factors for mixtures are defined by combining rules for the scaling temperature and volume. In contrast, combination of the third (or fourth) corresponding-states parameter in the perturbation expansion approach requires more arbitrary assumptions (Mentzer et al., 1980).

The most successful mixing rules in the shape factor method are the van der Waals one-fluid mixing rules (Leland et al., 1968a,b). The properties of the mixture are calculated as those of a pseudopure fluid whose Gibbs energy differs from that of the mixture only by the ideal change on mixing.

$$G_m(T, p, x) = G_x(T, p, x) + RT \sum_{\alpha} x_\alpha \ln x_\alpha \tag{4.3-15}$$

where

$$G_x = f_{x,0} G_0\left(\frac{T}{f_{x,0}}, \frac{p h_{x,0}}{f_{x,0}}\right) - RT \ln h_{x,0} \tag{4.3-16}$$

and the subscript x denotes the pseudofluid x. The mixing rules of Leland et al. for the conformal parameters are quadratic in composition:

$$h_{x,0} = \sum_{\alpha} \sum_{\beta} x_\alpha x_\beta h_{\alpha\beta,0} \tag{4.3-17}$$

$$f_{x,0} h_{x,0} = \sum_{\alpha} \sum_{\beta} x_\alpha x_\beta f_{\alpha\beta,0} h_{\alpha\beta,0} \tag{4.3-18}$$

The unlike-pair conformal parameters are

$$f_{\alpha\beta,0} = \zeta_{\alpha\beta}(f_{\alpha\alpha,0} f_{\beta\beta,0})^{1/2} \tag{4.3-19}$$

$$h_{\alpha\beta,0} = \tfrac{1}{8}\eta_{\alpha\beta}(h_{\alpha\alpha,0}^{1/3} + h_{\beta\beta,0}^{1/3})^3 \tag{4.3-20}$$

where $\zeta_{\alpha\beta}$ and $\eta_{\alpha\beta}$ are binary interaction parameters.

Leland et al. (1969) also developed a van der Waals two-fluid model in an attempt to account for effects of differences of size and intermolecular energy. Each component of a mixture is replaced by a pseudocomponent and then assumed to mix ideally:

$$G_m(T, p, x) = \sum_{\alpha} x_\alpha [G'_\alpha(T, p, x_\alpha) + RT \ln x_\alpha] \tag{4.3-21}$$

where

$$G'_\alpha(T, p, x_\alpha) = f_{\alpha,0} G_0\left(\frac{T}{f_{\alpha,0}}, \frac{p h_{\alpha,0}}{f_{\alpha,0}}\right) - RT \ln h_{\alpha,0} \tag{4.3-22}$$

The mixing rules then become

$$h_{\alpha,0} = \sum_{\beta} x_\beta h_{\alpha\beta,0} \tag{4.3-23}$$

$$f_{\alpha,0} h_{\alpha,0} = \sum_{\beta} x_\beta f_{\alpha\beta,0} h_{\alpha\beta,0} \tag{4.3-24}$$

Variations on the two-fluid model were presented by Leland and Fisher (1970) and Chapela-Castañares and Leland (1974). Other forms of mixing rules were proposed by Wheeler and Smith (1967). The working equations for computing vapor–liquid equilibria using the shape factor method were summarized by Mentzer et al. (1980). The method with the van der Waals one-fluid mixing rules was used extensively by Leach et al. (1968), Mollerup and Rowlinson (1974), Mollerup (1975, 1978) and Mollerup and Fredenslund (1976). The two-fluid model was used by Watson and Rowlinson (1969), Gunning and Rowlinson (1973) and Leland and Fisher (1970). Mollerup (1980) found that the accuracy of calculation using simpler shape factors derived from cubic equations of state is similar.

In general, the shape factor method enables most thermodynamic properties to be calculated within experimental accuracy. The primary region of applicability encompasses mixtures of low molecular weight nonpolar compounds although good results were obtained also for mixtures containing heavier hydrocarbons (Mentzer et al., 1980). However, as the components of the mixture become more dissimilar, the accuracy decreases.

Mixing rules for pseudocritical parameters used for the corresponding states perturbation expansion approach have the same justification as the quadratic mixing rules for equation-of-state parameters. The relationships between the potential parameters and the pure-component critical or mixture pseudocritical parameters.

$$\sigma^3 = k_1 V_c \tag{4.3-25}$$

$$\epsilon = k_2 T_c \tag{4.3-26}$$

enables the van der Waals one-fluid mixing rules for molecular parameters [Eqs. (4.3-3) and (4.3-4)] to be reexpressed in terms of critical constants:

$$V_c = \sum_i \sum_j x_i x_j V_{cij} \tag{4.3-27}$$

$$T_c V_c = \sum_i \sum_j x_i x_j T_{cij} V_{cij} \tag{4.3-28}$$

The cross coefficients T_{cij} and V_{cij} can be defined in various ways. The usual way is to introduce binary parameters as follows:

$$V_{cij} = \eta_{ij}[(V_{ci}^{1/3} + V_{cj}^{1/3})/2]^3 \tag{4.3-29}$$

$$T_{cij} = \xi_{ij}(T_{ci} T_{cj})^{1/2} \tag{4.3-30}$$

Extension of Eqs. (4.3-29) and (4.3-30) to three-parameter PCS requires an arbitrary mixing rule for the third parameter. Lee and Kesler (1975) and Plöcker et al. (1978) used a simple combining rule for the acentric factor

$$\omega = \sum_i x_i \omega_i \tag{4.3-31}$$

Plöcker et al. noted, however, that mixing rules (4.3-27)–(4.3-31) are successful only for mixtures containing small nonpolar molecules. In order to extend the method to wide-boiling nonpolar or weakly polar mixtures including components such as H_2S, CO_2 or H_2, they introduced an empirical exponent $\eta = 0.25$ into Eq. (4.3-28):

$$T_c V_c^\eta = \sum_i \sum_j x_i x_j T_{cij} V_{cij}^\eta \tag{4.3-32}$$

Equations (4.3-27), (4.3-29) and (4.3-30) with $\eta_{ij} = 1$ and a mixing rule for P_c

$$P_c = \frac{RT_c z_c}{V_c} \tag{4.3-33}$$

$$z_c = 0.2905 - 0.085\omega \tag{4.3-34}$$

constitute the Plöcker–Knapp–Prausnitz mixing rules for the Lee–Kesler method (4.2-73). Oellrich et al. (1981) evaluated the method in comparison with the HCBKS, Soave, Hamam et al. (1977) and Peng–Robinson equations using a very extensive database. Good results were obtained especially in view of the possibility of estimating the binary parameters by correlating them as a function of the quantity $T_{ci}V_{ci}/T_{cj}V_{cj}$.

Lee et al. (1979) extended the van der Waals one-fluid mixing rules by using an additional corresponding-states parameter, the anisotropic strength parameter δ^2. The mixture parameters were obtained from a perturbation technique as follows:

$$\delta^2 \epsilon^2 \sigma^2 = \sum \sum x_i x_j \delta_{ij}^2 \epsilon_{ij}^2 \sigma_{ij}^3 \tag{4.3-35}$$

while the equations for ϵ and σ remained unchanged [Eqs. (4.3-3) and (4.3-4)]. However, results of calculations for wide-boiling hydrocarbon mixtures led them to introduce empirical exponents into the mixing rules.

$$\sigma^{4.5} = \sum \sum x_i x_j \sigma_{ij}^{4.5} \tag{4.3-36}$$

$$\epsilon \sigma^{4.5} = \sum \sum x_i x_j \epsilon_{ij} \sigma_{ij}^{4.5} \tag{4.3-37}$$

$$\delta^2 \sigma^{4.5} = \sum \sum x_i x_j \delta_{ij}^2 \sigma_{ij}^{4.5} \tag{4.3-38}$$

Another method for improving mixing rules in the three-parameter PCS was proposed by Wong et al. (1983, 1984). The authors abandoned the linear mixing rule for ω [Eq. (4.3-31)], which implies also a linear mixing rule for the critical compressibility factor (Joffe, 1971). Their mixing rules

$$\frac{T_c}{P_c} = \frac{\sum_i \sum_j x_i x_j T_{cij}}{P_{cij}} \tag{4.3-39}$$

$$\frac{T_c^2}{P_c} = \frac{\sum_i \sum_j x_i x_j T_{cij}^2}{P_{cij}} \tag{4.3-40}$$

$$\omega \left(\frac{T_c}{P_c} \right)^{2/3} = \sum_i \sum_j x_i x_j \omega_{ij} \left(\frac{T_{cij}}{P_{cij}} \right)^{2/3} \tag{4.3-41}$$

are equivalent to the van der Waals mixing rules if we consider V_c to be proportional to $z_c T_c / p_c$ and z_c to be a quantity that changes slightly from compound to compound. These mixing rules were used with the variable-reference-fluid approach [Eq. (4.2-74)] to correlate VLE for a variety of mixtures including strongly nonideal ones. The model with two binary parameters appeared to yield satisfactory correlations although some limitations, typical to quadratic mixing rules, were also observed (e.g., phase equilibria for water + hydrocarbon systems could not be fit with unique values for binary parameters for both phases). The authors concluded that the method with cubic EOS as a reference fluid equation is, for nonpolar fluids, as good as conventional corresponding-states methods with fixed reference fluids. For polar mixtures it is clearly superior due to the introduction of polar fluids as references. In general, the method is as accurate as the direct use of cubic EOS but may be also used for compounds for which the parameters of cubic EOS are not available. Rosenthal et al. (1987) found the method to be applicable also to supercritical fluid phase equilibria.

Johnson and Rowley (1989) employed Eqs. (4.3-39)–(4.3-41) as mixing rules for their extension of the Lee–Kesler method (4.2-77) to polar mixtures. Equation 4.3-40 was modified by introducing an exponent η similarly with the work of Plöcker et al. (1978).

$$T_c = \left(\frac{T_c}{P_c} \right)^{-\eta} \sum_i \sum_j x_i x_j \left(\frac{T_{cij}}{P_{cij}} \right)^{\eta} \tag{4.3-42}$$

and the acentric factor in Eq. (4.3-41) was replaced by the anisotropy parameter α. For the polarity parameter the mixing rule was

$$\beta_H = K \sum x_i x_j (\beta_{Hij} + A_{ij}) \tag{4.3-43}$$

The combining rule for T_{cij} is given by Eq. (4.3-30) and that the P_{cij} is analogous to Eq. (4.3-29).

$$P_{cij} = \frac{8 T_{cij}}{[(T_{ci}/P_{ci})^{1/3} + (T_{cj}/P_{cj})^{1/3}]^3} \tag{4.3-44}$$

The α and β parameters are calculated as simple averages:

$$\alpha_{ij} = 0.5(\alpha_i + \alpha_j) \tag{4.3-45}$$

$$\beta_{ij} = 0.5(\beta_i + \beta_j) \tag{4.3-46}$$

To enhance the predictive capability of the method, the parameters η and A_{ij} were generalized by regression from a number of experimental data sets. The method can be thus used in a predictive manner, i.e., without adjusting binary parameters to yield a semiquantitative estimate of the data. Average deviations in pressure predicted in this way are 16.2 per cent for polar mixtures (60 systems) and 4.6 per cent for nonpolar systems (28 systems). If one binary parameter is fitted to individual data the deviation reduces to 4.8 per cent for polar mixtures. For comparison, deviations obtained from the Lee–Kesler method with the mixing rules of Plöcker et al. with one adjustable parameter are 7.6 per cent for the same polar mixtures and 2.2 per cent for nonpolar mixtures.

The mixing rules for pseudocritical parameters can be used not only with equations of state in the explicit corresponding-states format but also with simple equations of state. This type of mixing rule is advantageous for multiparameter EOS for which it is difficult to develop a prescription for all parameters with a reasonable number of adjustable parameters. An example is provided by the mixing rules of Arai et al. (1982) modified later by Orbey and Vera (1989) who used Eqs. (4.3-27)–(4.3-28), (4.3-31) and (4.3-33) with appropriate combining rules. In general, appropriate selection of reference fluids allows the above methods to be applicable to a wide variety of mixtures including highly asymmetric ones. However, their applicability to strongly nonideal mixtures containing polar and associating components is limited by the use of quadratic mixing rules for the pseudocritical parameters, which have essentially the same theoretical background as the classical quadratic mixing rules used for the parameters of the van der Waals-type EOS.

4.3.3. Mixing Rules from Excess Gibbs Energy Models

The inability of classical quadratic mixing rules to represent the phase behavior of strongly nonideal mixtures can be understood by examining the excess Gibbs energy calculated from simple EOS with these mixing rules. If an equation of state can describe the mixture and pure-component properties in the same physical state (e.g., liquid) at a given temperature and pressure, the excess Gibbs energy can be calculated according to (4.1-15) as

$$g^E(T, p, \mathbf{x}) = RT\left(\ln \phi_{\text{mix}} - \sum x_i \ln \phi_i\right) \tag{4.3-47}$$

where ϕ_{mix} and ϕ_i are the fugacity coefficients of the mixture and the ith pure component. Vidal (1978) derived the infinite-pressure limit of the excess Gibbs energy calculated from the Redlich–Kwong equation with quadratic mixing rules. To arrive at an explicit equation, Vidal assumed that $v = b$ at infinite pressure and $V^E = 0$.

$$g^E(P = \infty) = \ln 2\left(-\frac{a}{b} + \frac{\sum x_i a_i}{b_i}\right) \tag{4.3-48}$$

If the adjustable parameter k_{ij} in the mixing rule for the parameter a is set to zero, Eq. (4.3-48) for a binary system reduces to

$$g^E(P=\infty) = \ln 2\left(\frac{x_i b_{11}}{b}\right)\left(\frac{x_2 b_{22}}{b}\right) \times \left[\left(\frac{a_{11}/b_{11}}{b_{11}}\right)^{1/2} - \left(\frac{a_{22}/b_{22}}{b_{22}}\right)^{1/2}\right]^2 \tag{4.3-49}$$

where $x_i b_{ii}/b$ are the infinite-pressure volume fractions and the terms $(a_{ii}/b_{ii}) \ln 2/b_{ii}$ can be related to Hildebrand's solubility parameters. Therefore, Eq. (4.3-49) becomes essentially identical to the well-known Hildebrand–Scatchard regular solution theory.

$$g^E(P=\infty) = b\Phi_1\Phi_2(\delta_1 - \delta_2) \tag{4.3-50}$$

where

$$\delta_i = \left[(\ln 2)\,\frac{a_{ii}/b_{ii}}{b_{ii}}\right]^{1/2} \tag{4.3-51}$$

This result is quite general; for other equations of state a more general form of Eq. (4.3-48) is valid (Huron and Vidal, 1979):

$$g^E(P=\infty) = \Lambda\left(-\frac{a}{b} + \frac{\sum x_i a_i}{b_i}\right) \tag{4.3-52}$$

where Λ is a characteristic parameter for the equation of state used. It was implicitly assumed that $g^E(P=\infty)$ is similar to the excess Gibbs energy of the liquid at finite pressures. Equation (4.3-50) explains the inherent limitation of quadratic mixing rules: They are applicable only to mixtures whose excess properties can be approximated by the regular solution theory. While it is a good approximation for hydrocarbon mixtures, it cannot be applied to systems containing strongly polar or associating components. Equation (4.3-52) suggests a mixing rule for parameter a in cubic equations of state:

$$a = b\left(\frac{\sum x_i a_{ii}}{b_{ii}} + \frac{g^E(p=\infty)}{\Lambda}\right) \tag{4.3-53}$$

Any equation for the infinite-pressure excess Gibbs energy can be thus incorporated in the mixing rules. Appropriate expressions for $g^E(p=\infty)$ can be selected according to our experience in the correlation of phase equilibria by the γ–ϕ method. It should be kept in mind, however, that although the excess Gibbs energies of a liquid under finite and infinite pressure are similar, they are not identical and cannot be used interchangeably. Therefore, the parameters of the mixing rule (4.3-53) cannot be evaluated by standard low-pressure γ–ϕ regression but must be determined by the correlation of phase equilibria with the equation of state. For example, Pandit and Singh (1987) had to reevaluate the group contribution parameters for the ASOG method [(3.4-44)–(3.4-49)] coupled with cubic EOS. Sheng et al. (1989) reformulated the UNIFAC [Eqs. (3.4-44), (3.4-50)–(3.4-59)] combinatorial term for calculations with the Patel–Teja EOS. Alternatively, the components' fugacity coefficients can be determined from the actual activity coefficients taking into account Eq. (4.3-47). A detailed algorithm was given by Soave (1986) for finding a mixture's equation-of-state parameters, compressibility factor and fugacity coefficients when actual activity coefficients and pure component EOS parameters are given.

For the purpose of data smoothing and evaluation, Eq. (4.3-53) can be coupled with flexible multiparameter equations such as that of Redlich–Kister (1948)

$$g^E(P=\infty) = RTx_1x_2 \sum_{l=0}^{m} A_l(x_1 - x_2)^l \tag{4.3-54}$$

If only a limited amount of data are available, the local composition g^E models may be advantageous. Huron and Vidal (1979) used the NRTL equation (3.3-11) for $g^E(P=\infty)$:

$$g^E(P=\infty) = \sum_{l=1}^{n} x_i \frac{\sum_{j=1}^{n} x_j G_{ji} C_{ji}}{\sum_{k=1}^{n} x_k G_{k,i}} \tag{4.3-55}$$

with

$$C_{ji} = g_{ji} - g_{ii} \tag{4.3-56}$$

and

$$G_{ji} = b_j \exp\left(-\frac{\alpha_{ji} C_{ji}}{RT}\right) \tag{4.3-57}$$

The above expressions have been obtained from the NRTL model (3.3-11) modified by introducing infinite-pressure volume fractions in the local compositions. The purpose of the modification was to obtain a model that can be reduced to the conventional quadratic mixing rules for nonpolar mixtures. Huron and Vidal and Soave (1984) found that the mixing rules (4.3-53) and (4.3-55)–(4.3-57) represented a great improvement over the classical quadratic mixing rules and made it possible to correlate vapor–liquid equilibria for highly nonideal systems with very good accuracy. However, it should be noted that the improvement was obtained at the cost of increasing the number of adjustable constants from one to three.

Kurihara et al. (1987) proposed a modification of the Huron–Vidal mixing rule containing a quadratic term and a "residual" one that is calculated from the infinite-pressure Wilson (3.3-9) equation.

$$a = \sum \sum x_i x_j (a_{ii} a_{jj})^{1/2} - \left(\frac{b}{\ln 2} \right) g^{E(\mathrm{res})}(P = \infty) \tag{4.3-58}$$

$$g^{E(\mathrm{res})}(P = \infty) = -RT \sum x_i \ln \left(\sum x_j \Lambda_{ji} \right) \tag{4.3-59}$$

Tochigi et al. (1988) obtained good results of predicting ternary vapor–liquid equilibria from binary data using the Soave EOS with mixing rules (4.3-58) and (4.3-59).

Another approach inspired by the success of local composition models was proposed by Heyen (1981). He introduced local compositions directly into the mixing rules rather than utilizing existing G^E models.

$$a = \sum \sum x_i x_{ji} a_{ji} \tag{4.3-60}$$

where the local mole fractions x_{ji} are related to the bulk mole fractions through the empirical weighting factor τ_{ij}

$$x_{ji} = \frac{x_j \tau_{ij}}{\sum_k x_k \tau_{ik}} \tag{4.3-61}$$

Heyen fitted two parameters, τ_{ij} and τ_{ji}, for a binary system and found agreement between the model and data in most cases.

The more recent methods for coupling G^E models with equations of state are based on less restrictive assumptions than those needed to arrive at Huron and Vidal's mixing rules [(4.3-55)–(4.3-57)]. Mollerup (1986) retained the assumption that $V^E = 0$ but based his derivation on the zero pressure limit. This limit seems to be more reasonable in view of the applicability of g^E models to low and and moderate pressures. If such an assumption is made, $b \neq v$ and the mixing rule takes the form

$$\frac{a}{b} = \sum x_i \left(\frac{a_i}{b_i} \right) \left(\frac{f_i}{f} \right) - \frac{g^E}{f} + \left(\frac{RT}{f} \right) \sum x_i \ln \left(\frac{f_i b_i}{b} \right) \tag{4.3-62}$$

where the functions f_i and f depend on reduced density b/v for pure components and mixture, respectively. The mixing rule (4.3-62) depends thus on the liquid phase volume of the mixture and individual components. It should be noted, however, that such a procedure can be applied only if the equation can be solved in the zero pressure limit. For example, the zero pressure root of the van der Waals EOS exists if $a/bRT \geq 4$.

From the practical point of view it is convenient to use the same parameters in equations of state and g^E models. This is especially important if an elaborate group contribution method such as UNIFAC [(3.4-44) and (3.4-50)–(3.4-59)] is to be extended to high pressures. To achieve this, Gupte et al. (1986) developed the UNIWAALS EOS by forcing the g^E calculated from the van der Waals EOS to be identical with g^E calculated from UNIFAC at all pressures. Therefore, the UNIWAALS model employs the temperature and pressure of the mixture as the standard state for the mixing process in contrast with the infinite-pressure standard state in the Huron–Vidal method and the zero pressure standard state in Mollerup's proposal. The derived mixing rule

$$\frac{a}{RTb} = \frac{Pv^E}{fRT} - \frac{1}{f}\left[\ln\frac{P(v-b)}{RT} - \sum x_i \ln\frac{P(v_i - b_i)}{RT}\right] + \frac{1}{f}\sum\frac{x_i a_i f_i}{RTb_i} - \frac{1}{f}\frac{g^E}{RT} \tag{4.3-63}$$

where $f = b/v$ and $f_i = b_i/v_i$ reduces to the mixing rule of Mollerup (1986) by setting $v^E = 0$ and to that of Huron and Vidal by setting the first two terms equal to zero and $f_i = f = 1$. For the calculation of the a parameter the mixture and pure-component liquid volumes are required even if the liquid phase does not actually exist. Therefore, Gupte et al. (1986) proposed an appropriate extrapolation procedure. Gani et al. (1989) reformulated the computational procedure in a more consistent way and developed a procedure for supplying hypothetical liquidlike volumes.

Péneloux et al. (1989) coupled Guggenheim's (1952) quasi-lattice theory with equations of state. They analyzed the form of EOS mixing rules when a G^E model is imposed at constant "packing fraction" (i.e., the quantity b/v). The method served as a basis for developing a group contribution EOS (Abdoul, 1987). The method was restricted to nonpolar and weakly polar mixtures as Guggenheim's zeroth approximation was shown to be equivalent to classical van der Waals mixing rules.

Heidemann and Kokal (1990) and Michelsen (1989) returned to Mollerup's suggestion of using the zero pressure standard state for combining EOS and G^E models. Their purpose was identical with the purpose of developing UNIWAALS, i.e., to utilize the G^E model parameters in equation-of state calculations. When applying the zero pressure reference state, the volume root at zero pressure does not exist at higher reduced temperatures (above about 0.85 for pure compounds). A certain limiting value of a/bRT exists below which the root does not appear. Therefore, a reasonable although purely empirical extrapolation procedure is necessary. Heidemann and Kokal (1990) proposed that (1) the zero pressure standard state is used for pure components above a limiting value of a/bRT; (2) below the limiting a/bRT value, the reduced density ξ is extrapolated as

$$\xi_i = 1 + \beta\left(\frac{a_i}{b_i RT}\right)^2 + \delta\left(\frac{a_i}{b_i RT}\right)^3 \tag{4.3-64}$$

and (3) the standard state pressure for the mixture is set by the requirement that

$$z - \sum x_i z_i = 0 \tag{4.3-65}$$

To find the mixture ξ and a/bRT parameters, Eq. (4.3-65) is solved along with the condition of equality of the excess Gibbs energy calculated from a γ–ϕ

model with that calculated from the equation of state. Solution of these equations is necessarily iterative, but the a parameter is not density dependent in contrast to the UNIWAALS equation and the procedure is mathematically consistent. The method of Michelsen (1989) is based on the same principle, but the extrapolation technique is different.

The mixing rules derived from G^E models make it possible to incorporate the flexibility of G^E models into equations of state. Moreover, the predictive capability of group contribution methods can be extended to high pressures. Some limitations of the G^E models used are obviously unchanged, e.g., the nonuniqueness of adjustable parameters when calculated from VLE and LLE data.

4.3.4. Simple Composition-Dependent Combining Rules

The methods combining excess Gibbs energy models and equation-of-state parameters are equivalent to expressing the cross-parameters (a_{ij}) of an EOS by a suitable function of composition. The explicit use of a liquid phase G^E model is not essential to obtain a good accuracy of correlation. It is therefore possible to construct empirical composition-dependent combining rules that are sufficiently accurate yet simple. Stryjek and Vera (1986a,b,c), Adachi and Sugie (1986) and Panagiotopoulos and Reid (1986) proposed mixing rules containing two binary parameters. Stryjek and Vera called the mixing rule

$$a = \sum \sum x_i x_j (a_{ii} a_{jj})^{1/2} (1 - x_i k_{ij} - x_j k_{ji}) \tag{4.3-66}$$

a "Margules-type" mixing rule due to its analogy to the Margules equation for G^E. The forms of Adachi and Sugie

$$a = \sum \sum x_i x_j (a_{ii} a_{jj})^{1/2} [1 - l_{ij} + m_{ij}(x_i - x_j)] \tag{4.3-67}$$

and Panagiotopoulos and Reid

$$a = \sum \sum x_i x_j (a_{ii} a_{jj})^{1/2} [1 - k_{ij} + (k_{ij} - k_{ji}) x_i] \tag{4.3-68}$$

are very similar. Stryjek and Vera proposed also a Van Laar-type mixing rule

$$a = \sum \sum x_i x_j (a_{ii} a_{jj})^{1/2} \left(1 - \frac{k_{ij} k_{ji}}{x_i k_{ij} + x_j k_{ji}}\right) \tag{4.3-69}$$

and found it to give similar results as the activity coefficient methods in the correlation of vapor–liquid equilibrium data. Sandoval et al. (1989) tested the Stryjek–Vera and related mixing rules for the prediction of ternary VLE and found the results to compare favorably with those obtained with the Wilson and NRTL models.

Kabadi and Danner (1985) also proposed a two-parameter composition-dependent combining rule designed specifically for water–hydrocarbon mixtures.

$$a = x_1^2 a_{11} + x_2^2 a_{22} + 2x_1 x_2[(a_{11}a_{22})^{1/2}(1 - k_{12}) + x_1 G(1 - T_{r_1}^{0.8})] \quad (4.3\text{-}70)$$

where index 1 denotes water. Correlations were established for the parameters k_{12} and G within homologous series of hydrocarbons. Michel et al. (1989) evaluated this method and found large errors for the aqueous phase.

Michel et al. (1989) proposed an unconventional mixing rule for water–hydrocarbon systems that assigns a composition dependence to a_{11} rather than a_{12} with the boundary condition that $a \rightarrow a_{11}$ (pure) when $x_1 \rightarrow 1$.

$$a = x_1^2 a_{11}[1 + \tau T^n x_2 \exp(-\alpha x_2)] + 2x_1 x_2(1 - k_{12})(a_{11}a_{22})^{1/2} \quad (4.3\text{-}71)$$

The authors obtained a good correlation of mutual solubilities although attempts to correlate the parameters with molecular properties were only partially successful.

Schwartzentruber et al. (1987) introduced an additional third parameter to improve the flexibility of the composition-dependent combining rules.

$$a = \sum \sum x_i x_j (a_{ii}a_{jj})^{1/2}\left[1 - k_{ij} - \frac{l_{ij}(m_{ij}x_i - m_{ji}x_j)}{m_{ij}x_i + m_{ji}x_j}\right] \quad (4.3\text{-}72)$$

with

$$k_{ji} = k_{ij} \qquad l_{ji} = -l_{ij} \qquad m_{ji} = 1 - m_{ij} \quad \text{and} \quad k_{11} = l_{11} = 0$$

Equation (4.3-72) is a generalization of the Panagiotopoulos and Reid form of composition-dependent combining rules as it reduces to Eq. (4.3-68) when $m_{ij} = 0$. The equation was found to correlate VLE very accurately for the ternary system methanol–carbon dioxide–propane and its binaries. In particular, the equation does not predict a false miscibility gap for the methanol–propane system as the previous models did. Schwartzentruber et al. (1989) estimated the parameters of Eq. (4.3-72) by fitting them numerically to two isotherms generated from UNIFAC. The results obtained were comparable to those of Gani et al. (1989) who introduced the UNIFAC functional form directly into the equation of state. Margerum and Lu (1990) verified the mixing rules of Adachi and Schwartzentruber along with the Huron–Vidal rule, for 15 alcohol–hydrocarbon systems. Such systems provide a stringent test of the models due to their strong nonideality. Mean deviations in pressure were 3.1 per cent for Eq. (4.3-67), 1.5 per cent for Eq. (4.3-72) and 1.6 per cent for the Huron–Vidal model. However, all models incorrectly predicted phase splitting for the methanol–hexane mixture. It should be noted that the authors constrained their EOS to accurately represent pure component vapor pressures so that any deviations can be attributed only to the deficiency of the mixing rules.

In general, composition-dependent mixing rules constitute the simplest, yet successful, method to apply equations of state to complex mixtures. Their success stems from their greater flexibility in comparison with the classical quadratic mixing rules. However, the flexibility is obtained at the cost of introducing further adjustable constants. Therefore, problems such as strong correlation between adjustable parameters and nonuniqueness of parameter sets are likely to occur. Another, more fundamental disadvantage of this type of mixing rules has been pointed out by Michelsen and Kistenmacher (1990). These authors showed that the composition-dependent mixing rules are not invariant to dividing a component into a number of identical subcomponents. If a binary mixture (x_1, x_2) is treated as a ternary (x_1, x_2, x_3) where the ternary is formed by dividing component 2 into two pseudocomponents with identical properties, a different value for parameter a will result. Therefore, the calculated properties of complex mixtures will strongly depend on the number of pseudocomponents, which is in contrast to experimental evidence.

However, a good feature of this type of mixing rules is their simplicity. In particular, cubic equations of state do not lose their cubic volume dependence when composition-dependent mixing rules are applied. This feature, together with the numerical flexibility of the mixing rules, makes them attractive for engineering computations.

Lencka and Anderko (1991) analyzed the composition-dependent mixing rules with respect to their ability to predict phase equilibria in ternary systems using parameters determined from binary data. The predictive capability of the mixing rules is extremely important for most industrial applications where multicomponent mixtures are dealt with and the experimental information is limited to binary data. The results are shown in Table 4.3-3.

In the case of mixtures of not very high nonideality the mixing rules of Panagiotopoulos and Reid (1986) and Schwartzentruber et al. (1987) are very reliable. The results of prediction for ternary systems are of practically the same quality as the results of correlating binary data. The situation is less satisfactory for very strongly nonideal mixtures that either exhibit phase splitting (cyclohexane–hexane–methanol) or are close to phase separation (propane–methanol–carbon dioxide). In this case the mixing rules are incapable of predicting ternary equilibria with the same accuracy as the accuracy of binary correlation. The three-parameter mixing rule of Schwartzentruber et al. (1987) gives somewhat better results due to a better reproduction of the constituent binary systems.

It is worthwhile to examine the reasons of the failure of some mixing rules for selected systems. As the composition-dependent mixing rules necessitate the use of two or three parameters for each binary pair, it is probable that the parameters are not uniquely determined from binary data and are thus not always meaningful for the prediction of ternary data. To illustrate this problem, two ternary systems have been chosen for which only one binary pair (say $1+2$) shows appreciable nonideal behavior, and the remaining two pairs are easily correlated using the classical quadratic mixing rules. Therefore, both the correlation of the binary ($1+2$) and the prediction of the ternary ($1+2+3$) are influenced mainly by the binary parameters for the pair ($1+2$) and not by

Table 4.3-3
RESULTS OF CORRELATING BINARY VAPOR–LIQUID EQUILIBRIUM DATA AND PREDICTING TERNARY EQUILIBRIA USING PARAMETERS EVALUATED FROM BINARY DATA[a]

	CQ (*4.3-1*) (*1 param.*)		*PR* (*4.3-68*) (*2 param.*)		*SV* (*4.3-69*) (*2 param.*)		*SGR* (*4.3-72*) (*3 param.*)	
Mixture	$\delta P\%$	$\delta y\%$	$\delta P\%$	$\delta y\%$	$\delta P\%$	$\delta y\%$	$\delta P\%$	$\delta y\%$
Benzene + hexane	0.7	1.0	0.2	0.6	0.4	0.9	0.2	0.6
Hexane + cyclohexane	0.2	0.4	0.3	0.4	0.3	1.5	0.2	0.5
Benzene + cyclohexane	0.4	1.4	0.4	1.4	0.4	1.4	0.3	1.4
Benzene + hexane + cyclohexane	0.4	0.8	0.5	0.9	0.5	0.8	0.5	0.9
Water + methanol	2.2	1.1	0.9	0.4	0.8	0.4	0.5	0.2
Methanol + ethanol	0.3	0.4	0.4	0.5	0.5	0.5	0.3	0.4
Water + ethanol	6.4	3.5	1.8	1.1	1.3	0.6	1.2	0.5
Water + methanol + ethanol	1.9	1.4	1.2	0.7	7.5	3.6	1.4	0.8
Chloroform + methanol	5.9		0.7		1.2		0.7	
Acetone + chloroform	0.7		0.3		0.3		0.2	

Acetone + methanol	1.1		0.1		0.6		0.1	
Acetone + chloroform + methanol	2.7		0.8		4.5		0.5	
Pentane + methanol	10.1	5.9	3.0	2.3	2.0	1.8	1.0	1.0
Methanol + acetone	0.6	0.6	0.4	0.5	0.4	0.5	0.3	0.5
Pentane + acetone	1.4	1.2	0.4	0.7	0.3	0.6	0.2	0.6
Pentane + methanol + acetone	7.3	4.5	1.5	1.4	7.8	4.8	1.0	1.2
Propane + methanol	11.6	1.2	7.3	0.8	3.3	2.0	3.8	1.1
Methanol + carbon dioxide	3.9	0.3	2.9	0.2	2.8	0.2	1.6	0.2
Propane + carbon dioxide	2.1		1.4		1.4		0.9	
Propane + methanol + carbon dioxide	10.0	4.5	7.8	2.4	11.3	6.2	5.2	3.0
Cyclohexane + hexane	0.4		0.4		0.4		0.4	
Hexane + methanol	5.5		2.7		1.5		1.1	
Cyclohexane + methanol	8.4		3.6		2.9		1.5	
Cyclohexane + hexane + methanol	10.7		5.8		7.6		5.0	

[a]Using the classical quadratic (CQ) mixing rules and the composition dependent mixing rules of Panagiotopoulos and Reid (PR), Stryjek and Vera (SV) and Scwartzentruber et al. (SGR).

those for the remaining pairs (1 + 3) and (2 + 3). Then, the binary and ternary VLE were calculated for several pairs of binary parameters k_{12} and k_{21} in the vicinity of the optimum point determined from binary data. The results are shown in Figures 4.3-1 and 4.3-2 as lines of constant deviations on the k_{12} versus k_{21} plane.

In the figures the optimum point determined from binary data is marked with a solid triangle. It usually appears that the optimum point for the representation of ternary data does not coincide with the optimum point for the binary data. The optimum ternary point is marked by a solid circle. Additionally, deviation contours are shown for binary data (solid lines) and ternary data (dashed lines). The numbers along the deviation contours indicate by how many per cent the deviations are higher than those at the optimum, either binary or ternary, points. The regions enclosed by the 20 per cent deviation contours are shadowed. The results shown in Figures 4.3-1 and 4.3-2 are typical for systems that do not exhibit phase splitting. In the case of mixtures showing miscibility gaps the distance between the optimum points for

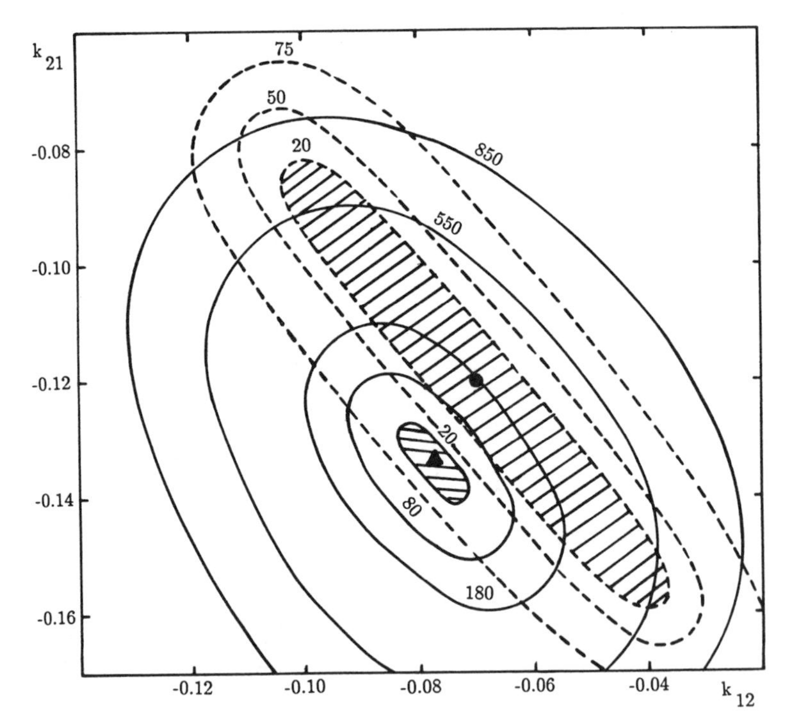

Figure 4.3-1
Deviation contours for the correlation of VLE for the binary system water–ethanol (solid lines) and the ternary system water–ethanol–methanol (dashed lines) using the Panagiotopoulos–Reid (1986) mixing rule. The optimum point for the binary system is denoted by a triangle and that for the ternary one by a circle. The numbers along the contours indicate by how many percent the reproduction of the data for a given pair of parameters (k_{12}, k_{21}) is worse than that at the optimum point.

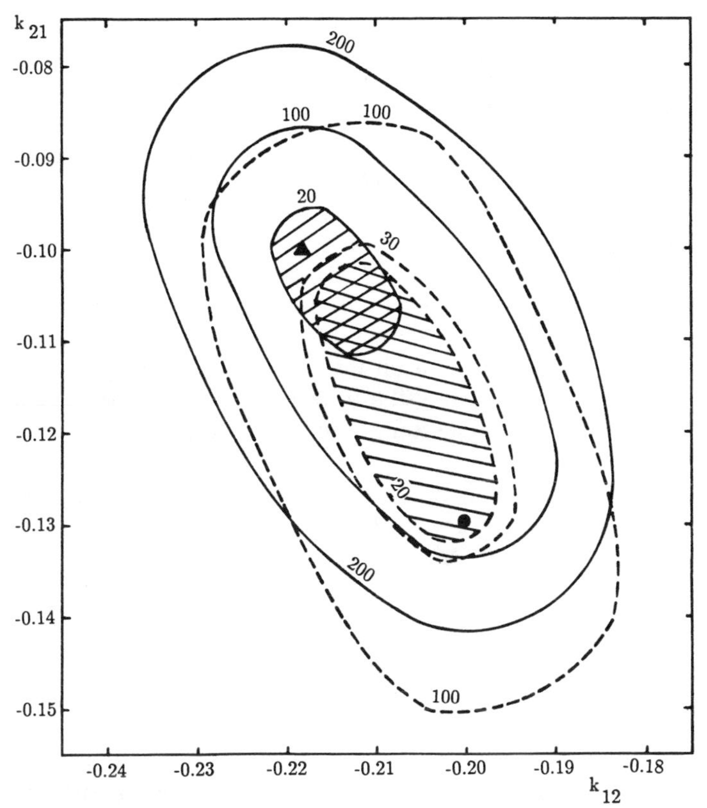

Figure 4.3-2
Deviation contours for the correlation of VLE for the binary system acetone–water (solid lines) and the ternary system acetone–water–methanol (dashed lines) using the Stryjek–Vera (1986a,b) mixing rule. The optimum point for the binary system is denoted by a triangle and that for the ternary one by a circle. The numbers along the contours indicate by how many percent the reproduction of the data for a given pair of parameters (k_{12}, k_{21}) is worse than that at the optimum point.

the binary and ternary mixtures is larger. This analysis shows that in some cases the two- or three-parameter composition-dependent mixing rules are inherently incapable of representing binary and ternary data with the same values of parameters. Thus, simultaneous correlation of binary and ternary data is not likely to improve the results in such cases. The above limitations should be kept in mind when applying any empirical mixing rules that have more than one binary parameter.

4.3.5. Density-Dependent Mixing Rules

Density-dependent mixing rules are formulated to obtain a consistent representation of mixture properties in the high- and low-density regions. The mixing rules described in Sections 4.3.3 and 4.3.4, while successful, suffer from a serious deficiency, i.e., an incorrect low-density limit. They do not reproduce the theoretically justified quadratic composition dependence of the second virial coefficient. This is indicative of their weak theoretical background. The density-dependent mixing rules are formulated in such a way that this boundary condition is satisfied. In principle, it seems probable that better results can be obtained if the constants of a chosen equation of state are allowed to depend not only on composition and, sometimes, on temperature (which is consistent with the van der Waals one-fluid theory underlying the quadratic mixing rules), but also on density. This concept, subject to the restriction that for a pure component no change is introduced into the equation of state leads to the formulation of density-dependent mixing rules. The first formulations of density-dependent mixing rules were based on the local composition concept, which had been earlier proved successful in the γ–ϕ methods. The application of local compositions toward the formulation of mixing rules was reviewed by Danner and Gupte (1986).

In general, the local composition concept assumes that the ratio of the mole fraction of j molecules around a central i molecule to the mole fraction of i molecules around the central i molecule is related to the ratio of the bulk mole fractions by Boltzmann factors

$$\frac{x_{ji}}{x_{ii}} = \left(\frac{x_j}{x_i}\right) \exp\left[\left(-\frac{\alpha E_{ji}}{RT}\right) - \left(\frac{\alpha E_{ii}}{RT}\right)\right] \tag{4.3-73}$$

where E_{ji} is the interaction energy between i and j molecules and α is a nonrandomness parameter. As a measure of local composition, local surface fractions are often considered to be more appropriate than local mole fractions (Abrams and Prausnitz, 1975; Maurer and Prausnitz, 1978).

Mollerup (1981) and Whiting and Prausnitz (1982) proposed a method for incorporating the local composition concept into an equation of state of the van der Waals type, that is, any equation of state that separates the repulsive and attractive contributions to the residual Helmholtz energy. As the local composition effects are temperature dependent according to Eq. (4.3-73), they influence only the attractive contribution. The mixing rules for the repulsive part are not density dependent because they are applicable to hard-core reference systems and as such are independent of temperature. Therefore, local compositions can be written for the internal energy, which is the only property being independent of the repulsive contribution to the Helmholtz energy $[U^{\text{rep}} = d(A^{\text{rep}}/T)/d(1/T) = 0]$. To incorporate local compositions, a two-fluid form was adopted for the residual internal energy U^r. For a binary mixture the following expression is postulated:

$$U^r = (\tfrac{1}{2})[N_1(N_{21}u_{21} + N_{11}u_{11}) + N_2(N_{12}u_{12} + N_{22}u_{22})] \tag{4.3-74}$$

where N_i is the total number of the molecules i, N_{ij} is the number of molecules of type j around the molecules of type i and u_{ij} is the molar attractive energy of a fluid whose molecules interact according to the ij potential. Defining the coordination number around a molecule of type i

$$N_{ci} = \sum N_{ji} \tag{4.3-75}$$

the local mole fraction becomes

$$x_{ji} = \frac{N_{ji}}{N_{ci}} \tag{4.3-76}$$

Thus, the residual energy U^r is expressed as

$$U^r = \left(\frac{N}{2}\right)[x_1 N_{c_1}(x_{21}u_{21} + x_{11}u_{11}) + x_2 N_{c_2}(x_{12}u_{12} + x_{22}u_{22})] \tag{4.3-77}$$

The coordination numbers around the molecules of the types 1 and 2 (N_{c_1} and N_{c_2}) are not necessarily the same and can be, in general, functions of composition (Danner and Gupte, 1986). If they are assumed to be independent of composition, they can be absorbed in the u parameters and

$$\frac{U^r}{N} = x_1(x_{21}u_{21} + x_{11}u_{11}) + x_2(x_{12}u_{12} + x_{22}u_{22}) \tag{4.3-78}$$

The next step is to assume a functional form for the attractive Helmholtz energy and internal energy from a suitable equation of state. Whiting and Prausnitz (1982) used a simple van der Waals attractive term that yields

$$A^{\text{attr}} = -\frac{4aN\xi}{b} \quad \text{where } \xi = b\rho/4 \tag{4.3-79}$$

$$U^{\text{attr}} = -Na\rho \tag{4.3-80}$$

Further, the authors identified the quantities E_{ji} from Eq. (4.3-73) with the Helmholtz energies [Eq. (4.3-79)]. The local compositions are now given by

$$\frac{x_{ji}}{x_{ii}} = \left(\frac{x_j}{x_i}\right)\exp\left[\alpha\left(\frac{a_{ji}}{b_{ji}} - \frac{a_{ii}}{b_{ii}}\right)\frac{4\xi_{ii}}{RT}\right] \tag{4.3-81}$$

where

$$\xi_i = \frac{b_{ii}\rho}{4}$$

Sandler and co-workers (Sandler, 1985; Lee et al., 1985) pointed out that Eq. (4.3-81) yields a wrong density dependence of local compositions. Notably, Eq. (4.3-81) predicts a random mixture in the zero density limit, i.e.,

$$\lim_{\rho\to 0}\left(\frac{x_{ji}}{x_{ii}}\right) = \left(\frac{x_j}{x_i}\right)$$

whereas the proper zero density limit is

$$\lim_{\rho\to 0}\left(\frac{x_{ji}}{x_{ii}}\right) = \left(\frac{x_j}{x_i}\right)\exp\left(-\frac{(u_{ji} - u_{ii})}{kT}\right)$$

Combining (4.3-77), (4.3-80) and (4.3-81), an expression for the residual internal energy is obtained. This expression, integrated with respect to reciprocal temperature, yields the attractive Helmholtz energy. In turn, the attractive Helmholtz energy differentiated with respect to volume gives the attractive contribution to pressure:

$$p^{\text{attr}} = -\rho^2 \sum x_i \left\{ \frac{\sum x_j (a_{ji} b_{ii}/b_{ji}) \exp(\alpha a_{ji} 4\xi_i / RTb_{ji})}{\sum x_j \exp(\alpha a_{ki} 4\xi_i / RTb_{ki})} \right\} \tag{4.3-82}$$

Equation (4.3-82), coupled with any repulsive term, gives the simplest density-dependent local composition equation of state. As shown by Whiting and Prausnitz, Eq. (4.3-82) with the van der Waals repulsive term (4.2-4) satisfies two boundary conditions: (1) in the low-pressure limit it reproduces the correct composition dependence of the second virial coefficient and (2) in the high-density limit the well-known Wilson (3.3-9) equation for Gibbs excess energy is recovered provided that both components are subcritical. The procedure outlined above can be refined by introducing local surface or volume fractions instead of mole fractions to make it more suitable for molecules of unequal sizes. Accordingly, the residual internal energy for an n-component mixture becomes (Mollerup, 1981)

$$U^r = \sum_{i=1}^{n} x_i q_i \frac{\sum_{j=1}^{n} \Psi_j E_{ji} U_{ji}}{\sum_{l=1}^{n} \Psi_l E_{li}} \tag{4.3-83}$$

where

$$E_{ji} = \exp\left[-\alpha_{ji}\left(\frac{V_{ji}}{RT}\right)\right] \tag{4.3-84}$$

q_1 is proportional to external surface area and Ψ_j is a measure of concentration. Furthermore, U_{ji} can be calculated from a more accurate equation of state. Thus, Mollerup derived local composition versions of the Redlich–Kwong EOS (4.2-12)

$$p = \frac{RT}{V-b} - \frac{1}{V(V+b)} \sum_{i=1}^{n} x_i q_i \frac{\sum_{j=1}^{n} x_j E_{ji} a_{ji}/q_{ji}}{\sum_{l=1}^{n} x_l E_{li}} \tag{4.3-85}$$

where

$$E_{ji} = \exp\left[\frac{a_{ji}}{q_{ji} bRT} \ln\left(1 + \frac{b}{V}\right)\right] \tag{4.3-86}$$

and of the Peng–Robinson EOS (4.2-25)

$$p = \frac{RT}{V-b} - \frac{1}{V(V+b)+b(V-b)} \sum_{i=1}^{n} x_i q_i \frac{\sum_{j=1}^{n} x_j E_{ji} a_{ji}/q_{ji}}{\sum_{l=1}^{n} x_l E_{li}} \tag{4.3-87}$$

where

$$E_{ji} = \exp\left[\frac{a_{ji}}{q_{ji} bRT^2 (2)^{1/2}} \ln \frac{V + b(1+2^{1/2})}{V + b(1-2^{1/2})}\right] \tag{4.3-88}$$

It is remarkable that the above local composition equations of state require only one binary-adjustable parameter. The binary parameter is introduced into a combining rule for the cross parameter a_{ji}/q_{ji} in Eq. (4.3-85) or (4.3-88)

$$\frac{a_{ji}}{q_{ji}} = (1 - k_{ij})\left(\frac{a_{ii}a_{jj}}{q_i q_j}\right)^{1/2} \tag{4.3-89}$$

Mathias and Copeman (1983) evaluated the Mollerup model for asymmetric but nonpolar mixtures. In this case the classical one-fluid quadratic mixing rules work very well, and it is extremely important that the density-dependent mixing rules retain this capability. As a test case the authors chose VLE data for the methane–decane system, which can be predicted by means of classical quadratic mixing rules with the default interaction parameter $k_{12} = 0$. Unfortunately, the density-dependent local composition (DDLC) model appeared to overpredict the asymmetric effects. Unreasonably large binary parameters were necessary to correlate this relatively simple system. This undesirable feature of Eqs. (4.3-85)–(4.3-87) is due to the implicit assumption that local composition inevitably occur in the $j - i$ pair interaction whenever $a_i/q_i = a_j/q_j$. To overcome this difficulty, Mathias and Copeman proposed to separate the internal energy into two parts: conformal $u_{ji}^{\mathrm{att},0}$ (analogous to classical mixing rules) and nonconformal ("excess") $u_{ji}^{\mathrm{att},E}$ ones:

$$u_{ji}^{\mathrm{att}} = u_{ji}^{\mathrm{att},0} + u_{ji}^{\mathrm{att},E} \tag{4.3-90}$$

$$u_{ji}^{\mathrm{att},0} = \frac{\partial}{\partial(1/T)}\left[(a_i a_j)^{1/2}(1 - k_{ij})\frac{F_v}{T}\right]_{V,\mathbf{N}} \tag{4.3-91}$$

$$u_{ji}^{\mathrm{att},E} = \frac{\partial}{\partial(1/T)}\left[(a_i a_j)^{1/2}(-d_{ij})\frac{F_v}{T}\right]_{V,\mathbf{N}} \tag{4.3-92}$$

where F_v is a function of volume and depends on the form of the adopted equation of state. The parameters $d_{ji} \neq d_{ij}$ determine the local composition effects. The total mixture molar energy is obtained by summing the conformal and nonconformal contributions using the quadratic and local composition mixing rules, respectively.

$$u^{\mathrm{att},0} = \sum\sum x_i x_j u_{ij}^{\mathrm{att},0} \tag{4.3-93}$$

$$u^{\text{att},E} = \sum \sum x_i q_i \sum x_{ji} U_{ji}^{\text{att},E} \tag{4.3-94}$$

Then, the equation of state is obtained similarly to the procedure outlined above. The authors also presented a simplified version of the equation to reduce the computation time. Finally, the EOS obtained contains two binary-adjustable parameters (d_{12} and d_{21}) instead of one in the original method of Mollerup. The resulting equation is, however, applicable to polar as well as nonpolar mixtures and is superior to classical quadratic mixing rules with respect to the correlation of VLE and LLE. Similarly to Mathias and Copeman, Mollerup (1983) modified his original model by dividing the excess internal energy into a *random* and a *nonrandom* contributions where the term *randomness* denotes the van der Waals approximation with classical quadratic mixing rules.

$$U^E = U^E_{\text{random}} + U^E_{\text{nonrandom}} \tag{4.3-95}$$

where U^E_{random} is analogous with the van der Waals approximation

$$U^E_{\text{random}} = \sum \sum x_i x_j (U^0_{ji} - U_{ji}) \quad \text{with } U^0_{ji} = -\frac{a^0_{ij}}{v} \tag{4.3-96}$$

where a^0_{ij} is a van der Waals attraction parameter for the $i-j$ pair. $U^E_{\text{nonrandom}}$ is represented by a local composition expression:

$$U^E_{\text{nonrandom}} = \sum \sum x_i x_j \frac{E_{ji}}{\sum x_l E_{li}} (U_{ji} - U^0_{ji}) \tag{4.3-97}$$

where $E_{ji} = \exp[-(U_{ji} - U^0_{ji})/RT]$ is the Boltzmann factor. If the pure-component Helmholtz energy has the general form

$$F_i = -N_i g_i(b_i, v_i) - N_i \left(\frac{a_{ii}}{T}\right) f_i(v_i) \tag{4.3-98}$$

where $g_i(b_i, v_i)$ and $f_i(v_i)$ are functions expressing the volume dependence of the repulsive and attractive terms, respectively, the expression for a mixture becomes

$$F = -Ng - \left(\frac{f}{TN}\right) \sum \sum N_i N_j a^0_{ji} - \sum N_i \ln\left(\frac{\sum N_j E_{ji}}{N}\right) \tag{4.3-99}$$

and contains two binary parameters k_{ji} and l_{ji} in the a_{ji} and b_{ji} terms:

$$\begin{aligned} a_{ji} &= a^0_{ji} k_{ji} \quad \text{with } a^0_{ji} = (a_{jj} a_{ii})^{1/2} \\ b_{ji} &= b^0_{ji} l_{ji} \quad \text{with } b^0_{ji} = (b_{jj} + b_{ii}) \end{aligned} \tag{4.3-100}$$

This model coupled with the Redlich–Kwong EOS (4.2-12) was shown to retain the capability of quadratic mixing rules to correlate nonpolar mixtures while improving the correlation for strongly nonideal mixtures. For example, the average deviations in P for seven alcohol–alkane systems were reduced from 13.3 to 4.3 per cent with the same number of adjustable parameters, i.e., two. The model was also shown to reproduce correctly the experimentally observed maxima in Henry's constants for gas solubility in water (Mollerup, 1985). Mathias and Copeman's and Mollerup's (1983) models are similar except for two differences. Namely, Mathias and Copeman, unlike Mollerup, retained the UNIQUAC surface area parameters and used unsymmetric binary parameters $d_{ij} \neq d_{ji}$. Won (1983) developed also a density-dependent local composition model with two unsymmetric binary parameters and applied it successfully to model supercritical phase equilibria. Adachi and Sugie (1985) evaluated the Whiting–Prausnitz and Won mixing rules along with those of Huron–Vidal and the classical two-parameter ones. For the five systems studied the authors found fairly similar average deviations: 2.48 per cent in pressure for Whiting–Prausnitz's, 3.47 per cent for Won's, 1.38 per cent for Huron–Vidal's and 2.10 per cent for classical mixing rules. However, the authors' conclusion that the functional form of the mixing rules is not very influential in VLE calculations is biased by their selection of mixtures that excluded strongly nonideal systems such as alcohol–hydrocarbon and aqueous systems.

Skjold-Jørgensen (1984, 1988) developed, following the work of Mollerup, a group contribution EOS for the prediction of gas solubilities in nonpolar and polar solvents. The equation is based on a Carnahan–Starling–van der Waals form with the attractive part of the residual Helmholtz energy calculated from a NRTL-type local composition model. The equation requires critical temperatures and pressures and group parameters as input data and predicts K factors ($K_i = y_i/x_i$) with the accuracy of about 15 per cent.

Density-dependent local composition models can be also derived from considerations based on statistical thermodynamics and computer-generated data. A concise review of developments along these lines has been given by Danner and Gupte (1986).

Lee et al. (1983) established a statistical groundwork connecting the local compositions with equation-of-state formalism using the radial distribution functions. The final expression for pressure for a binary system is

$$P = x_a[x_{aa}P_{aa} + x_{ba}P_{ba}] + x_b[x_{bb}P_{bb} + x_{ab}P_{ab}] \tag{4.3-101}$$

where the x_{ij} (i, $j = a, b$) denotes the local mole fraction of molecules of type j around molecules of type i. The local compositions x_{ij} are defined in terms of specific Helmholtz free energies $\alpha' A'_{ij}$ of interactions between isolated pairs of molecules of type i and j.

$$x_{ba} = \frac{x_b F_{ba} \exp(-\alpha' A'_{ba}/kT)}{x_a F_{aa} \exp(-\alpha' A'_{aa}/kT) + x_b F_{ba} \exp(-\alpha' A'_{ba}/kT)} \tag{4.3-102}$$

Li et al. (1985) applied this result to the correlation of VLE by calculating the quantities A'_{ji} and P_{ji}, which are the properties of a hypothetical system of molecules with $j-1$ interactions only, using a BWR-type EOS of Chung et al. (1984). The mixing rule for the reduced density

$$\rho^*_{ij} = \rho \sum \sum x_m x_n V^*_{mn} \quad \text{with } V^*_{ij} = \xi^3 (V^*_{ii} V^*_{jj})^{1/2} \tag{4.3-103}$$

is consistent with a one-fluid model, whereas that for reduced temperature

$$T^*_{ij} = \frac{kT}{(\epsilon^0_{ij}/k + D_{ij}/T)} \quad \text{with } \epsilon^0_{ij} = \zeta(\epsilon^0_{ii}\epsilon^0_{jj})^{1/2} \tag{4.3-104}$$

corresponds to a multifluid model. The model with three adjustable parameters was found to improve upon the conformal solution mixing rules.

Sandler and co-workers (Sandler, 1985; Lee et al., 1985, 1986, 1989; Lee and Sandler, 1987) presented a series of papers dealing with the development of density-dependent mixing rules on the basis of computer simulation data on square-well fluids. For practical calculations with van der Waals-type EOS the authors recommended to use the general mixing rule

$$P^{\text{att}} = \sum \sum x_i x_j P^{\text{att}}_{ij} \tag{4.3-105}$$

where P^{att} and P^{att}_{ij} are the attractive terms of an equation of state for mixture and pure fluid, respectively. Equation (4.3-105) was tested with the PR EOS (4.2-25) for nonpolar and weakly polar mixtures. The mixing rule was found to give more accurate predictions than the classical van der Waals mixing rule when no binary interaction parameters were used but did not improve the results when one binary parameter was used.

Lee and Chao (1986) introduced the density dependence, as determined from computer simulation results into local compositions

$$x_{ji} = \frac{x_j \exp[-\gamma_{ji}(\epsilon_{ji} - \epsilon^0_{ji})/RT]}{\sum x_k \exp[-\gamma_{ki}(\epsilon_{ki} - \epsilon^0_{ki})RT]} \tag{4.3-106}$$

with $\epsilon^0_{ji} = \alpha(\rho^*)\epsilon_{ii} + [1 - \alpha(\rho^*)]\epsilon_{jj}$ and $\gamma_{ji} = (\epsilon_{ii}\epsilon_{jj})^{1/2}/|\epsilon_{ji}|$ where ρ^* is the reduced density. Local compositions were then introduced directly into mixing rules [Eq. (4.3-60)]. The resulting model used with three binary parameters represents an improvement over two-parameter classical mixing rules for the majority of systems studied.

Adachi et al. (1989) determined the density dependence of the local mole fraction ratio and used it in conjunction with mixing rules derived analogously with those of Huron and Vidal. They found that the density dependence does not, in general, improve the correlation of VLE in comparison with the original density-independent Huron–Vidal mixing rules.

A different approach was proposed by Deiters (1987) who introduced density dependence into the van der Waals mixing rules for simple fluids by applying a variable exponent δ to account for nonequiform particle distribution in mixtures

$$\epsilon\sigma^{\delta} = \sum\sum x_i x_k \epsilon_{ik} \sigma_{ik}^{\delta} \qquad (4.3\text{-}107)$$

$$\sigma^{\delta} = \sum\sum x_i x_k \sigma_{ik}^{\delta} \qquad (4.3\text{-}108)$$

The exponent is not constant as in Eqs. (4.3-36) and (4.3-37) but is obtained by integration of radial distribution functions of rigid-sphere mixtures. The mixing rules were found to improve the reproduction of phase equilibria in cryogenic mixtures, especially in the vicinity of critical points. The method was extended later to nonspherical molecules (Deiters, 1989a).

While the development of density-dependent mixing rules has been based mostly on the local composition concept, it is possible to formulate them in a more empirical way without using the local composition formalism. The functional form of the mixing rules is then guided primarily by the physical boundary conditions. Along this line, Luedecke and Prausnitz (1985) constructed an equation of state containing an extension to mixtures (Mansoori et al., 1971) of the repulsive term of Carnahan–Starling (1969) and a simple van der Waals attractive term. For the constant a of the van der Waals term Luedecke and Prausnitz wrote

$$a = \sum\sum x_i x_j a_{ij} + a^{\text{nc}} \qquad (4.3\text{-}109)$$

where the first term is identical to the classical quadratic mixing rules and a^{nc} is a contribution of "noncentral" forces (due to differences in polarity, size and shape of molecules) to the van der Waals constant. As nothing is known a priori about this term, the safest way to construct an expression for a^{nc} is to look at the boundary conditions. First, as a^{nc} refers only to contributions from unlike pairs:

$$\lim_{x_i \to 0} a^{\text{nc}} = 0 \qquad (4.3\text{-}110)$$

Second, the authors assumed that as temperature rises or as density falls the importance of noncentral forces declines:

$$\lim_{\rho/RT \to 0} a^{\text{nc}} = 0 \qquad (4.3\text{-}111)$$

The simplest nonquadratic approximation consistent with the boundary conditions (4.3-110) and (4.3-111) introduces a cubic composition dependence

$$a^{\text{nc}} = \left(\frac{\rho}{RT}\right) \sum\sum x_i x_j (x_i c_{i(j)} + x_j c_{j(i)}) \qquad (4.3\text{-}112)$$

where $c_{i(j)}$ is a binary parameter that reflects noncentral forces when molecule j is infinitely dilute. The resulting mixing rule

$$a = \sum\sum x_i x_j (a_{ii} a_{jj})^{1/2} (1 - k_{ij}) + \left(\frac{\rho}{RT}\right) \sum\sum x_i x_j (x_i c_{i(j)} + x_j c_{j(I)}) \qquad (4.3\text{-}113)$$

contains three adjustable parameters k_{12}, $c_{1(2)}$ and $c_{2(1)}$ for a binary system. In the low-density limit Eq. (4.3-113) correctly reproduces the quadratic composition dependence of the second virial coefficient. With this three-parameter mixing rule the authors were able to correlate accurately vapor–liquid or liquid–liquid equilibria for strongly nonideal water-containing binaries. The model overpredicted, however, ternary LLE. Panagiotopoulos (1986) proposed a similar mixing rule to that reported by Luedecke and Prausnitz

$$a = \sum \sum x_i x_j a_{ij}(1 - k_{ij}) + \left(\frac{b}{vRT}\right) \sum \sum x_i x_j (x_i c_{ij} + x_j c_{ji}) \qquad (4.3\text{-}114)$$

with $k_{ij} = k_{ji}$ and $c_{ij} = -c_{ji}$. A cubic composition dependence of the mixing rule was also assumed by Cotterman and Prausnitz (1986) for the high-density contribution to the Helmholtz energy. As the method of Cotterman and Prausnitz separates the high- and low-density contributions in the pure-component equation of state, it is based on the same principle of interpolating between known boundary conditions as the above mixing rules.

Perhaps the simplest example of purely empirical density-dependent mixing rules is the mixing rule of Mohamed and Holder (1987). The authors adopted a linear density dependence for the binary parameter k_{ij} in the quadratic mixing rules

$$k_{ij} = a_{ij} + b_{ij}\rho \qquad (4.3\text{-}115)$$

Equation (4.3-115) was used together with the Peng–Robinson equation of state to represent vapor–liquid equilibria for systems containing carbon dioxide and aromatic hydrocarbons. On average, the density-dependent mixing rules reduced the deviations 2.2 times in relation to the classical one-parametric mixing rules (Mohamed and Holder, 1988). The improvement has been obtained, however, at the cost of increasing the number of binary parameters from one to two and complicating the volume dependence of the equation of state, which became quartic rather than cubic.

Wilczek-Vera and Vera (1987) proposed density-dependent mixing rules that preserve the cubic nature of the equation of state while recovering the quadratic composition dependence of the second virial coefficient. These mixing rules are inspired by the flexibility of the composition-dependent Margules-type and Van Laar-type mixing rules [Eqs. (4.3-66) and (4.3-69)]. The authors proposed two-parameter Margules-type and Van Laar-type forms

$$a = \sum \sum x_i x_j (a_{ii} a_{jj})^{1/2} \left[1 - \left(\frac{V^L}{V}\right)(x_i k_{ij} + x_j k_{ji})\right] \qquad (4.3\text{-}116)$$

and

$$a = \sum \sum x_i x_j (a_{ii} a_{jj})^{1/2} \left[1 - \frac{(V^L/V) k_{ij} k_{ji}}{x_i k_{ij} + x_j k_{ji}}\right] \qquad (4.3\text{-}117)$$

and their three-binary-parameter extensions

$$a = \sum \sum x_i x_j (a_{ii} a_{jj})^{1/2} \left\{1 - l_{ij} + \left(\frac{V^L}{V}\right)[l_{ij} - (x_i k_{ij} + x_j k_{ji})]\right\} \qquad (4.3\text{-}118)$$

and

$$a = \sum \sum x_i x_j (a_{ii} a_{jj})^{1/2} \left[1 - l_{ij} + \left(\frac{V^L}{V} \right) \left(l_{ij} - \frac{k_{ij} k_{ji}}{x_i k_{ij} + x_j k_{ji}} \right) \right] \qquad (4.3\text{-}119)$$

The above forms seem to be especially suitable for low-pressure vapor–liquid equilibrium calculations. When applied to the vapor phase for which $V \gg V^L$, they reduce to classical quadratic mixing rules. For the liquid phase ($V = V^L$) Eqs. (4.3-116)–(4.3-119) generate the density-independent mixing rules (4.3-66) and (4.3-69). The improvement due to the introduction of the density dependence was marginal for the systems studied because the density-independent forms are already sufficiently flexible to correlate VLE data. However, the three-parameter forms [Eqs. (4.3-118) and (4.3-119)] introduce additional flexibility into the mixing rules and lead to an improvement of results especially for polar–nonpolar systems and for nonpolar systems differing largely in size.

Melhem et al. (1989) presented a comprehensive evaluation of the density-dependent mixing rules of Luedecke–Prausnitz and Panagiotopoulos–Reid as well as the density-independent mixing rule of Panagiotopoulos–Reid [Eq. (4.3-68)]. Table 4.3-4 shows their results for selected types of mixtures containing polar components.

All the mixing rules studied yielded fairly good results when applied to correlate individual VLE data although larger deviations (about 4–6 per cent in pressure) appeared especially for very strongly nonideal systems such as alcohol–aliphatic hydrocarbons. According to Melhem et al., the three-parameter mixing rule of Luedecke and Prausnitz (4.3-113) failed when applied to the prediction of ternary VLE for binary data while the remaining two-parameter mixing rules gave reasonable results, although not significantly better than those obtained from one-parameter classical quadratic mixing rules.

When using phase equilibrium modelling methods, it is highly desirable to be able to correlate simultaneously vapor–liquid and liquid–liquid equilibria Melhem et al. (1989) also calculated several liquid–liquid equilibria using the binary parameters obtained from vapor–liquid equilibria. The results are summarized in Table 4.3-5 for five binary systems.

It is evident from Table 4.3-5 that the results of predicting liquid–liquid equilibria from vapor–liquid equilibria are very poor when the classical quadratic mixing rules are used and are only qualitatively accurate for the density-dependent and composition-dependent combining rules. The mixing rules discussed above share this drawback with the gamma–phi correlation methods. When the objective of calculations is to obtain a simultaneous correlation of VLE and LLE, these methods cannot be recommended.

In general, the concept of density-dependent mixing rules makes it possible to obtain a satisfactory correlation of *individual* phase equilibrium data. In comparison with the density-independent composition-dependent mixing rules, they are more realistic because they satisfy the necessary boundary conditions. The workable density-dependent mixing rules require two or three adjustable parameters for a binary system. Attempts to obtain a one-parameter representation do not seem to be satisfactory. Danner and Gupte (1986) evaluated a density-dependent local composition model of their own with

Table 4.3-4
ACCURACY OF CORRELATING INDIVIDUAL VLE ISOTHERMS OR ISOBARS FOR MIXTURES CONTAINING POLAR COMPOUNDS[a]

Type of Mixtures	Number of Data Sets	*CQ* (*4.3-1*) (*1 param.*)		*PRDI* (*4.3-68*) (*2 param.*)		*PRDD* (*4.3-114*) (*2 param.*)		*LPDD* (*4.3-113*) (*3 param.*)	
		δy%	δP%	δy%	δP%	δy%	δP%	δy%	δP%
Acid + hydrocarbon	6	3.3	2.1	3.3	1.4	2.8	3.1	3.1	2.2
Ketone + hydrocarbon	3	1.8	1.3	1.9	0.8	1.7	1.9	1.2	1.1
Alcohol + aromatic hydrocarbon	10	2.1	4.6	1.1	1.5	1.0	1.9	0.9	0.6
Alcohol + aliphatic hydrocarbon	8	6.4	14.7	2.2	5.2	3.7	5.6	2.1	3.6
Alcohol + ketone	15	2.0	4.2	1.5	2.5	1.3	3.1	1.2	2.6
Alcohol + alcohol	8	2.8	4.1	2.6	2.7	2.1	3.4	1.6	2.0
Alcohol + water	6	3.9	7.3	0.8	1.2	0.8	1.3	0.9	1.2
Ketone + water	5	5.3	16.3	1.3	2.2	1.1	2.4	0.9	2.7

After Melhem et al. (1989).

[a] Using the classical quadratic (CQ) mixing rules, the Panagiotopoulos–Reid density-independent composition-dependent (PRDI) and density-dependent (PRDD) mixing rules and the Luedecke–Prausnitz (LPDD) mixing rules.

Table 4.3-5
ACCURACY OF PREDICTING BINARY LIQUID–LIQUID EQUILIBRIA USING PARAMETERS FITTED TO VAPOR–LIQUID EQUILIBRIA[a]

		Compositions of Coexisting Phases x^{α} and x^{β}				
System	**T (K)**	**Experiment**	**CQ**	**PRDI**	**PRDD**	**LPDD**
Butanol + water	365	0.260 0.045	0.442 0.007	0.440 0.021	0.423 0.026	0.438 0.021
Isobutanol + water	364	0.335 0.055	0.418 0.008	0.385 0.028	0.400 0.033	0.437 0.030
2-Butanone water	346	0.640 0.045	0.416 0.008	0.564 0.034	0.640 0.031	0.646 0.065
Ethyl acetate + water	343	0.700 0.180	0.453 0.004	0.692 0.018	0.694 0.015	0.672 0.034
Methanol + hexane	323	0.885 0.155	no phase split	0.897 0.135	0.897 0.147	0.861 0.100

After Melhem et al. (1989).

[a]By means of the classical quadratic (CQ) mixing rules, the Panagiotopoulos–Reid density-independent composition-dependent (PRDI) and density dependent (PRDD) mixing rules and the Luedecke–Prausnitz (LPDD) mixing rules.

respect to the prediction of ternary equilibria from binary data. For most systems the improvement in the predictions with the DDLC model over the quadratic mixing rules was only marginal. The authors concluded that this result underscores the limited physical significance of the local composition parameters and indicates a need for more predictive models with fewer adjustable parameters. A drawback of the density-dependent mixing rules is the additional complexity they introduce into the functional form of equations of state. This price has to be paid for a more correct behavior of the EOS in the low- and high-density regions. From the point of view of practical phase equilibrium calculations, the density dependence does not improve markedly upon the density-independent, composition-dependent combining rules, which are deficient in the low-density region.

4.3.6. Improved Classical Mixing Rules for Strongly Nonideal Mixtures

The above methods for extending equations of state to strongly nonideal mixtures replace the classical quadratic mixing rules with those with more complicated composition and density dependence. It is possible, however, to enhance the flexibility of mixing rules without changing their simple quadratic composition dependence. Kwak and Mansoori (1986) and Benmekki and Mansoori (1987) noted that the conformal solution mixing rules of Leland et al. [Eqs. (4.3-3) and (4.3-4)] imply the quadratic composition dependence for EOS parameters provided that the dimensions of the parameters are (molecular volume) [Eqs. (4.3-3)] and (molecular volume) $\times$ (molecular energy) [Eq.

(4.3-4)]. On the other hand, the parameters of the widely used equations of state have different dimensions depending on their functional forms. Kwak and Mansoori (1986) postulated that the equations be reformulated so that the dimensions of EOS parameters be either (volume) or (volume) × (energy). The familiar Peng–Robinson equation of state reformulated by Kwak and Mansoori takes the form

$$z = \frac{V}{V-b} - \frac{c/RT + d - 2(cd/RT)^{1/2}}{(V+b) + (b/V)(V-b)} \tag{4.3-120}$$

where b and d are proportional to [volume] and c is proportional to (volume) × (energy). For parameters b, c and d the quadratic conformal solution mixing rules are employed [Eq. (4.3-1)]. The following combining rules are used together with the mixing rules:

$$b_{ij} = (1 - l_{ij})^3 \left(\frac{(b_{ii}^{1/3} + b_{jj}^{1/3})}{2} \right)^3 \tag{4.3-121}$$

$$d_{ij} = (1 - m_{ij})^3 \left(\frac{(d_{ii}^{1/3} + d_{jj}^{1/3})}{2} \right)^3 \tag{4.3-122}$$

$$c_{ij} = (1 - k_{ij})^3 \left(\frac{c_{ii} c_{jj}}{b_{ii} b_{jj}} \right)^{1/2} b_{ij} \tag{4.3-123}$$

A total of three parameters k_{ij}, l_{ij} and m_{ij} have to be adjusted from binary data. Benmekki and Mansoori (1987) applied Eqs. (4.3-120)–(4.3-123) to calculate binary VLE in strongly nonideal systems and obtained a good correlation. For the six systems studied by the authors the average deviation in pressure was 1.4 per cent. For comparison, the average deviation obtained from the one-parameter quadratic mixing rules was 6.8 per cent. Benmekki and Manssoori (1988) found that the mixing rules also made it possible to represent simultaneously vapor–liquid equilibrium (VLE) and vapor–liquid–liquid equilibrium (VLLE) data.

In the case of ternary systems Benmekki and Mansoori found it necessary to include the contribution of three-body interactions. The compressibility factor is then expressed as (Lan and Mansoori, 1977)

$$z = z^e + \sum_{i>j>k=1}^{n} \sum^{n} \sum^{n} x_i x_j x_k \frac{\eta \beta_{ijk}}{b^3 V} \left(\frac{f_1' f_2 - f_1 f_2'}{f_2^2} \right) \tag{4.3-124}$$

where z^e is the empirical equation of state [Eq. (4.3-120)], b is the co-volume, $\eta = b/4v$ is the reduced density, f_1 and f_2 are algebraic functions of h and b_{ijk} is given by

$$b_{ijk} = \epsilon \left(\tfrac{8}{27} \right) \pi N_0^4 \nu_{ijk} \tag{4.3-125}$$

and contains an adjustable parameter ϵ that has to be fitted to ternary data. The three-body interaction term with the ternary parameter ϵ improved the quality of correlating ternary data especially around the critical solution point of the mixture.

A somewhat similar concept based on dimensional analysis has been used by Vetere (1983) and Vetere et al. (1988) in conjunction with his EOS. The author noted that the three terms in Eq. (4.2-67) have the dimension of δ^2, the square of Hildebrand's solubility parameter. Therefore, quadratic mixing rules have been applied for them

$$(\text{term}) = \sum \sum x_i x_j \, (\text{term})_{ij} \tag{4.3-126}$$

where

$$\text{term} = a \frac{e^{-kT}}{v^2} \quad \text{or} \quad a_1 \frac{e^{-k_1 T}}{v^m} \quad \text{or} \quad b \frac{e^{-k_2 T}}{v^n}$$

The combining rules

$$(\text{term})_{ij} = [(\text{term})_i (\text{term})_j]^{1/2} (1 - k_{ij}^{(k)}) \qquad k = 1, 2, 3 \tag{4.3-127}$$

contain three binary parameters. These rules have the same justification as the Hildebrand–Scatchard mixing rules. Nevertheless, they are sufficiently flexible to correlate VLE in strongly nonideal mixtures with three adjustable parameters.

The above results show that a more flexible equation of state can be constructed using quadratic mixing rules provided that the EOS parameters are appropriately selected and a sufficiently large number of adjustable parameters (i.e., three per binary) is introduced.

4.3.7. Explicit Treatment of Association and Polarity

The group of methods explicitly treating association and polarity has to be distinguished due to their different character. These methods are fundamentally different from those described before in that they modify the form of the equation of state for pure substances. The performance of the equation of state for mixtures is a consequence of this modification. The possibility of representing phase equilibrium data by assuming that molecules associate or solvate to form new species has been recognized for a long time (Dolezalek, 1908). According to this concept, a binary mixture of compounds A and B is a multicomponent mixture containing, in addition to A and B, also self-associates containing one type of molecules A_n and B_m and cross-associates (solvates) $A_n B_m$ containing molecules of both types (m and n denote the numbers of molecules in an associate). Quantitative analysis of the behavior of such mixtures requires assigning equilibrium constants to each of the assumed association equilibria. The chemical theory has been shown by many authors to be particularly useful when applied to systems containing one associating substance and inert solvents. In this case the chemical approach made it possible to obtain an accurate representation of phase equilibria with a limited number of physically meaningful parameters. While chemical theories are useful for describing strongly nonideal mixtures, their application to the modelling of multicomponent mixtures containing any kind of chemical species encounters substantial difficulties. The major drawback of the chemical approach has been its specific character, restricted to particular classes of mixtures, and the difficulty of extending the models valid for binaries to multicomponent mixtures. In this section we shall show how these problems were attacked, in some cases successfully, by different investigators.

In general, there are two methods for incorporating association equilibria into an equation of state. The first one consists in simultaneously solving chemical equilibria between associated species and physical (phase) equilibria between all species present in a solution. A pure associated substance is assumed to be a mixture of an associated species A_i, each of them containing i monomeric units in a multimer. Once the pseudocomponents of the pure substance are assumed, the system of equations to be solved is as follows:

1. Chemical equilibria for the reaction $iA_i = A_i$:

$$i\mu_1(T, P, \mathbf{x}) = \mu_i(T, P, \mathbf{x}) \tag{4.3-128}$$

where μ_i is the chemical potential of the A_i species.

2. Phase equilibria:

$$\mu_i^{\text{phase 1}}(T, P, \mathbf{x}) = \mu_i^{\text{phase 2}}(T, P, \mathbf{x}) \tag{4.3-129}$$

where the superscripts stand for the liquid and vapor (or two liquid) phases in equilibrium. The standard chemical potentials μ_i^0 of the species are related to each other by equilibrium constants K_i of the reaction (4.3-128)

$$\mu_i^0(T) - i\mu_1^0(T) = \Delta G^0(T) = -RT \ln K_i(T) \tag{4.3-130}$$

The standard state chosen for the A_i species in both the liquid and gas phases is pure A_i in the ideal-gas state. This standard state is convenient for equation-of-state calculations because the association constant is then the same for all phases and depends only on temperature. It should be noted that the association constants used in the association models from the γ–ϕ family are defined on the basis of a different standard state for the associates, which is usually a pure liquid species A_i at the equilibrium temperature and pressure. In principle, it is possible to use these constants in the equation-of-state formalism but such an approach leads to much more complex expressions than the methods based on ideal-gas standard states (notably, density-dependent association constants are required).

Equations (4.3-129) and (4.3-130) can be solved numerically provided that the residual chemical potentials of associated species are calculated from an equation of state suitable for nonpolar mixtures. For this purpose, Wenzel et al. (1982) used the Schmidt–Wenzel (4.2-28) cubic equation of state whereas Gmehling et al. (1979) and Grenzheuser and Gmehling (1986) employed the PHCT EOS (4.2-46). The most difficult problem in the computational scheme [(4.3-129) and (4.3-130)] is the assignment of probable associated species. A priori statements concerning which chemical species are possible if independent information (e.g., spectroscopic data) is available. For example, there is abundant evidence that carboxylic acids form cyclic dimers and alcohols form continuous chainlike or cyclic associates. However, practical applications of the scheme [(4.3-129) and (4.3-130)] require that the number of assumed association equilibria be minimized. If we assumed a continuous association model, the number of equations to be solved would be infinitely large. Therefore, Gmehling et al. (1979) and Grenzheuser and Gmehling (1986) restrict their calculations to dimers that make their method applicable primarily to carboxylic acids. On the other hand, Wenzel et al. (1982) allow for the existence of higher multimers. For example, methanol is assumed to be a mixture of monomers, tetramers and dodecamers. The selection of multimers is arbitrary and results from a compromise between the requirements of reproducing the data and minimizing the number of adjustable pure-component parameters. It is not guided by the structure of associated compounds. Each of the associates is characterized by the critical temperature, critical pressure and acentric factor treated as adjustable parameters. Additionally, each of the assumed association reactions is described by the values of standard enthalpy and entropy of association, assumed to be independent of temperature

$$-RT \ln K_i(T) = \Delta H_i^0 - T\,\Delta S_i^0 \tag{4.3-131}$$

In some cases, especially when a wide temperature range has to be covered, standard heat capacity of association can be allowed for. This leads to

$$\ln K_i(T) = \frac{\Delta H_i^0(T_0) + \Delta C_{pi} T_0}{RT} + \frac{1}{R}\left(\Delta S_i^0(T_0) + \Delta C_{pi}^0 - \Delta C_{pi}^0 \ln T_0\right) + \frac{\Delta C_{pi}^0}{R} \ln T \tag{4.3-132}$$

or, in the notation of Grenzheuser and Gmehling,

$$\ln K_i(T) = \frac{\mathrm{HO}}{T} + \mathrm{SO} + \mathrm{CO} \ln T \tag{4.3-133}$$

Thus, the number of adjustable pure-component parameters is very large (three parameters for each mono- or multimer and two or three for each reaction) and their determination is by no means simple. In order to determine pure-component parameters, Wenzel et al. (1982) utilize not only pure-component data (i.e., vapor pressure, liquid and vapor volumes and critical properties) but also binary mixture phase equilibria.

Extension of the computational scheme [(4.3-129) and (4.3-130)] to mixtures is straightforward provided that the mixture contains only one associating component. In this case the chemical equilibria (4.3-128) and (4.3-130) are not changed and only the number of phase equilibrium equations

(4.3-129) increases by the number of inert components present in the solution. In the case of mixtures containing two or more associating components, a problem of cross-associating appears. As it is impossible to deduce all parameters characterizing the cross-associates from binary data, only arbitrarily selected cross-associates are allowed for. The mixing rules employed for the associates and inert components are analogous to those used for mixtures of nonpolar components. As the binary parameters for the (associate–inert species) pairs are interrelated, only one effective adjustable binary parameter is used. With one binary parameter Wenzel et al. (1982) obtained a very good correlation of VLE data. The equation is also able to correlate liquid–liquid equilibria (Kolasińska et al., 1983) provided that a suitable temperature dependence is adopted for the binary parameter. Prediction of ternary VLE from binary data is also very satisfactory (Peschel, 1986). Moreover, Lang and Wenzel (1989) found that if a sufficiently large aggregate is assumed, the equation can predict three phases, one of them being interpreted as a solid phase.

In the case of mixtures containing more than one associating species, a problem of formulating a cross-association scheme appears. Grenzheuser and Gmehling have circumvented it by employing simple mixing rules for the association parameters from Eq. (4.3-133). Using the same notation,

$$\mathrm{HO}_{ij} = \frac{\mathrm{HO}_{ii} + \mathrm{HO}_{jj}}{2}$$

$$\mathrm{CO}_{ij} = \frac{\mathrm{CO}_{ii} + \mathrm{CO}_{jj}}{2} \qquad (4.3\text{-}134)$$

$$\mathrm{SO}_{ij} = \frac{\mathrm{SO}_{ii} + \mathrm{SO}_{jj}}{2} + \ln 2$$

For models with more than one self-association equilibria the formation of specific cross-associates, usually restricted to cross-dimers, has to be assumed (Wenzel et al., 1982; Peschel, 1986).

The good features of this type of method have been achieved at a considerable cost. First, the determination of pure-component parameters is cumbersome and no clear-cut algorithm exists for guessing which associates should be assumed to obtain a good representation of phase equilibria. This can be done only by trial and error. Second, this approach cannot overcome the traditional deficiency of association models, that is the difficulty of formulating a self-consistent model for systems containing any number of associating and inert components.

An attempt to solve the above problems has been made in the development of the second method for incorporating association into an equation of state. This method consists in analytically solving the chemical equilibrium expressions for an assumed association model in order to derive an explicit equation of state with association built in. It is assumed that thermodynamic properties of pure polar compounds and their mixtures can be attributed to chemical equilibria between associated species and physical (nonspecific) interactions between all species present in a solution. The physical effects can be represented by an equation of state that is applicable to nonpolar fluids, whereas a realistic association model should be used for the chemical effects. A continuous association model seems to be a good choice for this purpose because such models can provide a satisfactory approximation of the thermodynamic behavior of associated systems with only one effective association constant.

The first method for combining a continuous linear model with a van der Waals-type equation of state was proposed by Heidemann and Prausnitz (1976). For the consecutive association reactions $A_i + A_1 = A_{i+1}$ the equilibrium constants have been assumed to be equal $K_{i,i+1} = K$. The authors used a simple equation of state expressed in a general form

$$P = \frac{RTz_{\mathrm{rep}}(\xi)}{v} - \frac{a}{b^2}\Pi_{\mathrm{att}}(\xi) \tag{4.3-135}$$

where a and b are the generalized van der Waals parameters and z_{rep} and Π_{att} are functions of reduced density ξ. The simple quadratic mixing rules

$$a = \sum\sum x_i x_j (a_i a_j)^{1/2} \tag{4.3-136}$$

with $a_j = j^2 a$ and

$$b = \sum x_i b_i \tag{4.3-137}$$

with $b_i = jb$ have been employed for the associated species. It appears that if such mixing rules are assumed, the reduced density ξ and, subsequently, the z_{rep} and Π_{att} functions are independent of association. The association effects enter the equation of state only through a factor n_T/n_0 where n_T is the overall number of moles of associated species and n_0 is the number of moles of the substance in absence of association. The van der Waals parameters a and b take then the form

$$a = \left(\frac{n_0}{n_T}\right)^2 a \tag{4.3-138}$$

$$b = \left(\frac{n_0}{n_T}\right) b \tag{4.3-139}$$

Subsequently, the equation of state becomes

$$P = \frac{n_T}{n_0}\frac{RT\xi}{b} z_{\mathrm{rep}} - \frac{a}{b^2}\Pi_{\mathrm{att}}(\xi) \tag{4.3-140}$$

After evaluating the n_T/n_0 factor and material balance equations, the equation of state takes the form

$$z = \frac{2}{1 + [1 + 4RTK\exp(g)/v]^{1/2}} z_{\mathrm{rep}} - \frac{a}{RTb\xi}\Pi_{\mathrm{att}} \tag{4.3-141}$$

where
$$g = \int_0^{\xi} \left(\frac{z_{\text{rep}} - 1}{\xi} \right) d\xi \tag{4.3-142}$$

and provides an explicit expression for the effect of association on the compressibility factor.

The Heidemann–Prausnitz technique in its original form is not applicable, however, to practical mixture calculations. It is necessary to introduce some modifications by changing the equation of state and/or the mixing and combining rules for the associated species. The mixing rules for the associated species, which are unavoidably hypothetical, have a strong effect on the mathematical form of the equation of state and, subsequently, its empirical effectiveness. In the following example we shall take a closer look at the Heidemann–Prausnitz mixing rules and their modifications that render the association-based equations of state practical. As a model system, we shall consider a mixture containing one associating and one inert component.

Example 4.3-1 The influence of mixing rules for associated species on the general form of the equation of state.

Assuming that a mixture contains an inert component B and a series of associated species A_i $(i = 1, 2, \ldots)$, the total value of an equation-of-state parameter q is

$$q_T = z_B^2 q_{BB} + 2 \sum_i z_B z_{A_i} q_{A_i B} + \sum_i \sum_j z_{A_i} z_{A_j} q_{A_i A_j} \tag{P4.3-1}$$

where z denotes the "true" mole fraction of a "real" component:

$$z_B = \frac{n_B}{n_T} \tag{P4.3-2}$$

$$z_{A_i} = \frac{n_{A_i}}{n_T} \tag{P4.3-3}$$

and n_T is the total number of moles of the associated mixture:

$$n_T = n_B + \sum_i n_{A_i} \tag{P4.3-4}$$

If association was not taken into account, parameter q would be expressed as an average of those for the analytical (apparent) components A and B:

$$q_0 = x_B^2 q_{BB} + 2 x_B x_A q_{BA} + x_A^2 q_{AA} \tag{P4.3-5}$$

where x denotes the analytical (apparent) mole fraction:

$$x_B = \frac{nB}{n_0} \tag{P4.3-6}$$

$$x_A = \frac{n_A}{n_0} = \frac{\sum i n_{A_i}}{n_0} \tag{P4.3-7}$$

and n_0 is the analytical number of moles of the mixture

$$n_0 = n_B + \sum_i i n_{A_i} \tag{P4.3-8}$$

The crucial step in the derivation is the assumption of combining rules for the parameters q_{A_iB} and $q_{A_iA_j}$. Heidemann and Prausnitz (1976) proposed to approximate the co-volume b_{A_i} and attractive parameter a_{A_i} of an i-mer represented by a van der Waals-type EOS by

$$b_{A_i} = i b_{AA} \tag{P4.3-9}$$

$$a_{A_i} = i^2 a_{AA} \tag{P4.3-10}$$

This approximation was tested by Wenzel and Krop (1990) who analyzed the b and a parameters calculated for a series of hydrocarbons using the Peng–Robinson EOS. They found the multipliers i and i^2 in Eqs. (P4.3-9) and (P4.3-10) to be rather satisfactory although a more detailed inspection of the b and a values would suggest the multipliers i^e where the exponent e is intermediate between 1 and 2 (closer to 1 for b and closer to 2 for a). However, it is not worthwhile to establish a more accurate approximation of parameters a and b for hydrocarbons because of two reasons. First, hydrocarbons provide only a very rough approximation to the properties of multimers and, second, any fractional exponent would preclude further analytical derivations.

In agreement with Eqs. (P4.3-9) and (P4.3-10), the combining rules for parameters a and b can be formulated as

$$b_{A_iB} = \frac{i b_{AA} + b_{BB}}{2} \tag{P4.3-11}$$

$$b_{A_iA_j} = \frac{b_{AA}(i+j)}{2} \tag{P4.3-12}$$

$$b_{A_iB} = i a_{AB} \tag{P4.3-13}$$

$$a_{A_iA_j} = ij a_{AA} \tag{P4.3-14}$$

If Eqs. (P4.3-11)–(P4.3-14) are inserted into Eq. (P4.3-1), the total values of the parameters can be related to the analytical ones by

$$n_T^2 a_T = n_0^2 a_0 \tag{P4.3-15}$$

where

$$a_0 = x_B^2 a_{BB} + 2 x_B x_A a_{BA} + x_A^2 a_{AA} \tag{P4.3-15a}$$

and

$$n_T b_T = n_0 b_0 \tag{P4.3-16}$$

where
$$b_0 = x_B b_{BB} + x_A b_{AA} \tag{P4.3-16a}$$

If a van der Waals-type equation is used in the general form

$$\frac{PV}{n_T RT} = 1 + G\left(\frac{n_T b_T}{V}\right) + \frac{a_T}{RTb_T} H\left(\frac{n_T b_T}{V}\right) \tag{P4.3-17}$$

where V is the total volume and G and H are some functions of reduced density, the resulting equation of state differs from the EOS for a nonassociating fluid only by the factor n_T/n_0:

$$\frac{Pv}{RT} = \left[1 + G\left(\frac{b_0}{v}\right)\right]\frac{n_T}{n_0} + \frac{a_0}{RTb_0} H\left(\frac{b_0}{v}\right) \tag{P4.3-18}$$

where v is the molar volume. The ratio n_T/n_0 can be evaluated by considering the equilibrium constant for the reaction

$$A_i + A_1 = A_{i+1} \tag{P4.3-19}$$

as well as material balances. If all consecutive equilibrium constants for Eq. (P4.3-19) are assumed to be equal, then

$$K_{i,i+1} = K = \frac{z_{A_{i+1}}}{z_{A_i} z_{A_i}} \frac{1}{P} \frac{\phi_{A_{i+1}}}{\phi_{A_i}\phi_{A_i}} \tag{P4.3-20}$$

The resulting expression is

$$\frac{n_T}{n_0} = \frac{2x_A}{1 + [1 + (4RTKx_A/v)e^g]^{1/2}} + 1 - x_A \tag{P4.3-21}$$

where g depends on the form of the repulsive term

$$g = \int_0^{b/v} \left[\left(G\left(\frac{b}{v}\right) - 1\right)\frac{v}{b}\right] d\left(\frac{b}{v}\right) \tag{P4.3-22}$$

While Heidemann and Prausnitz (1976) did not extend their calculations to mixtures, Wenzel and Krop (1990) have found that Eqs. (P4.3-18), (P4.3-21) and (P4.3-22) are not practicable when applied to mixtures. This surprising and unfavorable result prompted Wenzel and Krop (1990) to drop the factor e^g in Eq. (P4.3-21) while retaining the general form (P4.3-18). This approach, while inconsistent, was shown to give very good results.

Ikonomou and Donohue (1986) used the PACT equation of state and arrived at a different result due to a different mathematical form of their EOS. Their derivation led to the separation of the n_T/n_0 term in the equation of state. The e^g term was shown to be equal to 1.

The effect of the mathematical form of the EOS on the separability of the n_T/n_0 term was further studied by Anderko (1990) who considered a vdW-type EOS in a generalized form:

$$\frac{PV}{n_T RT} = 1 + c_T G\left(\frac{n_T b_T}{V}\right) + \frac{c_T a_T}{RTb_T} H\left(\frac{n_T b_T}{V}\right) \tag{P4.3-23}$$

where c is equal to one-third of the number of external degrees of freedom and accounts for the segmental nature of aggregates. If the mixing rule for c is analogous to that for b [Eqs. (P4.3-1), (P4.3-11), (P4.3-12), (P4.3-16)], i.e.,

$$n_T c_T = n_0 c_0 \tag{P4.3-24}$$

and, additionally,

$$n_T \frac{c_T a_T}{b_T} = n_0 \frac{c_0 a_0}{b_0} \tag{P4.3-25}$$

the term n_T/n_0 can be separated from an equation of state:

$$\frac{Pv}{RT} = \frac{n_T}{n_0} + \left[c_0 G\left(\frac{b_0}{v}\right) + \frac{c_0 a_0}{RTb_0} H\left(\frac{b_0}{v}\right)\right] \tag{P4.3-26}$$

The term in brackets is a generalized van der Waals-type EOS for a mixture of monomeric compounds minus 1.

Anderko (1990) heuristically postulated that the term in brackets on the right-hand side of Eq. (P4.3-26) be replaced by a cubic equation of state because the generalized van der Waals-type EOS [Eq. (P4.3-23)] can be, as a whole, approximated by a cubic EOS. This assumption has led to the equation

$$z = z^{(\mathrm{ph})} + z^{(\mathrm{ch})} - 1 \tag{P4.3-27}$$

where $z^{(\mathrm{ch})} = n_T/n_0 = 1/\chi$ where χ is the mean association number and $z^{(\mathrm{ph})}$ is expressed by the cubic EOS of Yu and Lu (1987):

$$z^{(\mathrm{ph})} = \frac{v}{v-b} - \frac{a(T)v}{RT[v(v+c) + b(3v+c)]} \tag{P4.3-28}$$

The assumptions leading to Eq. (P4.3-26) result in the cancellation of residual chemical potentials for associates undergoing the reaction (P4.3-19)

$$\mu^r_{A_{i+1}} - \mu^r_{A_i} - \mu^r_{A_i} = 0 \tag{P4.3-29}$$

Equivalently, the equation for the association constant (P4.3-20) simplifies to an expression for an ideal associated gas:

$$K = \frac{z_{A_{i+1}}}{z_{A_i} z_{A_i}} \frac{v}{RT} \chi = \frac{z_{A_{i+1}}}{z_{A_i} z_{A_1}} \frac{1}{P^{(\mathrm{ch})}} \tag{P4.3-30}$$

where $P^{(\mathrm{ch})} = RTz^{(\mathrm{ch})}/v$. Equation (P4.3-29) and (P4.3-30) are equivalent to the e^g term [Eq. (P4.3-21)] being equal to unity. Equation (P4.3-30) makes it possible to calculate $z^{(\mathrm{ch})}$ analogously with the compressibility factor of an ideal-gas mixture of associated species. The application of Eq. (P4.3-28) in conjunction with Eq. (P4.3-27) is not consistent with Heidemann–Prausnitz (1976) derivation starting from Eq. (P4.3-17) with Eqs. (P4.3-15) and (P4.3-16) for the a and b parameters, respectively. However, if different combining rules are chosen for the equation-of-state parameters for associated species, Eq. (P4.3-27) can be recovered. Notably, the cubic EOS [Eq. (P4.3-28)] can be rewritten as

$$\frac{PV}{n_T RT} = 1 + \frac{b'_T}{v - b_T} - \frac{a_T v}{RT[v(v + c_T) + b_T(3v + c_T)]} \tag{P4.3-31}$$

where the combining rules for a and b are given by Eqs. (P4.3-11)–(P4.3-14) and the combining rules for c are identical to those for b. Although the parameter b' is identical to b for a pure compound, different combining rules can be used for it. They are intermediate between those for a and b, i.e.,

$$b'_{A_i B} = \frac{i(b_{AA} + b_{BB})}{2} \tag{P4.3-32}$$

$$b'_{A_i A_j} = ijb_{AA} \tag{P4.3-33}$$

When inserted into Eq. (P4.3-1) the above combining rules yield the result:

$$n_T^2 b'_T = n_0^2 b'_0 \tag{P4.3-34}$$

and

$$b'_0 = x_B b_{BB} + x_A b_{AA} = b_0 \tag{P4.3-35}$$

Thus, parameter b'_0 is identical to b_0. After substituting Eqs. (P4.3-34), (P4.3-15) and (P4.3-16) we get

$$z = \frac{n_T}{n_0} + \frac{b_0}{v - b_0} - \frac{a_0 v}{RT[v(v + c_0) + b_0(3v + c_0)]} \tag{P4.3-36}$$

which is equivalent to Eqs. (P4.3-27) and (P4.3-28). Equation (P4.3-36) also satisfies Eq. (P4.3-29). Since b'_0 is equal to b_0 [Eq. (P4.3-35)], the combining rules [(P4.3-32) and (P4.3-33)] do not violate the functional form of the EOS or the mixing and combining rules for the analytical (apparent) components.

The above discussion shows that the form of an equation-of-state incorporating association is strongly dependent on the particular form of combining rules for associated species. The combining rules are only hypothetical because sufficient information about individual associated species is not available. Therefore, the choice of a particular form of the equation of state should be guided by its empirical effectiveness. ■

In the following discussion we shall present an overview of the practical equation-of-state methods incorporating association in an explicit form that can be used for several classes of mixtures.

Hu et al. (1984) extended the method of Heidemann and Prausnitz to mixtures containing one associating component and inert diluents. Their treatment, however, requires numerical solution of chemical equilibria and material balances to obtain the factor n_T/n_0 and numbers of moles of monomers. To express the repulsive and attractive contributions, the authors used the Carnahan–Starling (4.2-40) and van der Waals (4.2-4) terms, respectively. The method was applied to the correlation of Henry's constants [Eq. (P1.2-53)] in aqueous mixtures. The correlation appeared to be only partly successful, which was ascribed to the inability of the Carnahan–Starling–van der Waals model to account for differences of molecular size and shape. Hong and Hu (1989) used the method of Hu et al. (1984) restricted to considering dimerization equilibria for calculating VLE in mixtures.

Ikonomou and Donohue (1986, 1987) used the method of Heidemann and Prausnitz in conjunction with the PACT equation of state (4.2-48). Mixing rules assumed for the associated species are analogous to those used for nonassociating components. The combining rules for the parameters describing the associated species are somewhat similar to those used by Heidemann and Prausnitz:

$$\epsilon_{ij} = \epsilon_{11} \qquad \sigma_{ij}^3 = \sigma_{11}^3 \qquad c_j = jc_1 \qquad r_j = jr_1 \qquad q_j = jq_1 \tag{4.3-143}$$

However, the resulting expression for the compressibility factor (associated perturbed anisotropic chain theory, or APACT) is different from that obtained by Heidemann and Prausnitz in that it contains n_T/n_0 as a separate term. This is a consequence of a different equation of state that incorporates the external degrees of freedom through parameter c.

$$z = \frac{n_T}{n_0} + z_{\text{rep}}^{\text{PACT}} + z_{\text{att}}^{\text{PACT}} \tag{4.3-144}$$

The factor n_T/n_0 for pure associated compounds is similar to that obtained by Heidemann and Prausnitz except for the term $g(\xi)$ [Eq. (4.3-142)] that is equal to 0. For a mixture containing one associating component and any number of inert ones, the factor becomes

$$\frac{n_T}{n_0} = \frac{2x_A}{1 + (1 + 4RTKx_A/v)^{1/2}} + 1 - x_A \tag{4.3-145}$$

where x_A is the mole fraction of the associating component.

Anderko (1989, 1990, 1991a) arrived at the separation of the n_T/n_0 factor by using the techniques described in Example 4.3-1. Moreover, the author showed that the obtained equation of state satisfied the rigorous thermodynamic condition

$$\mu_A = \mu_{A_1} \tag{4.3-146}$$

where A refers to the (nominal) associated component and A_1 to the monomer. In general, Anderko (1989a) expressed the compressibility factor as

$$z = z^{(\mathrm{ch})} + z^{(\mathrm{ph})} - 1 \tag{4.3-147}$$

where $z^{(\mathrm{ch})}$ is a chemical contribution equal to the reciprocal mean association number n_T/n_0 and $z^{(\mathrm{ph})}$ is a physical contribution that is equivalent to an equation of state for the nonreacting monomers. In particular, the cubic equation of Yu and Lu (1987) was used for this purpose. For the Mecke–Kempter continuous linear association model (see Section 3.3-5) Eq. (4.3-145) was derived. This model was employed as a basis for further studies as it seems to be the best compromise between physical reality and mathematical simplicity. In the case of other association models (e.g., that of Kretschmer and Wiebe or monomer–dimer) different expressions for $z^{(\mathrm{ch})}$ were derived.

Although the continuous linear association models are quite successful for a variety of associating compounds, some difficulties have been encountered in their application to aqueous mixtures. This is not unexpected because the structures of water clusters are three-dimensional rather than linear. The existence of three-dimensional structures leads to an appreciable decrease of the n_T/n_0 term, which cannot be accounted for by Eq. (4.3-145). To overcome this difficulty, Anderko (1991) noted that the n_T/n_0 term is always an algebraic function of the product $(RTKx_A/v)$ for any one-constant association model:

$$\frac{n_T}{n_0} = x_A F\left(\frac{RTKx_A}{v}\right) + 1 - x_A \tag{4.3-148}$$

Equation (4.3-148) is valid irrespective of the association model characteristics. If the function F from Eq. (4.3-148) is given by

$$F(q) = \frac{2}{1 + (1 + 4q)^{1/2}} \tag{4.3-149}$$

Eq. (4.3-145) is recovered. For water, Anderko (1991) developed an empirical function:

$$F(q) = \frac{2}{1 + q + \alpha q^2} \tag{4.3-149a}$$

where $\alpha = 8.2$ is a numerical constant.

For a practically oriented thermodynamic model it is of utmost importance that it can be readily extended to systems with any number of components. Until recently, the serious weakness of the "chemical" theory was its inapplicability to systems with more than one associating component. Therefore, it was vital to construct a consistent extension of Eq. (4.3-145) to systems containing any number of associating and inert components. Ikonomou and Donohue (1988) assumed that the cross-association constant K_{ij} between the ith and jth associating compound is equal to $(K_{ii}K_{jj})^{1/2}$ and proposed the following approximation:

$$\frac{n_T}{n_0} = \sum_{i=1}^{m} \frac{2x_i}{1 + (1 + 4RTK_i/v)^{1/2}} + \sum_{i=m+1}^{n} x_i \tag{4.3-150}$$

where m is the number of associating components and n is the total number of components. Equation (4.3-150) is a very accurate approximation to the numerical solution of mass balance and chemical equilibrium expressions for self-association and cross-association with $K_{ij} = (K_{ii}K_{jj})^{1/2}$. Despite its accuracy, Eq. (4.3-150) suffers from some shortcomings. First, it does not readily reduce to the expression for a mixture with one self-associating component [Eq. (4.3-145)] when the number of associating components m is equal to 1. Second, Eq. (4.3-150) was derived on the assumption that $K_{ij} = (K_{ii}K_{jj})^{1/2}$, which restricts the equation to systems containing chemically similar compounds. There are numerous examples of systems in which the cross-association (solvation) of components is much stronger than their self-association (acetone–chloroform, phenols–pyridine bases). For such systems the cross-association constant cannot be approximated by an average of self-association constants. The third weakness of Eq. (4.3-150) is its behavior in the low-density limit. The chemical contribution to the second virial coefficient calculated from Eq. (4.3-150) is $\Sigma\,(-RTK_{ii})x_i$ and is equivalent to the theoretically justified expression $\Sigma\,\Sigma\,(-RTK_{ij})x_i x_j$ if the cross-association constant is taken to be the arithmetic mean $K_{ij} = (K_{ii} + K_{jj})/2$ of self-association constants. This is in contradiction with the starting assumption that $K_{ij} = (K_{ii}K_{jj})^{1/2}$.

Anderko (1989c) proposed a different expression for the chemical contribution to the compressibility factor of a mixture containing n associating components $A^{(1)}, \ldots, A^{(n)}$ and r inert ones $B^{(1)}, \ldots, B^{(r)}$. To extend Eq. (4.3-145) to such mixtures, it has been assumed that (1) only linear multimers of the type $A^{(1)}_{m_1}A^{(2)}_{m_2}, \ldots, A^{(n)}_{m_n}$ occur in the mixture. There is no upper limit on the size of the multimer and there is no restriction to the sequence of monomeric units within a multimer. (2) The multimers are formed in consecutive association reactions. For each of the reactions the equilibrium constant depends on the kind of monomers forming a bond but not on the number of monomers constituting the multimer. The latter assumption is analogous to that underlying the Mecke–Kempter model for self-associating molecules. In this scheme a system is characterized by n self-association constants K_{ii} and $n(n-1)/2$ cross-association (solvation) constants K_{ij} $(i \neq j)$.

The final result of the derivation is

$$z^{(\mathrm{ch})} = \frac{n_T}{n_0} = \sum_{i=1}^{n} \frac{2x_{A^{(i)}}}{1 + \left\{4RT\left[\sum_{j=1}^{n} K_{ij}x_{A^{(j)}}\right]\Big/ v\right\}^{1/2}} + \sum_{k=1}^{r} x_{B^{(k)}} \qquad (4.3\text{-}151)$$

and is called association + equation of state (AEOS). Equation (4.3-151) satisfies three important boundary conditions:

1. If the number of associating components is equal to 1, Eq. (4.3-151) reduces to the expression for a mixture with one self-associating component.
2. When all self- and cross-association constants are equal to each other, the associated species become indistinguishable. Then, the expression for z(ch) reduces to the pure-component case. It should be noted that

this condition is satisfied also by Eq. (4.3-150). The derivation of Eq. (4.3-150) was based, however, on an a priori requirement that the condition should be satisfied.

3. In the low-density limit the correct composition dependence of the chemical contribution to the second virial coefficient is reproduced:

$$B^{(\mathrm{ch})} = \sum \sum x_{A(i)} x_{A(j)} (-RTK_{ij}) \tag{4.3-152}$$

To apply Eqs. (4.3-145) and (4.3-150) or (4.3-151) to mixtures, the pure-component parameters (i.e., enthalpy and entropy of self-association and the parameters characterizing the physical contribution) must be first determined from experimental pure-compound data. The determination of pure-component parameters is the crucial step in the application of an equation of state incorporating association. It is not sufficient to obtain a good fit to pure-component data. It must be also ensured that the relative magnitude of the physical and chemical terms be correct, i.e., the effects of association and nonspecific interactions on the compressibility factor be correctly partitioned. This can be, in principle, accomplished by using physically meaningful values of the association enthalpy and entropy. However, although numerous data on thermodynamic functions of association have been published, they often disagree with each other and cannot be unambiguously used to estimate the equation-of-state parameters.

In both Ikonomou and Donohue's and Anderko's methods vapor pressure and liquid density data have been used to fit the pure-component parameters. Good initial estimates of the parameters are required to arrive at meaningful parameter sets. Anderko (1989a) used the properties of the associating compounds' homomorphs to obtain initial estimates for the parameters characterizing the physical contribution to the compressibility factor. Also, the enthalpies and entropies of association usually lie within narrow intervals for members of homologous series thus facilitating the selection of initial estimates for the fitting procedure. In general, the determination of pure-component parameters from vapor pressures and densities is reliable as long as good initial estimates for the parameters and, more obviously, extensive vapor pressure and density data are available.

Alternatively, the equation-of-state parameters can be evaluated locally for each temperature of interest using vapor pressure and liquid density data generated from another equation of state. This approach is based on the fact that while it is extremely useful to allow for association in mixture calculations (as will be shown below), the pure-component properties such as vapor pressure and liquid volume can be easily reproduced using simpler techniques. Therefore, it is convenient to use an equation of state (hereinafter called the "master" equation) that correctly reproduces the pure-component properties of association fluids without explicitly allowing for association (cf. Section 4.2). The simple algorithm for phase equilibrium calculations is then:

1. For the temperature of interest, calculate the vapor pressure and liquid volume of the pure associating fluid from the master equation.

2. Assume physically realistic values of the enthalpy and entropy of association.
3. Calculate the association constant from the values of the enthalpy and entropy of association. Thus, the chemical contribution to the compressibility factor is completely defined.
4. Use a two-parameter van der Waals-type equation of state for the physical contribution to the compressibility factor.
5. Calculate the two parameters a and b from the vapor pressure and liquid density calculated from the master equation. In this step the complete equation of state incorporating association should be used.
6. Thus, all parameters (i.e., a, b and K) of the complete equation of state are known at the temperature of interest. The pure-component vapor pressure and liquid density are exactly the same as those predicted by the master equation.

The key element in the above algorithm is the use of correct values of the association enthalpy and entropy. Fortunately, these values are similar in homologous series of associating components. Second, very precise values of the association enthalpy and entropy are not needed to obtain a good representation of phase equilibria in mixtures. There is usually a relatively large interval of association constant values that are meaningful for mixture calculations. For example, it is sufficient to assume that the standard enthalpy, entropy and heat capacity of association of alcohols [c.f. Eq. (4.3-132)] are equal to $-26.0\,\mathrm{kJ/mol}$, $-95.0\,\mathrm{J/mol\cdot K}$ and 0, respectively. Only the recommended value for the entropy of association of methanol is different, i.e., it is equal to $-91.0\,\mathrm{J/mol\cdot K}$ (Anderko, 1991b). Similar prescriptions can be developed for other classes of associating compounds. It should be noted that these values are recommended only if the above algorithm is used. If the parameters are determined from extensive data bases for pure components, their values are usually somewhat different.

It is also possible to estimate the parameters without any previous experience for a given class of compounds. For this purpose it is convenient to estimate the enthalpy of association using the homomorph concept according to a simple procedure proposed by Książczak and Anderko (1987, 1988). Then, the association constant at a single temperature can be estimated by calculating vapor–liquid equilibria for a mixture of the compound of interest with an inert component for several values of the association constant and minimizing the deviations from experimental data. The value of the entropy of association follows then from the values of the association enthalpy and association constant at the single temperature. The above algorithm is strictly valid when the associating compound is subcritical as parameters a and b are evaluated from vapor pressure and liquid density at a given temperature. However, the algorithm can be easily extended to supercritical conditions. For supercritical associating fluids parameters a and b should be evaluated from two PVT data points, preferably in the compressed fluid region and near the critical density. In this case, however, the effect of association on phase equilibria is small, and the method of evaluating the parameters is no longer crucial.

When applied to mixtures, the equations allowing for association require only one binary parameter as all specific effects are incorporated into the pure-component equation of state. In the case of the AEOS equation the classical quadratic mixing rules with one adjustable parameter k_{ij} [Eqs. (4.3-1) and (4.3-6)] are used for the physical term. In the case of mixtures with one associating component k_{ij} is the only adjustable binary parameter. With the binary parameter adjusted to experimental data, very good representation of phase equilibria is obtained. Moreover, the equilibria can be fairly well predicted from pure-component parameters alone, i.e., by setting the binary parameter equal to zero. The value of the binary parameter is usually small and rarely exceeds 0.02.

In the case of binaries whose components are able to form cross-associates, the cross-association constant K_{12} appears as the second binary parameter. However, it is sufficient to adjust the cross-association constant while leaving the default value of the nonspecific binary parameter. Therefore, only one binary parameter has to be adjusted for all types of binary systems. Moreover, in the case of systems containing two chemically similar associating compounds, the geometric mean rule for the cross-association constant provides satisfactory prediction of phase equilibria. The quality of correlating individual isothermal or isobaric binary low-pressure VLE data is similar to that obtained from activity coefficient models based on the local composition concept such as NRTL or UNIQUAC (Anderko, 1989a,c). The binary parameter has a simple temperature dependence that makes it possible to correlate phase equilibria over a wide temperature an pressure range with one parameter set. This feature makes AEOS advantageous when applied to the representation of VLE over wide temperature ranges. This is illustrated in Table 4.3-6, which compares the results obtained from the equations of state of Grenzheuser and Gmehling and of Anderko with the results of applying the NRTL model (Renon et al., 1971). It should be noted that Grenzheuser and Gmehling's equation was developed primarily for compounds whose association is restricted to dimerization, although Daumn et al. (1986) found it applicable also to mixtures containing such compounds as H_2O, NH_3, CO_2 and H_2S. The results obtained from the AEOS equation are very good, especially in view of the fact that only two effective mixture parameters (required to determine a linear temperature dependence of the binary parameters k_{12} [Eq. (4.3-6)] have been adjusted to binary data. In the case of high-pressure equilibria for which the activity coefficient methods are not applicable, the quality of correlation is very similar to that obtained in the low-pressure region. AEOS is also a reliable tool for predicting ternary vapor–liquid equilibria from binary data as shown in Table 4.3-7. Very good results are also obtained from the calculation of liquid–liquid and solid–liquid equilibria (Anderko and Malanowski, 1989). This is illustrated in Table 4.3-8, which shows deviations of the simultaneous correlation of different types of phase equilibria using the AEOS equation of state.

Table 4.3-6
REPRESENTATION OF VAPOR–LIQUID EQUILIBRIA OVER WIDE TEMPERATURE RANGES.[a,b]

System	Temperature Range (K)	*Grenzheuser* (*2 param.*)		*AEOS.* (*2 param.*)		*NRTL.* (*5 param.*)	
		δP	Δy	δP	Δy	δP	Δy
Methanol + ethanol	298–433	0.8	1.0	0.4	0.4	1.1	0.9
Methanol + 1-propanol	298–363			1.4	0.4	4.9	4.6
Methanol + propanone	298–473	1.4	1.6	0.9	1.2	1.3	2.3[c]
Methanol + hexane	298–348			1.8	2.5	4.7	2.0
Methanol + cyclohexane	298–328	11.7	5.9	2.4	2.0		
Ethanol + 1-propanol	298–353			0.6	0.7	0.5	0.6
Ethanol + propanone	305–348			0.9	0.8	0.5	0.8

[a]Using the equations of state of Grenzheuser and Gmehling (1986) and AEOS (Anderko, 1989a,c, 1990) and the NRTL activity coefficient model according to Renon et al. (1971).
[b]$\delta P = (100/N) \sum |(P_i^{\text{cal}} - P_i^{\text{exp}})/P_i^{\text{exp}}|$.
$\Delta Q = \sum |Q_i^{\text{cal}} - Q_i^{\text{exp}}|$ where $Q = x$ or y or T.
[c]Temperature range: 280–335.

Table 4.3-7
PREDICTION OF VAPOR–LIQUID EQUILIBRIA IN TERNARY MIXTURES FROM BINARY DATA.[a]

		prediction		*correlation*					
		AEOS		*Wilson*		*NRTL*		*UNIQUAC*	
System	***T* (K)**	δP	Δy	δP	Δy	δP	Δy	δP	Δy
Methanol + propanone + 2-propanol	328	0.6	0.8	1.5	0.6	1.2	0.5	1.3	0.6
Methanol + propanone + cyclohexane	328	1.9	1.6	0.6	0.6	0.8	0.7	0.9	0.7
Ethanol + propanone + hexane	328	1.3	1.1	1.2	0.6	1.3	1.0	1.2	1.0
Ethanol + benzene + hexane	328	1.5	2.1	2.6	1.5	2.1	1.9	2.1	1.9
1-Propanol + benzene + heptane	348	1.5	1.3	2.2	1.4	2.2	1.6	2.1	1.6
Mean 6 systems		1.4	1.3	1.6	0.9	1.5	1.1	1.4	1.1

[a]Using the AEOS equation of state (Anderko, 1989c) compared with correlation of ternary data using the local composition γ–ϕ models.

Table 4.3-8
SIMULTANEOUS CORRELATION OF VAPOR–LIQUID, LIQUID–LIQUID AND SOLID–LIQUID EQUILIBRIA.[a]

System	Type of Equilibrium	Temp. Range	Δx or Δy	ΔT (K)	δP
Phenol + 2-methylbutane	SLE + LLE	286–342	0.012 0.018	0.87	
Phenol + hexane	SLE + LLE	292–325	0.012 0.016	0.59	
Phenol + heptane	SLE + LLE	290–326	0.007 0.018	0.64	
Phenol + decane	SLE + LLE + VLE	291–393	0.028 0.014	0.56	2.5
Methanol + cyclohexane	LLE + VLE	277–316	0.044 0.020		2.4
Methanol + + methylcyclohexane	LLE + VLE	278–333	0.032 0.013		1.4
Methanol + hexane	LLE + VLE	255–348	0.036 0.025		1.8
Methanol + heptane	LLE + VLE	278–371	0.031 0.020	0.28	

After Anderko and Malanowski (1989).
[a]Using the AEOS equation of state.

The binary parameters and their temperature dependence obtained from different phase equilibrium data are the same within the experimental error. This makes it possible to obtain a simultaneous correlation of phase equilibria. In general, the possibility of correlating simultaneously VLE, LLE and solid–liquid equilibrium (SLE) is one of the finest features of the AEOS equation. This correlation can be extended to homologous series of compounds. For example, the binary parameter k_{12} [Eq. (4.3-6)] for systems containing phenol and hydrocarbons is given by

$$k_{12} = -0.094(\omega - 0.335) + 10^{-4}[290(\omega - 0.360)^{1/2} - 4.9](T - 300) \tag{4.3-153}$$

where ω is the acentric factor of the hydrocarbon. The average deviations obtained by calculating phase equilibria with the use of the generalized binary parameters are as follows:

$\Delta x = 0.021$ and $\Delta T = 1.41$ for SLE (8 data sets)

$\Delta x = 0.035$ for LLE (6 data sets)

$\Delta P = 1.8\%$ and $\Delta y = 0.014$ for VLE (3 data sets)

The chemical approach to the construction of accurate equations of state seems to be very attractive. It has a predictive capability that can be considered at several levels of prediction. For example:

1. Estimation of mixture properties from pure-component data alone. This is, in general, an extremely difficult task in the present state of knowledge, and still no satisfactory results can be obtained from first-principle methods even for relatively simple systems (e.g., Moser and Kistenmacher, 1987). However, the AEOS method gives very reasonable estimates of phase equilibria in strongly nonideal systems despite its semiempirical character. The predictive capability of the method is limited to systems containing one polar component or those containing more polar components of the same kind but, nevertheless, it may serve as a tool providing initial estimates of phase equilibria for a relatively wide group of systems.
2. Prediction of properties of a system in a wide temperature range from limited experimental information. Again, the method is very well suited for this purpose because of (a) its ability to represent VLE, LLE and SLE with the same values of the adjustable parameter, (b) a well-established temperature dependence of the binary parameter and (c) a good performance in predicting ternary equilibria from binary data.
3. Prediction of mixture properties using information on the behavior of systems that are similar to the system of interest. This kind of prediction can be made on the basis of the variation found for the binary parameter with an appropriate descriptor (e.g., acentric factor) of a component from a compound family.

From the points of view of prediction levels 2 and 3, it is important that the methods incorporating association are capable of representing mixture properties with only one binary parameter. Therefore, the problems of nonuniqueness of adjustable parameters that are common for empirical correlation methods (e.g., Prausnitz, 1977) are avoided. Subsequently, it is easy to find an unequivocal dependence of the adjustable parameter on temperature and properties of components in series of mixtures.

The idea of separating the compressibility factor into terms due to specific and nonspecific interactions is not necessarily restricted to the chemical approach. Notably, De Santis et al. (1974) decoupled a parameter of the Redlich–Kwong EOS for water into a polar and nonpolar terms, the latter being evaluated from cross second virial coefficient data. Their treatment was limited, however, to the gas phase. Won and Walker (1979) used a similar method to calculate VLE for hydrocarbon mixtures containing traces of polar solutes.

Lee and Chao (1988) developed a more elaborate method for decoupling the equation of state into polar and nonpolar contributions:

$$p = p_{\mathrm{rep}} + p_{\mathrm{att,np}} + p_{\mathrm{att,p}} \tag{4.3-154}$$

where the subscripts np and p stand for the nonpolar and polar contributions, respectively. The nonpolar contribution was evaluated for water from the BACK equation of state with parameters determined from the Lennard–Jones constants describing the nonpolar interactions of water (this is, in principle, similar to the homomorph concept). The polar contribution was obtained by forcing the equation of state (4.3-154) to be identical with the accurate EOS for pure water determined by Keenan et al. (1969). The polar pressure has been further generalized to other fluids by means of a corresponding-states approximation:

$$p_{\mathrm{att,np},i}[T, v] = \left(\frac{T_{c,i}}{T_{c,w}}\right)\left(\frac{v_w^{00}}{v_i^{00}}\right) p_{\mathrm{att},p,w}\left[\left(\frac{T_{c,w}}{T_{c,i}}\right)T, \left(\frac{v_w^{00}}{v_i^{00}}\right)v\right] \tag{4.3-155}$$

where $T_c = \mu^2/(kv^{00}/N_A)$ is a characteristic temperature in the formalism of dipole–dipole interactions, v^{00} is the close-packed volume at 0 K in the BACK EOS and the indices w and i refer to water (as the reference substance) and the fluid of interest, respectively. For mixtures the quadratic mixing rules with two binary parameters were used for the nonpolar contribution whereas the polar pressure was expressed through composition-averaged scaling volumes v^{00}. The equation was shown to represent VLE and LLE in water–hydrocarbon mixtures with two temperature-independent binary parameters having unique values for all phases.

4.4 CLOSING REMARKS

In spite of the great progress made in the field of equations of state, no equation can simultaneously represent all pure-component and mixture properties with satisfactory accuracy. They still have to be tailor-made to solve problems of a limited scope. The user has to consider the limitations of available methods and analyze them from the cost versus benefit point of view. Table 4.4-1 presents a comparison of the groups of methods discussed above with respect to their practical applications. When dealing with equations of state for phase equilibrium calculations, it is important that the equation represent accurately the properties of pure components in the saturation region with a special emphasis on vapor pressure. Second, extension of the equation to mixtures should be straightforward and not restricted to a particular class of mixtures. The large volume of publications on equations of state indicates that the difficulties encountered in solving the above problems are enormous. With respect to the representation of pure-component properties, the advantages and limitations of the currently available types of equations of state are now well established. On the other hand, the potential capabilities of equations of state for mixtures are far from being fully explored. A better representation of pure-component properties does not necessarily ensure an improvement for

Table 4.4-1
COMPARISON OF VARIOUS GROUPS OF METHODS FOR CALCULATING PHASE EQUILIBRIA IN MIXTURES FROM USER'S POINT OF VIEW

Group	Cost	Benefit	Limitations
1. Classical quadratic and related mixing rules.	One binary parameter for nonpolar mixtures. For polar mixtures: 3 parameters + suitable reformulation of the pure-component EOS.	With cubic EOS: very simple calculations, reliable primarily for nonpolar mixtures. With BWR-type and augmented vdW EOS: better reproduction of pure-component properties; accuracy of phase equilibria is similar. With augmented vdW EOS: may be better for asymmetric nonpolar mixtures	Applicable primarily to nonpolar and weakly polar mixtures. With cubic EOS: some pure-component properties must be sacrificed (tailormade EOS). With BWR-type EOS: more computational difficulties, esp. in the vicinity of critical loci. With augmented vdW EOS: more computer time needed
2. Mixing rules in explicit corresponding states format	One binary parameter; computation time depends on the reference fluid EOS.	Can be used for compounds for which parameters of EOS (e.g., cubic) are not available; accuracy can be enhanced by using accurate reference fluid EOS for diverse fluids.	Applicability limited primarily to nonpolar and weakly polar mixtures; extensions to polar mixtures are of limited usefulness due to limitations of quadratic mixing rules for pseudocritical parameters.

3. Mixing rules from G^E models.	Two to three binary parameters needed. If a G^E model is used at $P \rightarrow \infty$, the parameters have to be reevaluated. The use of parameters from $\gamma-\phi$ correlations requires a rather complicated algorithm.	Flexibility of activity coefficient models is transferred to EOS methods. Applicability of low-pressure group contribution methods can be extended to high pressures.	Incorrect composition dependence of virial coefficients; limitations of G^E models are retained (e.g., nonuniqueness of adjustable parameter sets); incorrect phase splitting is frequently predicted.
4. Simple composition-dependent combining rules.	Two to three temperature-dependent binary parameters needed.	As flexible as G^E models, very easy to use.	The same as for group 3; may be less accurate for very strongly nonideal mixtures.
5. Density dependent mixing rules.	Two to three temperature dependent binary parameters needed; the mixing rules complicate the volume dependence of EOS (e.g., cubic form can not be retained).	Thermodynamically consistent representation of low- and high-density regions.	Quality of correlating phase equilibria is not improved upon groups 3 and 4.
6. Explicit treatment of association	Only one binary parameter for strongly nonideal mixtures; additional complexity introduced into the pure component EOS.	Accurate simultaneous reproduction of VLE and LLE over wide temp. and pressure ranges; small amount of binary input data required to predict phase behavior also within homologous series.	The maximum accuracy limit is set by the accuracy of the nonpolar fluid mixing rules used for the physical contribution. Only for methods with numerical solution of chemical and physical equilibria: difficult extension to systems with any number of associating compounds.

mixtures and, what is less obvious, a better theoretical background of a semiempirical EOS does not always lead to better results. At present, almost everybody is convinced that future progress will come from the developments of statistical thermodynamics, but almost all practical calculations are performed with empirical or at most semitheoretical models. From the practical

Table 4.2-2
SELECTED PAPERS DEALING WITH THE COMPUTATIONAL ASPECTS OF MODELLING PHASE EQUILIBRIA USING EQUATIONS OF STATE

Heidemann (1983)	Review of numerical methods for phase equilibrium calculations
Anderson and Prausnitz (1980)	Simple algorithms for bubble- and dew-point and flash calculations
Topliss et al. (1988)	A general procedure for solving noncubic EOS for density
Assalineau et al. (1980)	Bubble- and dew-point calculations
Michelsen (1980)	Calculation of critical points and phase envelopes
Michelsen (1982)	Flash calculations including stability analysis
Gundersen (1982)	Flash calculations using cubic EOS
Ziervogel et al. (1983)	Phase envelope calculations
Michelsen (1985)	Bubble- and dew-point calculations
Michelsen (1985)	Bubble- and dew-point calculations
Nghiem and Li (1984)	Computation of multiphase equilibria
Lu et al. (1974)	Three-phase equilibrium computations
Heidemann (1974)	Three-phase equilibrium computations
Heidemann and Khalil (1980)	Calculation of critical points
Michelsen (1984)	Calculation of critical points
Michelsen and Heidemann (1981)	Critical points from cubic EOS
Michelsen and Heidemann (1988)	Calculation of tri-critical points
Mathias et al. (1984)	Computation strategy in the absence of appropriate density roots
Poling et al. (1981)	Computation strategy in the absence of appropriate density roots
Deiters (1985)	Computation of phase equilibria using numerical differentiation of the Gibbs energy
Campbell (1988)	Initial estimates of vapor pressures
Paunović et al. (1981)	Objective functions for parameter optimization
Aim and Boublik (1986)	Maximum likelihood methods applied to EOS calculations

point of view, the picture is much clearer for mixtures containing nonpolar and weakly polar components for which simple mixing rules are recommended. For strongly nonideal mixtures a multitude of methods have been proposed that follow roughly two trends of research.

The more popular one consists in leaving the pure-component EOS unchanged and introducing additional flexibility into the mixing rules. These methods improve the performance of equations of state at the cost of introducing further binary parameters. While this is fully justified when the objective of the method is to correlate individual data, one cannot expect much predictive capability of the method. It is possible, however, to construct a predictive method by using the group contribution technique.

The second trend of research consists in reformulating the pure-component equation of state in order to obtain a better representation of phase equilibria in mixtures. This approach necessitates a clear separation of the effects of various types of intermolecular interactions on the properties of mixtures. As shown above, a semiempirical treatment of association and polarity can be very accurate and predictive.

When applied to phase equilibria, equations of state require nontrivial numerical methods to find the saturation points, phase envelopes, critical points etc. It is beyond the scope of this work to discuss the various numerical methods relevant to phase equilibrium computations. Nevertheless, Table 4.4-2 is provided to refer the reader to some useful sources of information about the computational problems associated with the use of equations of state.

As the main thrust of this review was to evaluate various equation-of-state methods for the modelling of experimentally measured phase equilibria, only empirical or semiempirical methods have been dealt with. Therefore, perturbation and related theories, although very promising, have not been mentioned here. The equations based on lattice-gas and related models have been also considered to be beyond the scope of this chapter. Although the "first-principle" methods are still unable to serve as a basis for the representation of experimental phase equilibrium data, several molecularly inspired concepts have been shown to be very useful in constructing semiempirical models. It can be hoped that future progress in the field of equations of state will arise from the synergistic interactions between molecularly based concepts and requirements of the representation of empirical data.

REFERENCES

Abbott, M. M., 1973. *AIChE. J.* 19: 596.
Abbott, M. M., 1979. *Adv. Chem. Ser.* 182: 47–70.
Abbott, M. M. and Prausnitz, J. M., 1987. *Fluid Phase Equilibria* 37: 29–62.
Abdoul, W., 1987. Une méthode de contributions de groupes applicable à la correlation et la prediction des propriétés thermodynamiques des fluides petroliers, Ph.D. Thesis, Marseille.
Abrams, D. S. and Prausnitz, J. M., 1975. *AIChE J.* 21: 116–128.

Adachi, Y. and Sugie, H., 1985. *Fluid Phase Equilibria* 24: 353–362.

Adachi, Y. and Sugie, H., 1986. *Fluid Phase Equilibria* 28: 103–118.

Adachi, Y., Lu, B. C.-Y. and Sugie, H., 1983a. *Fluid Phase Equilibria* 13: 133–142.

Adachi, Y., Lu, B. C.-Y. and Sugie, H., 1983b. *Fluid Phase Equilibria* 11: 29–48.

Adachi, Y., Sugie, H. and Lu, B. C.-Y., 1986. *Fluid Phase Equilibria* 28: 119–136.

Adachi, Y., Sugie, H., Nakanishi, K. and Lu, B. C.-Y., 1989. *Fluid Phase Equilibria* 52: 83–90.

Aim, K. V. and Boublik, T., 1986. *Fluid Phase Equilibria* 29: 583.

Albright, P. C., Edwards, T. J., Chen, Z. Y. and Sengers, J. V., 1987. *J. Chem. Phys.* 87: 1717.

Alder, B. J., Young, D. A. and Mark, M. A., 1972. *J. Chem. Phys.* 56: 3013.

Anderko, A., 1989a. *Fluid Phase Equilibria* 45: 39–67.

Anderko, A., 1989b. *Chem. Eng. Sci.* 44: 713–725.

Anderko, A., 1989c. *Fluid Phase Equilibria* 50: 21–52.

Anderko, A., 1990. *J. Chem. Soc., Faraday Trans.* 86: 2823.

Anderko, A., 1991. *Fluid Phase Equilibria*, 65: 89.

Anderko, A., 1992. *Fluid Phase Equilibria*, in press.

Anderko, A. and Malanowski, S., 1989. *Fluid Phase Equilibria* 48: 223–241.

Anderson, T. F. and Prausnitz, J. M., 1980. *Ind. Eng. Chem. Proc. Des. Dev.* 19: 1–8, 19: 9–14.

Androulakis, I. P., Kalospiros, N. S. and Tassios, D. P., 1989. *Fluid Phase Equilibria* 45: 135–163.

Arai, K., Inomata, H. and Saito, S., 1982. *J. Chem. Eng. Japan* 4: 1–5.

Assalineau, L., Bogdanic, G. and Vidal, J., 1980. *Fluid Phase Equilibria* 3: 273.

Barker, J. A., 1952. *J. Chem. Phys.* 20: 1526.

Barker, J. A. and Henderson, D., 1972. *Adv. Phys. Chem.* 23: 233–244.

Beattie, J. A. and Bridgeman, O. C.., 1927. *J. Am. Chem. Soc.* 63: 1665.

Bender, E., 1975. *Cryogenics* 15: 667–673.

Benedict, M., Webb, G. R. and Rubin, L. C., 1940. *J. Chem. Phys.* 8: 334–345.

Benedict, M., Webb, G. R. and Rubin, L. C., 1942. *J. Chem. Phys.* 10: 747–758.

Benmekki, E. H. and Mansoori, G. A., 1987. *Fluid Phase Equilibria* 32: 139–149.

Benmekki, E.-H. and Mansoori, G. A., 1988. *Fluid Phase Equilibria* 41: 43–57.

Beret, S. and Prausnitz, J. M., 1975. *AIChE J.* 21: 1123.

Berthelot, D. J., 1899. *J. Phys.* 8: 263.

Bienkowski, P. R. and Chao, K. C., 1975. *J. Chem. Phys.* 62: 615–619.

Bienkowski, P. R., Denenholz, H. S. and Chao, K. C., 1973. *AIChE J.* 19: 167–173.

Bondi, A., 1968. *Physical Properties of Molecular Crystals Liquids and Glasses*, Wiley, New York.

Boublík, T., 1970. *J. Chem. Phys.* 53: 471–472.
Boublík, T. and Nezbeda, I., 1977. *Chem. Phys. Lett.* 46: 315.
Boyle, R., 1660. *New Experiments Physico-Mechanical. Touching the Spring of the Air* (cf. Partington, J. R., *A History of Chemistry*, St Martin's Press, New York, 1961–1970).
Brennecke, J. F. and Eckert, C. A., 1989. *AIChE J.* 35: 1409–1427.
Brulé, M. R. and Starling, K. E., 1984. *Ind. Eng. Chem. Proc. Des. Dev.* 23: 833–845.
Brulé, M. R., Lin. C. T., Lee, L. L. and Starling, K. E., 1982. *AIChE J.* 28: 616–625.
Bryan, P. F. and Prausnitz, J. M., 1987. *Fluid Phase Equilibria* 38: 201–216.
Campbell, S. W., 1988. *Ind. Eng. Chem. Res.* 27: 1333.
Carnahan, N. F. and Starling, K. E., 1969. *J. Chem. Phys.* 51: 635–636.
Carnahan, N. F. and Starling, K. E., 1972. *AIChE J.* 18: 1184–1189.
Chang, S. D. and Lu, B. C.-Y., 1970. *Can. J. Chem. Eng.* 48: 261.
Chao, K. C. and Lin, H. M., 1985. Presented at the 2nd Int. Symp. on Critical Evaluation and Prediction of Phase Equilibria in Multicomponent Systems, Paris, France.
Chapela, G. A. and Rowlinson, J. S., 1974. *Trans. Faraday Soc.* 70: 584.
Chapela-Castañares, G. A. and Leland, T. W., 1974. *AIChE Symp. Ser.* 70: 48–55.
Chen, S. S. and Kreglewski, A., 1977. *Ber. Bunsenges. Phys. Chem.* 81: 1048–1052.
Chien, C. H., Greenkorn, R. A. and Chao, K. C., 1983. *AIChE J.* 29: 560–571.
Chorn, L. G. and Mansoori, G. A. (Eds.), 1989. *C_{7+} Fraction Characterization, Adv. Thermodynamics*, Vol. 1., Taylor & Francis.
Chou, G. F. and Prausnitz, J. M., 1989. *AIChE J.* 35: 1487–1496.
Chung, T. H., Khan, M. M., Lee, L. L. and Starling, K. E., 1984. *Fluid Phase Equilibria* 17: 351.
Clausius, R., 1857. *Ann. Phys.* 100: 353.
Clausius, R., 1881. *Ann. Phys. Chem.* 9: 337.
Cooper, H. W. and Goldfrank, J. C., 1967. *Hydrocarbon Process* 46: 141–146.
Cotteman, R. L. and Prausnitz, J. M., 1986. *AIChE J.* 32: 1799–1812.
Cotterman, R. L., Schwarz, B. J. and Prausnitz, J. M., 1986. *AIChE J.* 32: 1787–1798.
Cox, K. W., Bono, J. L., Kwok, Y. C. and Starling, K. E., 1971. *Ind. Eng. Chem. Fundam.* 10: 245–250.
Czerwienski, G. J., Tomasula, P. and Tassios, D., 1988. *Fluid Phase Equilibria* 42: 63–83.
Danner, R. P. and Gupte, P. A., 1986. *Fluid Phase Equilibria* 29: 415–430.
Daumn, K. J., Harrison, B. K., Manley, D. B. and Poling, B. E., 1986. *Fluid Phase Equilibria* 30: 197–212.
Deiters, U. K., 1981a. *Chem. Eng. Sci.* 36: 1139–1146;
Deiters, U. K., 1981b. *Chem. Eng. Sci.* 36: 1147–1151.
Deiters, U. K., 1982. *Chem. Eng. Sci.* 37: 855–861.
Deiters, U. K., 1985. *Fluid Phase Equilibria* 19: 287.

Deiters, U. K., 1987. *Fluid Phase Equilibria* 33: 267–293.
De Santis, R., Breedveld, G. J. F. and Prausnitz, J. M., 1974. *Ind. Eng. Chem., Proc. Des. Dev.* 13: 374–377.
Dimitrelis, D. and Prausnitz, J. M., 1986. *Fluid Phase Equilibria* 31: 1–21.
Dolezalek, F., 1908. *Z. Phys. Chem.* 64: 727.
Donohue, M. D. and Prausnitz, J. M., 1978. *AIChE J.* 24: 849.
Edmister, W. C., Vairogs, J. and Klekers, A. J., 1968. *AIChE J.* 14: 479.
Erickson, D. D. and Leland, T. W., 1986. *Int. J. Thermophys.* 7: 911–922.
Erickson, D. F., Leland, T. W. and Ely, J. F., 1987. *Fluid Phase Equilibria* 37: 185–205.
Fischer, J. and Lustig, R., 1985. *Fluid Phase Equilibria* 22: 245–251.
Fox, J. R., 1983. *Fluid Phase Equilibria* 14: 45–53.
Fuller, G. G., 1976. *Ind. Eng. Chem. Fundam.* 15: 254–257.
Gallagher, J. S., and Haar, L., 1978. *J. Phys. Chem. Ref. Data*, 7: 635–677.
Gani, R., Tzouvaras, N., Rasmussen, P. and Fredenslund, A., 1989. *Fluid Phase Equilibria* 47: 133–152.
Georgeton, G. K. and Teja, A. S., 1988. *Ind. Eng. Chem. Res.* 27: 657–664.
Georgeton, G. K., Smith, R. L., Jr. and Teja, A. S., 1986. *ACS Symp. Ser.* 300: 434–451.
Gmehling, J., Liu, D. D. and Prausnitz, J. M., 1979. *Chem. Eng. Sci.* 34: 951.
Goodwin, R. D., 1974. National Bureau of Standards, Technical Note 653.
Graboski, M. S. and Daubert, T. E., 1978. *Ind. Eng. Chem.Proc. Des. Dev.* 17: 443–448.
Graboski, M. S. and Daubert, T. E., 1979. *Ind. Eng. Chem. Proc. Des. Dev.* 18: 300–306.
Gray, R. D., Jr., Heidman, J. L., Hwang, S. C. and Tsonopoulos, C., 1983. *Fluid Phase Equilibria* 13: 59–76.
Gregorowicz, J., 1990. Jednolity opis termodynamiczny objętości faz i równowagi ciecz-para w układach węglowodorów cyklicznych (in Polish), Ph.D. Thesis, Institute of Physical Chemistry, Warszawa.
Grenzheuser, P. and Gmehling, J., 1986. *Fluid Phase Equilibria* 25: 1–29.
Grigull, U., Straub, J. and Schiebener, P., 1984. *Steam Tables in SI Units*. Springer-Verlag, New York.
Gubbins, K. E. and Twu, C. H., 1978. *Chem. Eng. Sci.* 33: 3863.
Guggenheim, E. A., 1952. *Mixtures*, Clarendon Press, Oxford, UK.
Guggenheim, E. A., 1965. *Mol. Phys.* 9: 199–200.
Gundersen, T., 1982. *Comput. Chem. Eng.* 3: 245.
Gunning, A. J. and Rowlinson, J. S., 1973. *Chem. Eng. Sci.* 28: 521–527.
Guo, T. M., Kim, H., Lin, H. M. and Chao, K.-C. 1985a. *Ind. Eng. Chem. Proc. Des. Dev.*, 24: 764–767.
Guo, T. M., Lin, H. M. and Chao, K.-C., 1985b. *Ind. Eng. Chem. Proc. Des. Dev.*, 24: 768–773.
Gupte, P. A., Rasmussen, P. and Fredenslund, A., 1986. *Ind. Eng. Chem. Fundam.*, 25: 636–645.
Halm, R. L. and Stiel, L. I., 1967. *AIChE J.* 13: 351–355.
Hamam, S. E. M., Chang, W. K., Elshayal, I. M. and Lu, B. C. Y., 1977. *Ind. Eng. Chem. Proc. Des. Dev.*, 16: 51–59.

Han, S. J., Lin, M. M. and Chao, K. C., 198. *Chem. Eng. Sci.* 43: 2327–2367.
Harmens, A., 1977. *Cryogenics* 17: 519–522.
Harmens, A. and Knapp, H., 1980. *Ind. Eng. Chem. Fundam.* 19: 291–294.
Heidemann, R. A., 1974. *AIChE J.* 20: 847.
Heidemann, R. A., 1983. *Fluid Phase Equilibria*, 14: 55.
Heidemann, R. A. and Khalil, A. M., 1980. *AIChE J.* 26: 769.
Heidemann, R. A. and Kokal, S. L., 1990. *Fluid Phase Equilibria* 56: 17–37.
Heidemann, R. A. and Prausnitz, J. M., 1976. *Proc. Nat. Acad. Sci.* 73: 1773.
Hemmer, P. C., Kac, M. and Uhlenbeck, G. E., 1964. *J. Math. Phys.* 5: 60–74.
Henderson, D., 1974. *J. Chem. Phys.* 61: 926.
Henderson, D., 1979. *ACS Symp. Ser.* 182: 1–30.
Heyen, G., 1980. 2nd. Int. Conf. Phase Eq. Fluid Prop. Chem. Ind., West Berlin.
Heyen, G., 1981. 2nd. World Congress on Chemical Engineering, Montreal.
Hildebrand, J. H., Prausnitz, J. M. and Scott, R. L., 1970. *Regular and Related Solutions*, Van Nostrand, Reinhold, New York.
Hirn, J., 1863. *Cosmos (Paris)* 22: 283, 413.
Hong, J. and Hu, Y., 1989. *Fluid Phase Equilibria* 51: 37–52.
Horvath, A. L., 1974. *Chem. Eng. Sci.* 29: 1334.
Hu, Y., Azevedo, E., Luedecke, D. and Prausnitz, J., 1984. *Fluid Phase Equilibria*, 17: 303–321.
Huron, M.-J. and Vidal, J., 1979. *Fluid Phase Equilibria*, 3: 255–271.
Ikonomou, G. D. and Donohue, M. D., 1986. *AIChE J.* 32: 1716–1725.
Ikonomou, G. D. and Donohue, M. D., 1987. *Fluid Phase Equilibria* 33: 61–90.
Ikonomou, G. D. and Donohue, M. D., 1988. *Fluid Phase Equilibria* 39: 129–159.
Ishikawa, T., Chung, W. K. and Lu, B. C.-Y., 1980. *AIChE J.* 26: 372–378.
IUPAC Thermodynamic Tables of Fluid State, Vol. 1, *Argon* (1972), Vol. 2, *Ethylene* (1974), Vol. 3, *Carbon Dioxide* (1976), Vol. 4, *Helium* (1977), Vol. 5, *Methane* (1978), Vol. 6, *Nitrogen* (1979), Vol. 7, *Propylene* (1980), Vol. 8, *Chlorine* (1985), Vol. 9, *Oxygen* (1987), Vol. 10, *Ethylene* (1988), IUPAC Thermodynamic Tables Project Centre, Imperial College, London.
Iwai, Y., Margerum, M. R. and Lu, B. C.-Y., 1988. *Fluid Phase Equilibria* 42: 21–41.
Jain, R. K. and Simha, R., 1979. *J. Chem. Phys.* 70: 2792; 70: 5329.
Jin, G., Walsh, J. M. and Donohue, M. D., 1986. *Fluid Phase Equilibria* 31: 123–146.
Joffe, J., 1949. *Chem. Eng. Prog.* 45: 160–166.
Joffe, J., 1971. *Ind. Eng. Chem. Fundam.* 10: 532–533.
Joffe, J., 1981. *Ind. Eng. Chem. Proc. Des. Dev.* 20: 168–172.
Joffe, J., Schroeder, G. M. and Zudkevitch, D., 1970. *AIChE J.* 16: 496–498.
Johnson, J. K. and Rowley, R. L., 1989. *Fluid Phase Equilibria* 44: 255–272.
Kabadi, V. N. and Danner, R. P., 1985. *Ind. Eng. Chem. Proc. Des. Dev.* 24: 537–541.

Keenan, J. H., Keyes, F., Hill, P. and Moore, J., 1969. *Steam Tables: Thermodynamic Properties of Water*, Wiley, New York.
Kim, C.-H., Vimalchand, P., Donohue, M. D. and Sandler, S. I., 1986. *AIChE J.* 32: 1726–1733.
Kim, H., Wang, W. C., Lin, H. M. and Chao, K. C., 1986. *Ind. Eng. Chem. Fundam.* 25: 75–84.
Kolasinska, G. 1986. *Fluid Phase Equilibria* 27: 289–308.
Kolasinska, G., Moorwood, R. A. S. and Wenzel, H., 1983. *Fluid Phase Equilibria* 13: 121–132.
Kreglewski, A. and Chen, S. S., 1978. *J. Chim. Phys.* 75: 347–352.
Książczak, A. and Anderko, A., 1987. *Ber. Bunsenges. Phys. Chem.* 91: 1048.
Książczak, A. and Anderko, A., 1988. *Ber. Bunsenges. Phys. Chem.* 92: 496.
Kubic, W. L., 1982. *Fluid Phase Equilibria* 9: 79–97.
Kubic, W. L., 1986. *Fluid Phase Equilibria* 31: 35–56.
Kumar, K. H. and Starling, K. E., 1982. *Ind. Eng. Chem. Fundam.* 21: 255–262.
Kwak, T. Y. and Mansoori, G. A., 1986. *Chem. Eng. Sci.* 41: 1303–1309.
Kurihara, K., Tochigi, K. and Kojima, K., 1987. *J. Chem. Eng. Japan*, 20: 227–231.
Lan, S. S. and Mansoori, G. A., 1977. *Int. J. Eng. Sci.* 15: 323–341.
Lang, E. and Wenzel, H., 1989. *Fluid Phase Equilibria* 51: 101–118.
Leach, J. W., Chappelear, P. S. and Leland, T. W., 1968. *AIChE J.* 14: 568–576.
Lee, M. J. and Chao, K. C., 1988. *AIChE J.* 34: 825–833.
Lee, R. J. and Chao, K. C., 1986. *Fluid Phase Equilibria* 29: 475–484.
Lee, B. I. and Kesler, M. G., 1975. *AIChE J.* 21: 510–527.
Lee, K. H. and Sandler, S. I., 1987. *Fluid Phase Equilibria* 34: 113–147.
Lee, K. H., Dodd, L. and Sandler, S. I., 1989. *Fluid Phase Equilibria* 50: 53–77.
Lee, K. H., Lombardo, M. and Sandler, S. I., 1985. *Fluid Phase Equilibria* 21: 177–196.
Lee, K. H., Sandler, S. I. and Patel, N. C., 1986. *Fluid Phase Equilibria* 25: 31–49.
Lee, L. L., Chung, T. H. and Starling, K. E., 1983. *Fluid Phase Equilibria* 12: 105–124.
Lee, T. J., Lee, L. L. and Starling, K. E., 1979. *ACS Symp. Ser.* 182: 125–141.
Leland, T. W. and Fisher, G. D., 1970. *Ind. Eng. Chem. Fundam.* 9: 537–544.
Leland, T. W. and Chappelear, P. S., 1968a. *Ind. Eng. Chem.* 60: 15–24.
Leland, T. W., Rowlinson, J. S. and Sather, G. A., 1968b. *Trans. Faraday Soc.* 64: 1447–1460.
Leland, T. W., Rowlinson, J. S., Sather, G. A. and Watson, I. D., 1969. *Trans. Faraday Soc.* 65: 2034–2043.
Lencka, M. and Anderko, A., 1991. *Chem. Eng. Commun.*, 107: 173–188.
Levelt-Sengers, J. M. H., Kamagar Parsi, B. and Sengers, J. V., 1983a. *J. Chem. Eng. Data* 28: 354.
Levelt-Sengers, J. M. H., Morrison, G. and Chang, R. F., 1983b. *Fluid Phase Equilibria* 14: 19–44.

Li, M. H., Chung, F. T. H., Lee, L. L. and Starling, K.E., 1985. *Fluid Phase Equilibria* 24: 221–240.
Luedecke, D. and Prausnitz, J. M., 1985. *Fluid Phase Equilibria* 22: 1–19.
Machát, V. and Boublík, T., 1985. *Fluid Phase Equilibria* 21: 1–9, 21: 11–24.
Mansoori, G. A., Carnahan, N. F., Starling, K. E. and Leland, T. W., 1971. *J. Chem. Phys.* 54: 1523–1525.
Margerum, M. R. and Lu, B. C. Y., 1990. *Fluid Phase Equilibria* 56: 105–118.
Martin, J. J., 1967. *Ind. Eng. Chem.* 59: 34.
Martin, J. J., 1979. *Ind. Eng. Chem. Fundam.* 18: 81–97.
Masuoka, H. and Chao, K. C., 1984. *Ind. Eng. Chem. Fundam.* 23: 24–29.
Mathias, P. M. and Copeman, T. W., 1983. *Fluid Phase Equilibria* 13: 91–108.
Mathias, P. M., Boston, J. F. and Watanasiri, S., 1984. *AIChE J.* 30: 182.
Mathias, P. M., Naheiri, T. and Oh, E. M., 1989. *Fluid Phase Equilibria* 47: 77–87.
Maurer, G. and Prausnitz, J. M., 1978. *Fluid Phase Equilibria* 2: 91–99.
Maxwell, J. C., 1874. *Nature* 44: 499, 597.
Melhem, G. A., Saini, R. and Goodwin, B. M., 1989. *Fluid Phase Equilibria* 47: 189–237.
Mentzer, R. A., Greenkorn, R. A. and Chao, K. C., 1980. *Sep. Sci. Tech.* 15: 1613–1678.
Michel, S., Hooper, H. H. and Prausnitz, J. M., 1989. *Fluid Phase Equilibria* 45: 173–189.
Michelsen, M. L., 1980. *Fluid Phase Equilibria* 4: 1.
Michelsen, M. L., 1982. *Fluid Phase Equilibria* 9: 1, 9: 21.
Michelsen, M. L., 1984. *Fluid Phase Equilibria* 16: 57.
Michelsen, M. L., 1985. *Fluid Phase Equilibria* 23: 181.
Michelsen, M. L., 1989. Report SEP 8919, Instituttet for Kemiteknik, DTH, Lyngby, Denmark.
Michelsen, M. L. and Heidemann, R. A., 1981. *AIChE J.* 27: 521.
Michelsen, M. L. and Heidemann, R. A., 1988. *Fluid Phase Equilibria* 39: 53.
Michelsen, M. L. and Kistenmacher, H., 1990. *Fluid Phase Equilibria* 58: 229.
Mohamed, R. S. and Holder, G. D., 1987. *Fluid Phase Equilibria* 32: 295–317; Errata, 1988. *Fluid Phase Equilibria* 43: 359–360.
Mollerup, J., 1975. *Adv. Cryog. Eng.* 20: 172–194.
Mollerup, J., 1977. *Ber. Bunsenges. Phys. Chem.* 81: 1015–1020.
Mollerup, J., 1978. *Adv. Cryog. Eng.* 23: 550–560.
Mollerup, J., 1980. *Fluid Phase Equilibria* 4: 11–34.
Mollerup, J., 1981. *Fluid Phase Equilibria* 7: 121–138.
Mollerup, J., 1983. *Fluid Phase Equilibria* 15: 189–207.
Mollerup, J., 1985. *Fluid Phase Equilibria* 22: 139–154.
Mollerup, J., 1986. *Fluid Phase Equilibria* 25: 323–327.
Mollerup, J. and Fredenslund, A., 1976. *Inst. Chem. Eng. Symp. Ser.* 44: 18–28.
Mollerup, J. and Rowlinson, J. S., 1974. *Chem. Eng. Sci.* 29: 1373–1381.
Morris, W. O., Vimalchand, P. and Donohue, M. D., 1987. *Fluid Phase Equilibria* 32: 103–115.
Moser, B. and Kistenmacher, H., 1987. *Fluid Phase Equilibria* 34: 189–201.

Naumann, K.-H., Chen, Y. P. and Leland, T. W., 1981. *Ber. Bunsenges. Phys. Chem.* 85: 1029–1033.
Nghiem, L. X. and Li, Y.-K., 1984. *Fluid Phase Equilibria* 17: 77.
Nies, E., Kleintjens, L. A., Koningsveld, R., Simha, R. and Jain, R. K., 1983. *Fluid Phase Equilibria* 12: 11.
Nishiumi, H., 1980. *J. Chem. Eng. Japan* 13: 178–183.
Nishiumi, H. and Saito, S., 1975. *J. Chem. Eng. Japan* 8: 356–360.
Oellrich, L., Plöcker, U., Prausnitz, J. M. and Knapp, H., 1981. *Int. Chem. Eng.* 21: 1–16.
Opfell, J. B., Sage, B. H. and Pitzer, K. S., 1956. *Ind. Eng. Chem.* 48: 2069–2076.
Orbey, H. and Vera, J. H., 1986. *Chem. Eng. Commun.* 44: 95–106.
Orbey, H. and Vera, J. H., 1989. *Pure Appl. Chem.* 61: 1413–1418.
Orwoll, R. A. and Flory, P. J., 1967. *J. Am. Chem. Soc.* 89: 6814, 89: 6822.
Panagiotopoulos, A. Z., 1986. High pressure phase equilibria: experimental and Monte Carlo simulation studies, Ph.D. Thesis, Massachusetts Institute of Technology, Cambridge, MA.
Panagiotopoulos, A. Z. and Reid, R. C., 1986. In, K. C. Chao and R. L. Robinson (Eds.), *Equations of State—Theories and Applications*, ACS Symp. Ser., 300: 571–582.
Panayiotou, C. and Vera, J. H., 1981. *Can. J. Chem. Eng.* 59: 501.
Panayiotou, C. and Vera, J. H., 1982. *Polymer J.* 14: 681.
Pandit, A. and Singh, R. P., 1987. *Fluid Phase Equilibria* 33: 1–12.
Partington, J. A., 1949. *An Advanced Treatise on Physical Chemistry*, Vol. I: *Fundamental Principles. The Properties of Gases*, Longmans, London.
Patel, N. C. and Teja, A. S., 1982. *Chem. Eng. Sci.* 37: 463–473.
Paunović, R., Jovanović, S. and Mihajlov, A., 1981. *Fluid Phase Equilibria* 6: 141.
Pedersen, K. S., Thomassen, P. and Fredenslund, A., 1989. *The Properties of Oils and Natural Gases*, Gulf Publishing Inc., Houston.
Péneloux, A., Rauzy, E. and Fréze, R, 1982, *Fluid Phase Equilibria* 8: 7–23.
Péneloux, A., Abdoul, W. and Rauzy, E., 1989. *Fluid Phase Equilibria* 47: 115–132.
Peng, D. Y. and Robinson, D. B., 1976. *Ind. Eng. Chem. Fundam.* 15: 59–64.
Peng, D. Y. and Robinson, D. B., 1980. *ACS Symp. Ser.* 133: 393–414.
Peschel, W., 1986. Die Berechnung von Phasengleichgewichten und Mischungswärmen mit Hilfe einer Zustandsgleichung in Verbindung mit der Chemischen Theorie: Methanolsysteme, Ph. D. Thesis, Erlangen.
Peters, C. J., de Swaan Arons, J., Levelt-Sengers, J. M. H. and Gallagher, J. S., 1988. *AIChE J.* 34: 834–839.
Pitzer, K. S., Lippman, D. Z., Curl, R. F., Huggins, C. M. and Peterson, D. E., 1955. *J. Am. Chem. Soc.* 77: 3433–3440.
Pitzer, K. S., and Schreiber, D. R., 1988. *Fluid Phase Equilibria* 41: 1–17.
Platzer, B. and Maurer, G., 1989. *Fluid Phase Equilibria* 51: 223–236.
Plöcker, U., Knapp, H. and Prausnitz, J. M., 1978. *Ind. Eng. Chem. Proc. Des. Dev.* 17: 324–332.

Poling, B. E., Grens, E. A. and Prausnitz, J. M., 1981. *Ind. Eng. Chem. Proc. Des. Dev.* 20: 127.
Prausnitz, J. M., 1977. In, Storvick, T. S. and Sandler, S. I., *Phase Equilibria and Fluid Properties in the Chemical Industry, Estimation and Correlation*, ACS Symp. Ser. 60, Washington, D.C.
Prigogine, I., 1957. *The Molecular Theory of Solutiions*, North Holland, Amsterdam.
Prigogine, I., Bellemans, A. and Naar-Colin, C., 1957. *J. Chem. Phys.* 26: 751–755.
Redlich, O. and Kister, A. T., 1948. *Ind. Eng. Chem.* 40: 345–348.
Redlich, O. and Kwong, J. N. S., 1949. *Chem. Rev.* 44: 233–244.
Renon, H., Assalineau, L., Cohen, G. and Raimbault, C., 1971. *Calcul sur ordinateur des équilibres liquide-vapeur et liquide-liquide*, Technip, Paris.
Robinson, D. B., Peng, D.-Y. and Chung, S. Y.-K., 1985. *Fluid Phase Equilibria* 24: 25–41.
Rosenthal, D., Tawfik, W. and Teja, A. S., 1987. *Fluid Phase Equilibria* 37: 85–104.
Rowlinson, J. S. and Swinton, F. L., 1982. *Liquids and Liquid Mixtures*, 3rd ed., Butterworths, London.
Saito, S. and Arai, Y., 1986. In, *Physico-chemical Properties for Chemical Engineering*, Vol. 8, Kagaku Kogyosha Co., Tokyo.
Salerno, S., Cascella, M., May, D., Watson, P. and Tassios, D., 1986. *Fluid Phase Equilibria* 27: 15–34.
Sanchez, I. C. and Lacombe, R. H., 1976. *J. Phys. Chem.* 80: 2352, 80: 2568.
Sandler, S. I., 1985. *Fluid Phase Equilibria* 19: 233–257.
Sandoval, R., Wilczek-Vera, G. and Vera, J. H., 1989. *Fluid Phase Equilibria* 52: 119–126.
Schmidt, G. and Wenzel, H., 1980. *Chem. Eng. Sci.* 35: 1503–1512.
Schwartzentruber, J., Ponce-Ramirez, L. and Renon, H., 1986. *Ind. Eng. Chem. Proc. Des. Dev.* 25: 804–809.
Schwarzentruber, J., Galivel-Solastiouk, F. and Renon, H., 1987. *Fluid Phase Equilibria* 38: 217–226.
Schwartzentruber, J., Renon, H. and Watanasiri, S., 1989. *Fluid Phase Equilibria* 52: 127–134.
Scott, R. L., 1971. In, D. Henderson (ed.), *Physical Chemistry, An Advanced Treatise*, Vol. 8a, Chapter 1, Academic Press, New York.
Sengers, J. V. and Levelt-Sengers, J. M. H., 1986. *Ann. Rev. Phys. Chem.* 37: 189–222.
Sheng, Y. J., Chen, Y. P. and Wong, D. S. H., 1989. *Fluid Phase Equilibria* 46: 197–210.
Simnick, J. J., Lin, H. M. and Chao, K. C., 1979. *ACS Symp. Ser.* 182: 209–233.
Singh, A. and Teja, A. S., 1976. *Chem. Eng. Sci.* 31: 404–405.
Skjold-Jørgensen, S., 1984. *Fluid Phase Equilibria* 16: 317–351.
Skjold-Jørgensen, S., 1988. *Ind. Eng. Chem. Res.*, 27: 110–118.
Smirnova, N. A. and Victorov, A. I., 1987. *Fluid Phase Equilibria* 34: 235.

Smith, J. M. and H. C. Van Ness, 1975. *Introduction to Chemical Engineering Thermodynamics*, 3rd ed., McGraw-Hill, New York.
Smoluchowski, M., 1915. *Ann. Phys.* 48: 1098.
Soave, G., 1972. *Chem. Eng. Sci.* 27: 1197–1203.
Soave, G., 1984. *Chem. Eng. Sci.* 39: 357–369.
Soave, G., 1986. *Fluid Phase Equilibria* 31: 147–152.
Starling, K. E. and Han, M. S., 1972. *Hydrocarb. Process.* 51: 129–132.
Strobridge, T. R., 1962. NBS Technical Note No. 129.
Stryjek, R. and Vera, J. H., 1986a. *Can. J. Chem. Eng.* 64: 334–340.
Stryjek, R. and Vera, J. H., 1986b. *Can. J. Chem. Eng.* 64: 820–826.
Stryjek, R. and Vera, J. H., 1986c. *Fluid Phase Equilibria* 25: 279–290.
Sugie, H., Iwahori, Y. and Lu, B. C.-Y., 1989. *Fluid Phase Equilibria* 50: 1–29.
Teja, A. S., 1980. *AIChE J.* 26: 337–341.
Teja, A. S., Sandler, S. I. and Patel, N. C., 1981. *Chem. Eng. J.* 21: 21–28.
Tochigi, K., Kurihara, K. and Kojima, K., 1988. *Fluid Phase Equilibria* 42: 105–115.
Topliss, R. J., Dimitrelis, D. and Prausnitz, J. M., 1988. *Comp. Chem. Eng.* 12: 483–489.
Trebble, M. A., 1988. *Fluid Phase Equilibria* 42: 117–128.
Trebble, M. A. and Bishnoi, P. R., 1986. *Fluid Phase Equilibria*, 29: 465.
Trebble, M. A. and Bishnoi, P. R., 1987a. *Fluid Phase Equilibria* 35: 1–18.
Trebble, M. A. and Bishnoi, P. R., 1987b. *Fluid Phase Equilibria* 40: 1–21.
Tsonopoulos, C. and Heidman, J. L., 1985. *Fluid Phase Equilibria* 24: 1–23.
Tsonopoulos, C. and Heidman, J. L., 1986. *Fluid Phase Equilibria* 29: 391–414.
Tsonopoulos, C. and Prausnitz, J. M., 1969. *Cryogenics*, 315–327.
Twu, C. H., 1983. *Fluid Phase Equilibria* 13: 189–194.
Twu, C. H., Lee, L. L. and Starling, K. E., 1980. *Fluid Phase Equilibria* 4: 35–44.
Van der Waals, J. D., 1873. Over de Continuitet van den Gas- en Vloeistoftoestand, Doctoral Dissertation, Leiden.
Van der Waals, J. D., 1890. *Z. Phys. Chem.* 5: 133–173.
Vennix, A. J. and Kobayashi, R., 1969. *AIChE J.* 15: 926–931.
Vera, J. H., Huron, M. J. and Vidal, J., 1984. *Chem. Eng. Commun.* 26: 311–318.
Vetere, A., 1982. *Chem. Eng. Sci.* 37: 601–610.
Vetere, A., 1983. *Chem. Eng. Sci.* 38: 1281–1291.
Vetere, A., Sguera, O., Paggini, A., Sanfilippo, D., 1988. *Comp. Chem. Eng.* 12: 491–502.
Vidal, J., 1978. *Chem. Eng. Sci.* 33: 787–791.
Vidal, J., 1983. *Fluid Phase Equilibria* 13: 15–33.
Vidal, J., 1984. *Ber. Bunsenges. Phys. Chem.* 88: 784–791.
Vimalchand, P., Donohue, M. D., 1985. *Ind. Eng. Chem. Fundam.* 24: 246.
Vimalchand, P., Celmins, I. and Donohue, M. D., 1986. *AIChE J.* 32: 1735–1738.
Watson, I. D. and Rowlinson, J. S., 1969. *Chem. Eng. Sci.* 24: 1575–1580.

Watson, P., Cascella, M., May, D., Salerno, S. and Tassios, D., 1986. *Fluid Phase Equilibria* 27: 35–52.

Wenzel, H. and Krop, E., 1990. *Fluid Phase Equilibria* 59: 147.

Wenzel, H., Moorwood, R. A. S. and Baumgartner, M., 1982. *Fluid Phase Equilibria* 9: 225–266.

Wheeler, J. D. and Smith, B. D., 1967. *AIChE J.* 13: 303–311.

Whiting, W. B. and Prausnitz, J. M., 1982. *Fluid Phase Equilibria* 9: 119–147.

Wichterle, I., 1978. *Fluid Phase Equilibria* 2: 59–78.

Wilczek-Vera, G. and Vera, J. H., 1987. *Fluid Phase Equilibria* 37: 241–253.

Wilding, W. V. and Rowley, R. L. 1986. *Int. J. Thermophys.* 7: 525–539.

Wilding, W. V., Johnson, J. K. and Rowley, R. L., 1987. *Int. J. Thermophys.* 8: 717–735.

Wilson, G. M., 1964. *Adv. Cryog. Eng.* 9: 168–176.

Won, K. W., 1983. *Fluid Phase Equilibria* 10: 191–120.

Won, K. W. and Walker, C. K., 1979. *Adv. Chem. Ser.* 182: 235–251.

Wong, D. S. H., Sandler, S. I. and Teja, A. S., 1983. *Fluid Phase Equilibria* 14: 79–90.

Wong, D. S. H., Sandler, S. I. and Teja, A. S., 1984. *Ind. Eng. Chem. Fundam.* 23: 38–44, 23: 45–49.

Wong, J. O. and Prausnitz, J. M., 1985. *Chem. Eng. Commun.* 37: 41–53.

Wu, G. Z. A. and Stiel, L. I., 1985. *AIChE J.* 31: 1632–1644.

Yao, J., Greenkorn, R. A. and Chao, K. C., 1982. *J. Chem. Phys.* 76: 4657–4664.

Yu, J. M., Adachi, Y. and Lu, B. C.-Y., 1986. *ACS Symp. Ser.* 300: 537–559.

Yu, J. M. and Lu, B. C.-Y., 1987. *Fluid Phase Equilibria* 34: 1–19.

Yu, J. M., Lu, B. C.-Y. and Iwai, Y., 1987. *Fluid Phase Equilibria* 37: 207–222.

Ziervogel, R. G. and Poling, B.E., 1983. *Fluid Phase Equilibria* 11: 127.

Zwanzig, R. W., 1954. *J. Chem. Phys.* 22: 1420–1426.

5

Thermodynamic Consistency

5.1 INTRODUCTION

In Chapters 3 and 4 we focused on empirical and semitheoretical models that can be used to represent experimental data. In this chapter we shall concentrate on experimental data. The problem to be dealt with is the quality of data. Experience shows that a considerable amount of data are of low quality. The question is how to verify which data are of low or high or reasonable quality. Various methods are used for this purpose, including the analysis of laboratories, equipment, methods and skills of experimenters. The results of such an analysis depend on the evaluator's experience. To find an independent and free of the human factor method of evaluation we consider the problem of thermodynamic consistency of the data, i.e., their conformity with general rules of thermodynamics.

The problem of thermodynamic consistency is encountered when more data have been measured than are needed for the complete determination of the state of a system. This is often the case for vapor–liquid equilibrium measurements when all equilibrium parameters, i.e., temperature, pressure and liquid and vapor composition are determined. For example, a binary two-phase system has, according to the phase rule [Eq. (1.2-123)], two degrees of freedom. Thus, only two of the experimental variables (temperature, pressure and composition of the liquid and gas phases) can be regarded as independent. The remaining two variables can be calculated according to the rules given by thermodynamics. It is completely at our discretion which of the variables we choose as independent. If the data are isothermal, a usual choice is to use T rather than P as an independent variable. As the experimental measurement of vapor phase composition is usually more uncertain than the measurement of liquid composition, it is advisable to use the liquid phase (x) rather than vapor phase composition (y) as independent variables. Of course, the results of calculations can have the same or lower accuracy than the accuracy of the input data. It should be remembered that no computational methods exist that can improve the accuracy of experiment. The amount of information contained in experimental data cannot be increased by a correlation.

Among the many thermodynamic relations that have to be satisfied, two are of great value for testing thermodynamic consistency: the Gibbs–Duhem [Eq. (1.2-91)] and Gibbs–Helmholtz [Eq. (1.2-82)] equations. It should be noted that all consistency tests can be satisfied only in relation to the accuracy of the original experiment. A statement that a test has been "passed" or "not passed" should be always accompanied by the acceptable error threshold.

5.2 CONSISTENCY WITH RESPECT TO THE GIBBS–DUHEM EQUATION

The Gibbs–Duhem equation in its general form for an n-component mixture

$$S\,dT - V\,dP + \sum_{i=1}^{n} N_i\,d\mu_i = 0 \tag{1.2-91}$$

can be analogously written for excess functions. If the definition of the activity coefficient [Eq. (1.2-192)] is used and molar properties are introduced, the Gibbs–Duhem equation takes the form

$$s^E\,dT - v^E\,dP + RT\sum_{i=1}^{n} x_i d\ln\gamma_i = 0 \tag{5.2-1}$$

or, equivalently,

$$\frac{-h^E}{RT^2}\,dT - \frac{v^E}{RT}\,dP + \sum_{i=1}^{n} x_i d\ln\gamma_i = 0 \tag{5.2-2}$$

The Gibbs–Duhem equation provides a relation between activity coefficients of all components in a mixture. Therefore, correct data for the activity coefficients should obey the Gibbs–Duhem equation. A converse statement is not necessarily true; it is possible that an incorrect data set fortuitously satisfies the Gibbs–Duhem equation.

The obviously simplest technique is to utilize Eq. (5.2-2) directly. At isothermal conditions $dT = 0$, and well below the critical region we can neglect the term involving v^E, which is orders of magnitude smaller than the term involving activity coefficients. For a binary system containing components 1 and 2 we obtain then

$$x_1\frac{d\ln\gamma_1}{dx_1} = x_2\frac{d\ln\gamma_2}{dx_2} \tag{5.2-3}$$

The slopes of $\ln\gamma_1$ versus the mole fraction x_1 and $\ln\gamma_2$ versus x_2 can be measured and substituted into Eq. (5.2-3) at various compositions to check if the Gibbs–Duhem equation is satisfied. Such a procedure cannot be quantitatively accurate due to the inaccuracies of determining the slopes from experimental data alone. It can be applied only in a qualitative manner to detect rough errors in the data (e.g., if $d \ln \gamma_1/dx_1$ is zero, then $d \ln \gamma_2/dx_2$ must also be zero, etc.).

5.2.1. Tests Based on Integral Forms of the Gibbs–Duhem Equation

To test the data in a quantitative manner, integral rather than differential forms of the Gibbs–Duhem equation must be used. The integral tests were proposed first by Herington (1947), Redlich and Kister (1948) and Ibl and Dodge (1953).

Writing the molar Gibbs energy for a binary system according to Eq. (3.1-13) as

$$\frac{g^E}{RT} = x_1 \ln \gamma_1 + x_2 \ln \gamma_2 \tag{5.2-4}$$

and differentiating it at constant temperature and pressure we obtain

$$\frac{d(g^E/RT)}{dx_1} = x_1 \frac{\partial \ln \gamma_1}{\partial x_1} + \ln \gamma_1 + x_2 \frac{\partial \ln \gamma_2}{\partial x_2} + \ln \gamma_2 \frac{dx_2}{dx_1} \tag{5.2-5}$$

Substituting the Gibbs–Duhem equation (5.2-2) and noting that $dx_1 = -dx_2$ we get

$$\frac{d(g^E/RT)}{dx_1} = \ln \frac{\gamma_1}{\gamma_2} - \frac{h^E}{RT^2}\frac{dT}{dx_1} - \frac{v^E}{RT}\frac{dP}{dx_1} \tag{5.2-6}$$

Integration from $x_1 = 0$ to $x_1 = 1$ yields

$$\int_0^1 \ln \frac{\gamma_1}{\gamma_2}\, dx_1 - \int_0^1 \frac{h^E}{RT^2}\frac{dT}{dx_1}\, dx_1 - \int_0^1 \frac{v^E}{RT}\frac{dP}{dx_1}\, dx_1$$

$$= \frac{g^E(x_1 = 1)}{RT} - \frac{g^E(x_1 = 0)}{RT} = 0 \tag{5.2-7}$$

Several practical tests have been proposed on the basis of Eq. (5.2-7). Before we proceed to them, we should estimate the relative significance of the terms on the left-hand side of Eq. (5.2-5)

Example 5.2-1 Orders of magnitude of the terms involving excess enthalpy and excess volume in Eq. (5.2-7).

SOLUTION Let us approximate the integrals involving h^E and v^E using mean values of h^E and v^E. For the whole concentration from 0 to 1 we obtain:

$$I_v = \int_0^1 \frac{v^E}{RT}\, dP \cong \frac{\overline{v^E}}{RT} |P_1^0 - P_2^0| \tag{P5.2-1}$$

$$I_h = \int_0^1 \frac{h^E}{RT^2}\, dT \cong \frac{\overline{h^E}}{R}\left(\frac{1}{T_{b_1}^0} - \frac{1}{T_{b_2}^0}\right) \tag{P5.2-2}$$

where $\overline{v^E}$ and $\overline{h^E}$ denote mean values of v^E and h^E, respectively, and $T_{b_i}^0$ are the boiling point temperatures of components i ($i = 1, 2$).

To estimate the order of magnitude of I_v, we insert some typical values into Eq. (P5.2-1), i.e., $\overline{v^E} = 0.3\ \text{cm}^3/\text{mol}$, $|P_1^0 - P_2^0| = 50\ \text{kPa}$ and $T = 300\ \text{K}$. The result is

$$I_v \cong 6 \times 10^{-6}$$

This is a very small value when compared with the value of $I_\gamma = \int_0^1 \ln \gamma_1/\gamma_2$, which can be of the order of 10^{-3} to 10^0. The I_h value can be estimated using typical values $\overline{h^E} = 1000\ \text{J/mol}$, $T_{b_2}^0 = 300\ \text{K}$ and $T_{b_1}^0 = 350\ \text{K}$: $I_h \cong 5.7 \times 10^{-2}$. This quantity is with no doubt important for testing nonisothermal data. ■

Therefore, it is of utmost importance to use excess enthalpy data when testing nonisothermal data. The pressure correction can be safely neglected. Moreover, the activity coefficients should be always properly computed, i.e., if the activity coefficients are computed from vapor–liquid equilibria, the vapor phase nonideality should be taken into account in the data reduction. This means that the fugacity coefficients of the vapor phase should be computed from independent measurement of vapor phase *PVT* properties. If this is not the case, the consistency tests described below can give totally misleading results.

THE OVERALL AREA TESTS

The oldest consistency test in this group is the so-called overall area test (Herington, 1947; Redlich and Kister, 1948). The test consists in plotting the values of

$$\frac{d(g^E/RT)}{dx_1} = \ln \frac{\gamma_1}{\gamma_2} - \frac{h^E}{RT^2}\frac{dT}{dx_1} - \frac{v^E}{RT}\frac{dP}{dx_1}$$

against mole fraction x_1. The test requires that the integral from $x_1 = 0$ to $x_1 = 1$ be zero as indicated by Eq. (5.2-7). This can be visualized as a requirement that the area above the x_1 axis (the "plus area") be equal to the area below the axis (the "minus area"). In the case of real data the areas are not necessarily equal. Therefore, an arbitrary threshold value for acceptable values of the integral is needed. If x_1^0 is the mole fraction for which $d(g^E/RT)/dx_1) = 0$, the usually accepted criterion for the data to be consistent is

$$c_1 = \frac{\int_0^{x_1^0} [d(g^E/RT)/dx_1)]\, dx_1 - \int_{x_1^0}^1 [d(g^E/RT)/dx_1)]\, dx_1}{\int_0^{x_1^0} [d(g^E/RT)/dx_1)]\, dx_1 + \int_{x_1^0}^1 [d(g^E/RT)/dx_1)]\, dx_1} \tag{5.2-8}$$

or, equivalently,

$$c_2 = \left| \frac{\text{(plus area)} - \text{(minus area)}}{\text{(plus area)} + \text{(minus area}} \right| \tag{5.2-9}$$

This is the so-called *relative area test*. The integrations should be performed numerically using appropriate interpolation functions. For example, spline functions can be used for this purpose (cf. Klaus and Van Ness, 1967).

Equations (5.2-8) and (5.2-9) are reasonable if the errors in activity coefficients [and, subsequently, in the quantity $\ln(\gamma_1/\gamma_2)$] are proportional to the γ_i values. This is not always the case as one should not expect small activity coefficients to be determined more precisely than large ones. Therefore, the difference between the plus area and minus area can be used as a complimentary test (the *absolute area test*):

$$c_2 = |\text{(plus area)} - \text{(minus area)}| \tag{5.2-10}$$

In this case the threshold value should be estimated by the user on the basis of the experimental uncertainty of activity coefficients. The absolute difference between the plus and minus area can be compared with the standard error of $\ln(\gamma_1/\gamma_2)$ estimated from the experimental uncertainties of the data. Similarly, the plot of $d(g^E/RT)/dx_1)$ obtained from Eq. (5.2-6) can be analyzed with marked confidence intervals, calculated at some reasonable confidence level (Van Ness et al., 1973). Some modifications of the area test, based on the error propagation law, have been also developed by Ulrichson and Stevenson (1972) and Samuels et al. (1972).

When using such consistency tests, it should be noted that the value of the function being integrated is not determined at both ends of the integration interval (i.e., $x_1 = 0$ or 1). The values of the function are simply not determined when the activity coefficients are not determined. This affects the accuracy of integration at the concentration limits in the vicinity of $x_1 = 0$ or 1. The accuracy can be improved by taking into account infinite-dilution activity coefficient data γ_1^∞ for $x_1 = 0$ and γ_2^∞ for $x_2 = 0$.

THE POINT-TO-POINT TEST

The overall area tests suffer from their integral nature, providing thus only averaged results. Some data can pass the overall area tests even with severe local inconsistencies if their effect on the whole test is balanced. To overcome this deficiency, the point-to-point test was introduced (Stevenson and Sater, 1966). This test consists of integrating Eq. (5.2-6) from one point x_1' to another x_1'' rather than from 0 to 1. In this case the right-hand side of the integrated form is no longer zero and the criterion becomes

$$\int_{x_1'}^{x_1''} \left[\ln \frac{\gamma_1}{\gamma_2} - \frac{h^E}{RT^2} \frac{dT}{dx_1} - \frac{v^E}{RT} \frac{dP}{dx_1} \right] dx_1 = \frac{g^E(x_1'')}{RT} - \frac{g^E(x_1')}{RT} + \sigma\left(\frac{g^E}{RT}\right) \tag{5.2-11}$$

where $\sigma(g^E/RT)$ is estimated as the sum of standard errors of g^E/RT at mole fractions x_1' and x_1''. To apply this test the number of experimental points must be relatively large and good estimates of errors must be available.

All the tests described above (i.e., the relative area, absolute area and point-to-point tests) suffer from a serious deficiency: They do not take into account the experimental values of pressure. In fact, pressure cancels out in the ratio of activity coefficients calculated according to a standard γ–ϕ method:

$$\frac{\gamma_1}{\gamma_2} = \frac{\Phi_1(y_1/x_1)P_2^0}{\Phi_2(y_2/x_2)P_1^0} \tag{5.2-12}$$

Strictly speaking, pressure enters into the calculation of correction factors Φ_i. This effect is, however, of only minor importance especially at low pressures. The only data used in the construction of area tests are the liquid and vapor compositions, whereas the most valuable (and usually the most accurate) information is provided by the pressure measurements. Therefore, the area tests are of little value for deciding whether the activity coefficient data are accurate or not.

EXTENDED (PRESSURE-DEPENDENT) AREA TEST

When only three among the usually measured four thermodynamic variables (T, P, x, y) are measured, the value of the fourth one can be calculated and compared with the measured value. This obvious fact can be utilized to complement the area tests by taking into account the measured values of pressure.

Instead of integrating the derivative of the excess Gibbs energy [Eq. (5.2-6)] for the whole concentration interval, we can integrate it from 0 to x_1

$$\int_0^{x_1} \left[\ln \frac{\gamma_1}{\gamma_2} - \frac{h^E}{RT^2}\frac{dT}{dx_1} - \frac{v^E}{RT}\frac{dP}{dx_1} \right] dx_1 = \frac{g^E(x)}{RT} \tag{5.2-13}$$

Following Eq. (3.1-14), the activity coefficients for a binary system are related to the excess Gibbs energy by

$$\ln \gamma_1 = \left(\frac{g^E}{RT}\right) + x_2\left(\frac{\partial(g^E/RT)}{\partial x_1}\right) \tag{5.2-14}$$

$$\ln \gamma_2 = \left(\frac{g^E}{RT}\right) + x_1\left(\frac{\partial(g^E/RT)}{\partial x_1}\right) \tag{5.2-15}$$

The total pressure can be calculated by substituting Eqs. (5.2-6) and (5.2-13) into Eqs. (5.2-14) and (5.2-15) and, subsequently, into Eq. (3.1-29) for the total pressure:

$$P = \frac{x_1\gamma_1P_1^0}{\Phi_1} + \frac{x_2\gamma_2P_2^0}{\Phi_2} \tag{5.2-16}$$

where Φ_i are correction factors involving vapor phase nonideality and the Poynting factors. In this manner the area test can be performed simultaneously with calculating the total vapor pressure at each experimental point. Two

Table 5.2-1
RESULTS OF TESTING CONSISTENCY OF ISOTHERMAL VAPOR-LIQUID EQUILIBRIUM DATA FOR THE ETHANOL–HEPTANE SYSTEM ACCORDING TO THE PRESSURE-DEPENDENT AREA TEST

Data	T (K)	Relative Area Defect c_1 [Eq. (5.2-8)]	Absolute Area Defect c_2 [Eq. (5.2-9)]	ΔP^a (kPa)	δPa (%)
Berro et al. (1982)	303.17	0.004	0.0049	0.085	0.6
	343.17	0.001	0.0015	0.091	0.1
Diaz Peña and Rodriguez Cheda (1970)	313.15	0.006	0.0068	0.309	1.2
	333.15	0.006	0.0066	0.699	1.2
Ferguson et al. (1933)	303.15	0.035	0.0402	0.209	1.4
Ramalho and Delmas (1968)	343.15	0.008	0.0094	2.997	3.3
	353.15	0.035	0.0386	4.545	3.2
	363.15	0.042	0.0470	4.745	2.3
Ratcliff and Chao (1969)	313.15	0.046	0.0516	0.693	2.7

After Oracz (1989).
[a] $\Delta P = [\Sigma (P_i^{\text{cal}} - P_i^{\text{exp}})^2/N]^{1/2}$.
[b] $\delta P = [\Sigma [(P_i^{\text{cal}} - P_i^{\text{exp}})/P_i^{\text{exp}}]^2/N]^{1/2}$.

criteria have to be satisfied simultaneously: the area test criterion and the requirement that the pressure residuals be less than an acceptable threshold value. Another refined area test based on this principle was proposed by Dohnal and Fenclová (1985).

To illustrate the performance of the above tests, Table 5.2-1 compares the results obtained by Oracz (1989) by testing consistency of several isothermal data sets for the ethanol–heptane system.

According to the pressure residuals, only the data sets of Berro et al. (1982) are consistent. On the other hand, the results of the area tests are satisfactory for some data for which the residuals of pressure are unacceptably large. This underscores the limited reliability of the area tests if they are used without additional tests.

5.2.2. Testing Consistency by Data Reduction

This is the oldest method for testing thermodynamic consistency. It goes back to the works of Margules (1895). Since then, the data reduction methods have been well established. The most significant contribution to establishing the data reduction techniques was due to Barker (1953) who introduced an iterative least-squares technique to fit expressions for g^E to vapor–liquid equilibrium data. The problem of testing consistency can be, in principle, solved simultaneously with data reduction. For this purpose a thermodynamic model for g^E has to be used that is assumed to perfectly represent the system under investigation. After fitting the adjustable parameters to the experimental data set (or subset), the user has to analyze the deviations obtained for all measured variables. The data set is called consistent if the deviations are random and within experimental error. The method of testing consistency by data reduction has been elaborated on by Van Ness and his school and is described in detail in Van Ness and Abbott's (1982) textbook.

This approach to consistency testing started in 1959 with a paper by Van Ness in which it was shown that a comparison of activity coefficients calculated directly from vapor–liquid equilibrium data and those obtained from a plot of g^E/RTx_1x_2 against x_1 constitutes an effective consistency test. For this purpose the quantity g^E/RTx_1x_2 was correlated using a suitable polynomial expression and the activity coefficients were calculated by means of Eq. (3.1-14). The activity coefficient values generated from the correlation of g^E are inherently consistent with the Gibbs–Duhem equation, whereas the activity coefficients generated from the raw data are not so constrained. The deviations between both sets of activity coefficients are thus a measure of the consistency of the data.

It should be noted that tests of this type, unlike the tests based on the integral forms of the Gibbs–Duhem equation, are equally applicable when the data are reduced by means of the gamma--phi or equation-of-state (EOS) methods. It is only prerequisite that the correlation equation have sufficient flexibility to produce a random distribution of the objective function residuals. If the distribution is not random, then the equation has introduced systematic deviations into the correlation. The error introduced by a model equation will be undetectable by thermodynamic consistency checks because results of correlations using either the γ–ϕ or EOS methods are always thermodynamically consistent. Inappropriate correlation equations introduce systematic deviations and, therefore, part of the experimental information is lost.

The methods contemporarily recommended employ the analysis of residuals of measurable variables. They will be described below for some important cases.

ISOTHERMAL DATA

As discussed in Section 3.1 we can perform a data reduction procedure that consists of calculating P and y for given T and x. If we use a sufficiently flexible equation (such as the polynomial equations described in Section 3.3.2), we can minimize the sum of squares of pressure residuals $(P_i^{\text{cal}} - P_i^{\text{exp}})$:

$$\sum_i (P_i^{\text{cal}} - P_i^{\text{exp}})^2 \rightarrow \min \tag{5.2-17}$$

to obtain a correlation for which the pressure residuals scatter randomly around zero. A consequence of this data reduction procedure with an equation of adequate flexibility is that all systematic errors in the data are transferred to the residuals of vapor phase composition $(y_i^{\text{cal}} - y_i^{\text{exp}})$. Thus, a systematic deviation from zero of the vapor phase composition residuals is a consequence of systematic errors in one or more variables. This method provides a clear test of whether the data are thermodynamically consistent or not.

The second procedure for testing isothermal data minimizes the sum of squares of vapor phase composition residuals

$$\sum_i (y_i^{\text{cal}} - y_i^{\text{exp}})^2 \rightarrow \text{min} \tag{5.2-18}$$

This minimization, however, is subject to a constraint that manifests itself in a systematic departure from zero of the obtained residuals of y. This constraint can be explained by considering the usual γ–ϕ data reduction procedure for which the activity coefficient follows from the equation

$$\gamma_i = \frac{y_i P \Phi_i}{x_i P_i^0} \tag{5.2-19}$$

where Φ_i is the vapor phase nonideality correction factor [Eq. (3.2-8)]. The residual of the quantity $\ln(\gamma_1/\gamma_2)$ can be defined as

$$\delta \ln \frac{\gamma_1}{\gamma_2} = \ln \frac{\gamma_1^{\text{cal}}}{\gamma_2^{\text{cal}}} - \ln \frac{\gamma_1}{\gamma_2} \tag{5.2-20}$$

Substituting Eq. (5.2-19) into Eq. (5.2-20) we get

$$\delta \ln \frac{\gamma_1}{\gamma_2} = \ln \frac{\gamma_1^{\text{cal}}}{\gamma_2^{\text{cal}}} + \ln \frac{y_2 x_1 \Phi_2}{y_1 x_2 \Phi_2} + \ln \frac{P_1^0}{P_2^0} \tag{5.2-21}$$

The function defined by Eq. (5.2-21) is a derivative [cf. Eqs. (5.2-14) and (5.2-15)] and can be integrated from $x_1 = 0$ to $x_1 = 1$ to yield

$$\int_0^1 \delta \ln \frac{\gamma_1}{\gamma_2}\, dx_1 = \int_0^1 \ln \frac{\gamma_1^{\text{cal}}}{\gamma_2^{\text{cal}}}\, dx_1 + \int_0^1 \ln \frac{y_2 x_1 \Phi_2}{y_1 x_2 \Phi_2}\, dx_1 + \int_0^1 \ln \frac{P_1^0}{P_2^0}\, dx_1 \tag{5.2-22}$$

The first term on the right is necessarily zero at isothermal conditions because activity coefficients calculated from a correlation equation must satisfy the Gibbs–Duhem equation. The residuals $\delta \ln(\gamma_1/\gamma_2)$ can be evaluated by dividing the ratios $\gamma_1^{\text{cal}}/\gamma_2^{\text{cal}}$ and γ_1/γ_2 calculated according to Eq. (5.2-19):

$$\frac{\gamma_1^{\text{cal}}/\gamma_2^{\text{cal}}}{\gamma_1/\gamma_2} = \frac{y_1^{\text{cal}}/y_2^{\text{cal}}}{y_1/y_2}\, \frac{\Phi_1^{\text{cal}}/\Phi_2^{\text{cal}}}{\Phi_1/\Phi_2} \tag{5.2-23}$$

In logarithmic form this becomes

$$\ln \frac{\gamma_1^{\text{cal}}}{\gamma_2^{\text{cal}}} - \ln \frac{\gamma_1}{\gamma_2} = \ln \frac{y_1^{\text{cal}}}{y_2^{\text{cal}}} - \ln \frac{y_1}{y_2} + \ln \frac{\Phi_1^{\text{cal}}}{\Phi_2^{\text{cal}}} - \ln \frac{\Phi_1}{\Phi_2} \tag{5.2-24}$$

or

$$\delta \ln \frac{\gamma_1}{\gamma_2} = \delta \ln \frac{y_1}{y_2} + \delta \ln \frac{\Phi_1}{\Phi_2} \tag{5.2-25}$$

The second term on the right is of lower order than the remaining two and we get

$$\delta \ln \frac{\gamma_1}{\gamma_2} \cong \delta \ln \frac{y_1}{y_2} \cong \frac{\delta y_1}{y_1 y_2} \tag{5.2-26}$$

Substituting Eq. (5.2-26) into Eq. (5.2-22), we obtain the constraint equation

$$\int_0^1 \ln \frac{\gamma_1}{\gamma_2}\, dx_1 \cong \int_0^1 \ln \frac{y_2 x_1 \Phi_2}{y_1 x_2 \Phi_2}\, dx_1 + \ln \frac{P_1^0}{P_2^0} \tag{5.2-27}$$

When the terms on the right do not cancel, the integral on the left does not vanish and neither does the integral:

$$\int_0^1 \delta y_1\, dx_1 \tag{5.2-28}$$

which approximates the net area on a plot of δy_1 versus x_1. If this value is not zero, no data reduction procedure can yield unbiased δy_1 residuals.

Therefore, minimization of vapor phase composition residuals may lead to systematic deviations from zero for both P and y residuals. However, systematic errors in the equilibrium pressure P are reflected only in the pressure residuals because the δy residuals do not depend on P. A bias in vapor composition residuals reflects systematic errors in x or y or pure-component saturation pressures P_i^0. In general, data that yield δy residuals from minimization of $\Sigma\,(\delta y_i)^2$ that scatter regularly around zero are not necessarily consistent unless they also yield δP residuals that scatter around zero, or, alternatively, δy residuals from minimization of $\Sigma\,(\delta P_i)^2$ that scatter around zero.

To illustrate the above general features of consistency testing by data reduction, Figure 5.2-1 shows the results obtained by Rogalski and Malanowski (1977) for the system carbon tetrachloride–acetonitrile (Brown and Smith, 1954) using the sum of symmetric functions (SSF) equation (3.3-37). The data were analyzed by minimizing the pressure residuals (A) or vapor composition residuals (B). Figure 5.2-1 illustrates that the equation used for data analysis must be sufficiently flexible. If a four-parameter SSF equation (3.3-37) is used (version I in the figure), the systematic errors cannot be separated due to the inadequacy of the model with four parameters. On the other hand, the SSF equation with six parameters provides a very good reproduction of the data within the limits of experimental random errors as shown by the dashed lines. In this case the systematic errors are clearly seen in the figure, pointing to the inconsistency of the data. To underline the crucial role of the flexibility of the equation, Figure 5.2-2 is also provided, which shows the quality of fit as a function of the number of parameters for the Redlich–Kister [Eq. (3.3-3)], SSF [Eq. (3.3-37)] and Wilson [Eq. (3.3-9)] equations. The best correlation equation should be used for a meaningful consistency test. In general, the correlation equation for consistency testing can be chosen from among polynomial

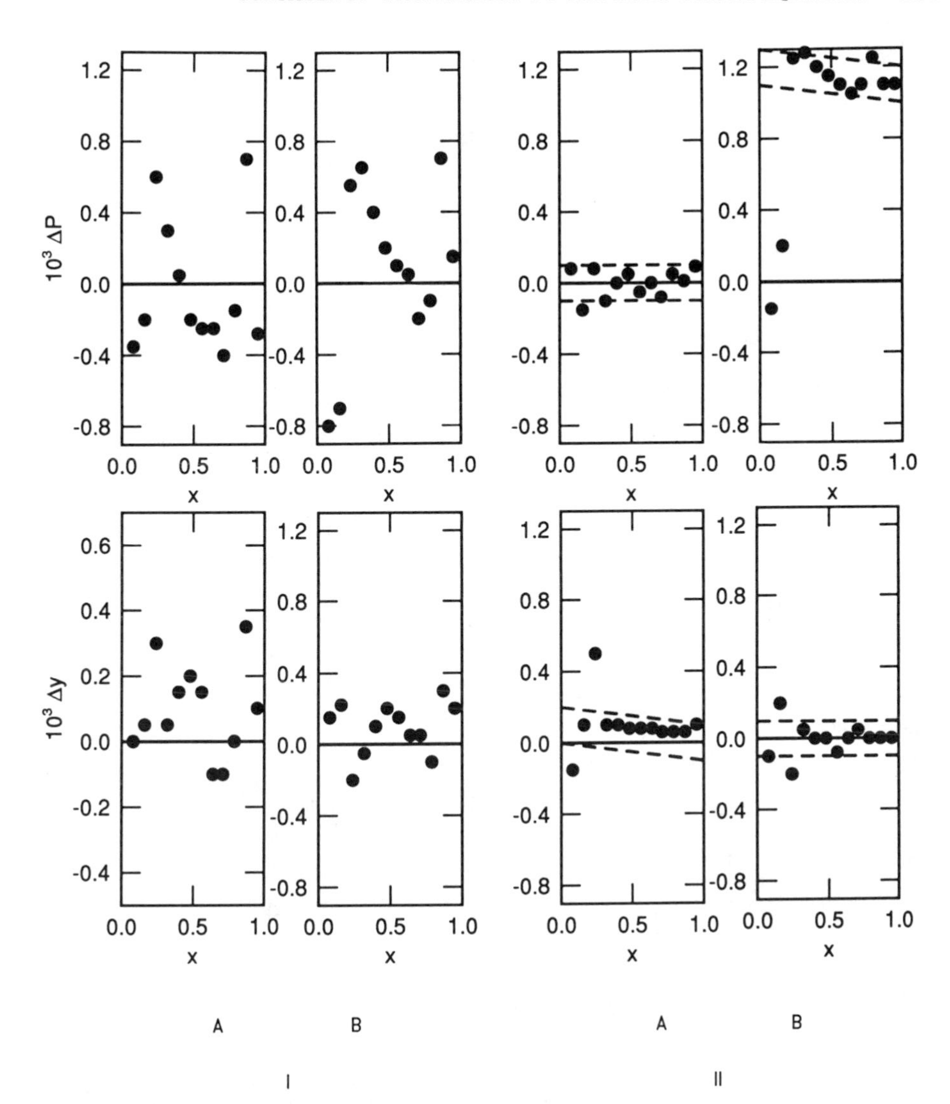

Figure 5.2-1
Deviations between the calculated and experimental values of pressure ($10^3 \times \Delta P$) and vapor composition ($10^3 \times \Delta y$) as a function of liquid composition (x) obtained by correlating the $P - x$ (column A) and $y - x$ (column B) data using the four-parameter (I) and six-parameter (II) SSF equation for the carbon tetrachloride–acetonitrile system at 318.15 K (Brown and Smith, 1954). Dashed lines represent the estimated errors of the data.

equations (Section 3.3.2), SSF (Section 3.3.4) or spline functions (Klaus and Van Ness, 1967).

To make a decision whether a data set is consistent or not, it may be convenient to use a numerical criterion rather than to observe qualitatively the residuals. Such a criterion was proposed by Christensen and Fredenslund (1975) for the mean residual of y:

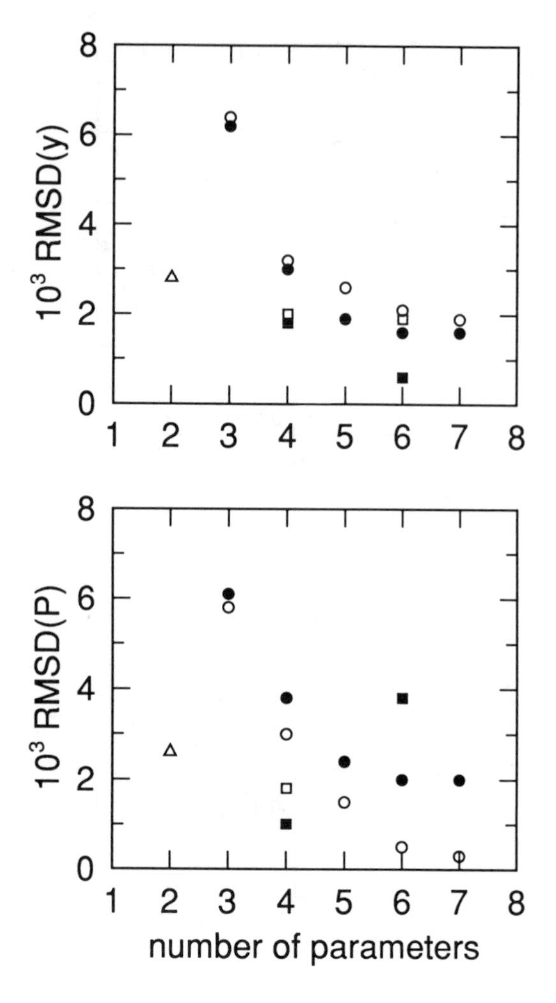

Figure 5.2-2
Root-mean-square deviations of pressure and vapor composition for the system carbon tetrachloride–acetonitrile at 318.15 K calculated with the use of the Redlich–Kister (○), SSF (□) and Wilson (△) equations. Hollow signs denote the values obtained from the reduction of $P-x$ data and the solid lines from $y-x$ data.

$$|\overline{\delta y}| \leq \Delta x + \Delta y \tag{5.2-29}$$

where Δx and Δy are the uncertainties of the x and y measurements, respectively. The right side is practically approximated by 0.01.

The existence of systematic deviations can be also checked using statistical tests (cf. Kémeny et al., 1982; Kollár-Hunek et al., 1986).

CONSISTENCY OF PURE-COMPONENT VAPOR PRESSURE

The consistency of pure-component vapor pressure values P_i^0 with the rest of the data set is important because the P_i^0 values have a special significance in the gamma–phi methods as they are used to define the reference state. Thus, uncertainty in P_i^0 greatly affects the results for all other data points.

The consistency of pure-component vapor pressures can be tested using a similar data reduction technique. The P_i^0 values are evaluated along with the constants of the correlating equation. The values obtained are then compared with the experimental pure-component vapor pressures. Such a procedure makes sense if the (isothermal) data are not far removed $x_1 = 0$ or $x_1 = 1$.

ISOBARIC DATA

In the case of isobaric data the customarily chosen independent variables are P and x. While the choice of x rather than y is guided, similarly as for isothermal data, by the usually better accuracy of x values, there is no advantage of using P instead of T. The data reduction can be performed by minimizing the temperature residuals:

$$\sum_i (T_i^{\text{cal}} - T_i^{\text{exp}})^2 \rightarrow \text{min} \quad \text{at } P = P^{\text{exp}} = \text{const} \tag{5.2-30}$$

or the pressure residuals

$$\sum_i (P_i^{\text{cal}} \rightarrow P_i^{\text{exp}})^2 \rightarrow \text{min} \quad \text{at each } T_i^{\text{exp}} \ (T_i^{\text{exp}} \neq T_j^{\text{exp}}) \tag{5.2-31}$$

The second data reduction procedure consists of minimizing the vapor mole fraction residuals similarly as in the isothermal case [Eq. (5.2-18). When the γ–ϕ method is used and the temperature residuals are minimized, the procedure is inherently more time consuming than the corresponding procedure for finding the pressure residuals. To calculate the equilibrium temperature, each step of the regression procedure must incorporate an iterative scheme to yield the value of T_i^{cal}. However, solution to this problem is very simple and requires the use of standard methods such as the Newton or secant method. If the equation-of-state technique is employed, the same computational effort is required to solve for P or T as the calculations always involve an iterative scheme. In practical applications both procedures [Eqs. (5.2-30) and (5.2-31)] yield identical results. Therefore, the use of Eq. (5.2-30) is not recommended as it leads only to a more complicated algorithm in the case of the γ–ϕ methods.

The only difference between the isobaric and isothermal cases is the necessity to allow for the temperature dependence of parameters. The pressure dependence is practically insignificant due to the small influence of v^E on the data (cf. Example 5.2-1). The temperature dependence must conform to the Gibbs–Helmholtz equation [Eq. (1.2-82)] interrelating the excess Gibbs energy and excess enthalpy at constant composition:

$$\frac{h^E}{RT} = -T\left(\frac{\partial(g^E/RT)}{\partial T}\right)_{P,x} \tag{5.2-32}$$

Equation (5.2-32) makes it possible to express the temperature dependence of parameters using experimentally determined excess enthalpy data. For example, polythermal heat of mixing data can be correlated using the Redlich–Kister equation (3.3-3):

$$\frac{h^E}{RT} = x_i x_j \sum_{k=0}^{m} A_k (x_i - x_j)^k \tag{5.2-33}$$

with temperature-dependent parameters

$$A_k = \frac{A_{k,1}}{T} + A_{k,3} \tag{5.2-34}$$

The corresponding equation for the excess Gibbs energy is

$$\frac{g^E}{RT} = x_i x_j \sum_{k=0}^{m} B_k (x_i - x_j)^k \tag{5.2-35}$$

with

$$B_k = \frac{A_{k,1}}{T} + A_{k,2} + A_{k,3} \ln T \tag{5.2-36}$$

If eq. (5.2-35) is differentiated according to Eq. (5.2-32), Eq. (5.2-33) is recovered.

The isobaric data reduction procedure should be performed with the values of $A_{k,1}$ and $A_{k,3}$ fixed using the excess enthalpy data. Moreover, a good equation should be used for correlating pure-component vapor pressures with temperature in the temperature range of experimental data. For example, the Antoine equation can be used for this purpose. Thus, only the parameters $A_{k,2}$ have to be adjusted to the isobaric vapor–liquid equilibrium data.

After performing the data reduction in the manner described above, the residuals of the calculated dependent variables T (or P) and y should be analyzed analogously as in the isothermal case. If the excess enthalpy data are not included in the data reduction procedure, the consistency testing is much less reliable.

The method involving the use of polynomial equations to fit both g^E and h^E data may be dangerous. The polynomial equations frequently lead to unexpected and experimentally unjustified inflection points. Moreover, the simultaneous correlation of phase equilibria and caloric data may lead to an excessive number of adjustable parameters (cf. Van Ness et al., 1967).

TERNARY AND MULTICOMPONENT DATA

To obtain a flexible correlation equation for testing the consistency of ternary and multicomponent data, it is often necessary to include ternary and higher-order parameters. The structure of such a correlation is given by the general equation (3.3-1). For example, a convenient correlating expression for g^E in ternary mixtures is based on an equation proposed by Wohl (1953):

$$g_{123}^E = g_{12}^E + g_{13}^E + g_{23}^E + (C_0 + C_1 x_1 + C_2 x_2) x_1 x_2 x_3 \tag{5.2-37}$$

The quantities g_{12}^E, g_{13}^E and g_{23}^E are the contributions of binary pairs, established on the basis of binary data. The ternary parameters C_0, C_1 and C_2 are determined by regressing the data for ternary mixtures. This procedure has been analyzed by Abbott et al. (1975). Once the data reduction is performed, the residuals of dependent variables can be evaluated as in the case of binary mixtures.

OTHER METHODS

While the above methods are deemed most reliable and straightforward, other methods also exist that deserve some comments. Some of them are called *model-free* methods in that they do not utilize any model for the excess Gibbs energy. This name is, however, rather inappropriate because such methods require some expressions for fitting the experimental $P - x$ or $T - x$ data. The problem of finding a suitable expression is not simpler, or even more difficult, than the problem of finding a good correlation equation for g^E.

These methods involve integration of some thermodynamic equations that can be derived from the Gibbs–Duhem equation. The Gibbs–Duhem equation is expressed in the form of a first-order differential equation that is integrated numerically. This equation can be solved for activity coefficients, excess Gibbs energy, vapor phase composition or relative volatility. The underlying idea of these methods is to calculate one of the usually measured four variables (T, P, x, y) and to compare it (or some derived quantity) with experimental data. Among these methods, the method of Mixon et al. (1965) consists of simultaneously solving for the excess Gibbs energy at discreet, equally spaced grid points. Highly elaborate methods have been proposed by Tao (1961, 1969, 1980). The model-free methods have been reviewed by Sayegh and Vera (1980).

In general, the model-free methods are more complicated and cumbersome than those directly employing correlation equations.

COMPUTATIONAL ASPECTS

The functions defined by Eqs. (5.2-17), (5.2-18), (5.2-30) and (5.2-31) are usually minimized using the simple unconstrained optimization technique. However, more sophisticated techniques can also be employed. These techniques involve using weighting factors w_j in the objective function F_0:

$$F_0 = \sum_j \frac{(Y_j^{\text{cal}} - Y_j^{\text{cal}})^2}{w_j} \tag{5.2-38}$$

where Y denotes P or T or y. The weighting factors can be deduced from a statistical theory called the principle of maximum likelihood. When using the maximum-likelihood principle, the following conditions must be satisfied:

1. The data set contains randomly distributed experimental errors. No systematic errors exist.
2. The model equation perfectly reproduces the data, i.e., it is capable of representing the true values of all thermodynamic variables.

3. The experimental random errors (i.e., the variances of measured variables) can be estimated from experimental information.

In practice, all these conditions are difficult to meet. The first two conditions reflect the most important problems of any branch of applied science: uncertain data and inadequate correlation equations. As a matter of fact, if perfect thermodynamic data and models existed, it would be pointless to write this book. Moreover, we are interested here in methods to *detect* experimental inconsistencies and cannot presume that they do not exist.

According to Van Ness and Abbott (1982, p. 327) "the contribution [of statistical methods] to the accuracy of correlation is relatively minor. When the preconditions to statistical treatment are not closely met, and they seldom are, one can do little further harm by setting $w_j = 1$ [Eq. (5.2-38)]. Moreover, improper application of statistical methods produces misleading results. Since these methods distribute systematic error over all measured variables, the error is divided among the possible sources, giving the appearance of minimal error without increasing the reliability of correlation".

While we deem it unnecessary to go into details of the maximum-likelihood methods, Table 5.2-2 is given to provide the interested reader with

Table 5.2-2
SELECTED PAPERS ON THE APPLICATION OF STATISTICAL METHODS TO THE REDUCTION OF PHASE EQUILIBRIUM DATA

Author(s)	Subject
Péneloux (1990)	A review of variance analysis and statistical methods for data reduction
Skjold-Jørgensen (1983)	A critical review of statistical principles in the reduction of experimental data
Kémeny et al. (1982)	Evaluation of some maximum-likelihood methods
Neau and Péneloux (1981)	Comparison of methods based on the maximum-likelihood principle
Sutton and McGregor (1977)	Parameter estimation and consistency tests using the maximum-likelihood principle
Van Ness et al. (1978)	Evaluation of the maximum-likelihood methods in relation to the least squares method
Anderson et al. (1978)	A maximum-likelihood method for the evaluation of the adequacy of models and accuracy of data
Péneloux et al. (1976)	A maximum-likelihood method including the estimation of experimental inaccuracies
Neau and Péneloux (1982)	A maximum-likelihood method involving reduction of subsets rather than complete data sets
Fabriès and Renon (1975)	A maximum-likelihood method
Figurski et al. (1980)	The orthogonal regression method for VLE and h^E calculations

references to some more important papers dealing with the application of statistics to the reduction of phase equilibrium data.

5.3 CONSISTENCY WITH RESPECT TO THE GIBBS–HELMHOLTZ EQUATION

The methods based on the Gibbs–Duhem equation were designed to examine the consistency of individual data. The Gibbs–Helmholtz equation [Eq. (1.2-82) makes it possible to verify the mutual consistency of several data sets. The Gibbs–Duhem equation interrelates the excess Gibbs energy and excess enthalpy:

$$h^E = -RT^2\left(\frac{\partial(g^E/RT)}{\partial T}\right)_{P,x} \tag{5.3-1}$$

Thus, the excess enthalpy can be computed from experimental values of the excess Gibbs energy at two or more temperatures and compared with experimental h^E values. As the experimental values of g^E are usually available for only a few temperatures, the derivative on the right-hand side of Eq. (5.3-1) has to be approximated by finite differences of g^E at constant composition. To arrive at accurate constant composition values of g^E, suitable correlation equations must be employed.

To apply this test, it is prerequisite to have experimental excess enthalpy data measured at similar temperatures as the g^E data. The h^E data need not be of very good accuracy. This is due to the fact that the numerical values calculated from Eq. (5.3-1) are very sensitive to the temperature dependence of g^E, which is usually not very strong, but are much less sensitive to the errors in absolute h^E values.

Although very large h^E data collections have been published (Christensen et al., 1982, 1984–), the vast majority of the measurements have been taken at 298 K and similar temperatures. The experimental h^E data can be recalculated to the temperature of interest if experimental c_p^E data are known:

$$h^E(T) = h^E(T_0) + \int_{T_0}^{T} c_p^E \, dT \tag{5.3-2}$$

However, c_p^E data rarely exist for systems of interest. If excess enthalpies are available only at different temperatures, the test can be performed only in a qualitative way.

To illustrate the application of this test when all necessary data are available, Figure 5.3-1 compares the excess enthalpies of the butan-1-ol + n–decane system at 378.15 K calculated by Gierycz et al. (1988) from vapor–liquid equilibria and those obtained calorimetrically. As the vapor–liquid equilibria were measured at 373.15 and 383.15 K, h^E was computed in the middle of the temperature range, i.e., at 378.15 K. The calorimetric h^E data

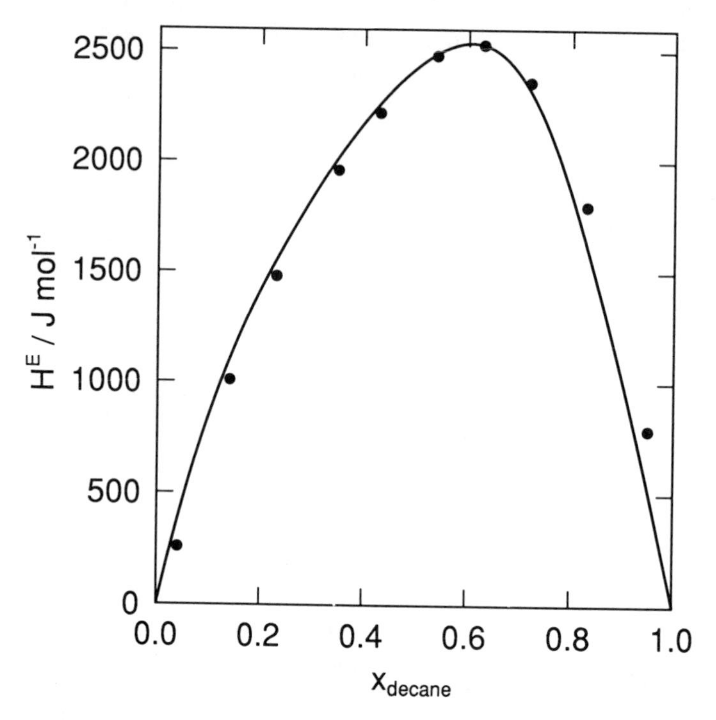

Figure 5.3-1
Excess molar enthalpy for the butan-1-ol–decane system at 378.15 K. The line represents the results calculated from VLE (Gierycz et al., 1988) and the circles denote the values obtained from excess enthalpies at 288.15 K (Nguyen and Ratcliff, 1975) and excess heat capacities (Gates et al., 1986).

obtained by Nguyen and Ratcliff (1975) at 288.15 K were extrapolated to 378.15 K using excess heat capacity data reported by Gates et al. (1986) for the temperature range 298.15 to 368.15 K.

The agreement is very satisfactory, all the more so since the small deviations observed in the vicinity of the pure substances can be explained by the 10 K gap between the measured values of c_p^E (highest value at 368.15 K) and the computed h^E.

It should be noted that such results are by no means typical for the majority of vapor–liquid equilibrium data. They can be obtained only from very accurate measurements.

The second example of the application of the Gibbs–Helmholtz equation is provided in Figures 5.3-2 and 5.3-3 for the *N*-methylpyrrolidone–water system (Malanowski, unpublished data). This case is rather unusual in that the excess Gibbs energy is small and *s*-shaped whereas the excess enthalpy is very large and negative. Due to the small values of g^E, the calculations of h^E according to Eq. (5.3-1) are very sensitive to errors in experimental g^E values. If all experimental vapor–liquid equilibrium (VLE) data points are correlated using a polynomial equation as illustrated in Figure 5.3-2, the predicted excess

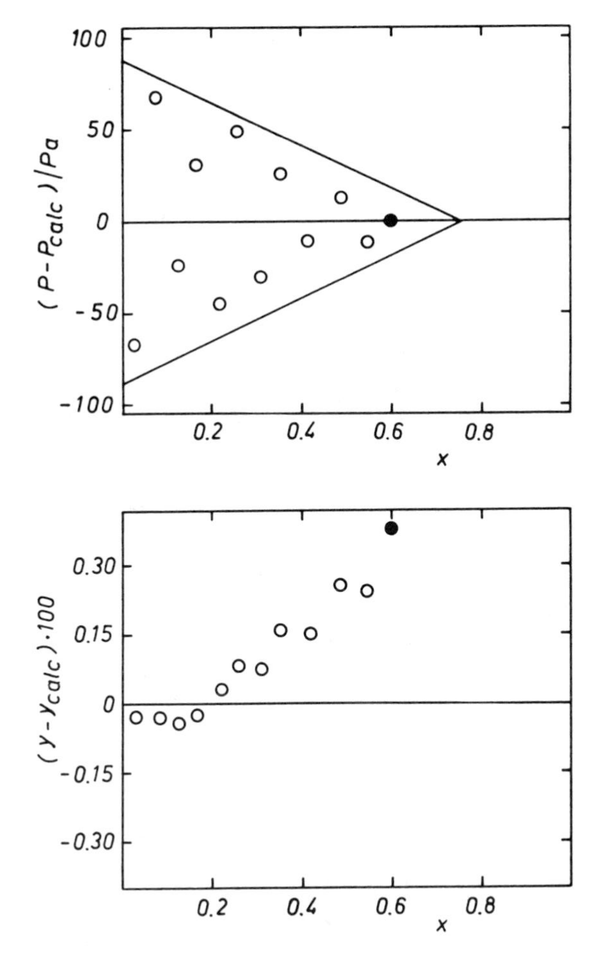

Figure 5.3-2
Deviations between the calculated and experimental values of pressure and vapor composition obtained by correlating all vapor–liquid equilibrium data points for the *N*-methylpyrrolidone–water system (Malanowski, unpublished data). The lines represent the estimated errors of the data.

enthalpy is grossly underestimated (Fig. 5.3-4, solid line). After removing one experimental VLE point that was less accurate (marked with a solid circle in Fig. 5.3-2), the obtained correlation is much better as illustrated in Figure 5.3-3. The pressure and vapor composition residuals scatter then around zero. What is more important, the calculated h^E values are in perfect agreement with experiment (Fig. 5.3-4, broken line).

This example shows that great care must be taken when correlating experimental data for consistency testing. A good correlation of the data is prerequisite for a meaningful consistency test.

5.4. CONCLUDING REMARKS

The most reliable and straightforward method of consistency testing is the analysis of residuals of phase equilibrium parameters obtained by means of

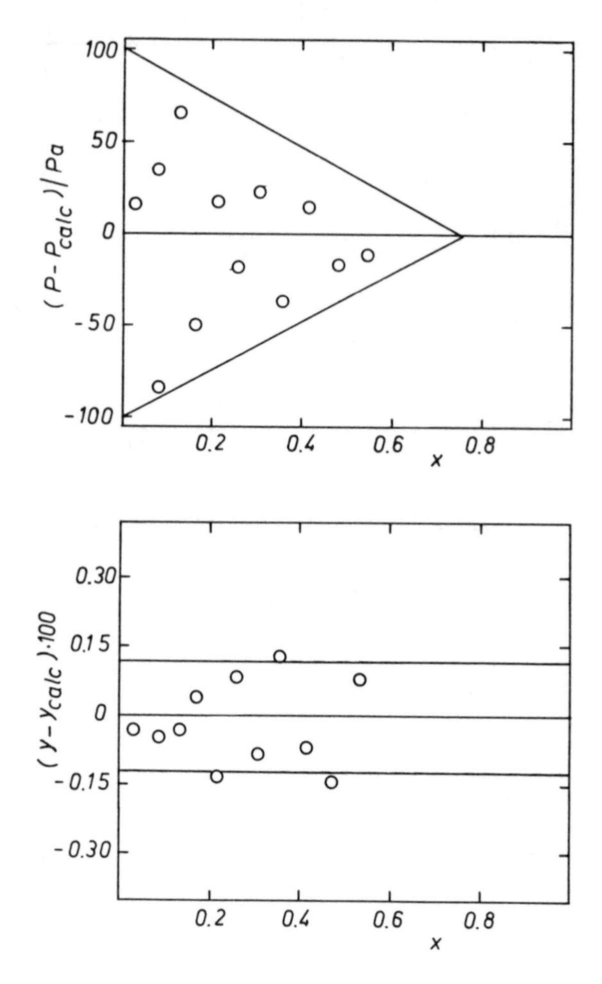

Figure 5.3-3
Deviations between the calculated and experimental pressure and vapor phase composition values for the same system as in Figure 5.3-2 after removing the experimental point denoted by a solid circle in Figure 5.3-3. The lines represent the estimated errors of the data.

data reduction procedures. Unlike the area tests, this procedure is equally applicable to binary and multicomponent data. However, implementation of data reduction methods requires a careful choice of suitable correlation equations.

Reliability of phase equilibrium data can be checked by comparing some derived thermodynamic quantities with their experimental values. The best choice is to use excess enthalpies. If thermodynamic data at various pressures are available, one could be tempted to verify the pressure dependence of the excess Gibbs energy using excess volume data according to the relation

$$\left(\frac{\partial g^E}{\partial P}\right)_{T,x} = v^E$$

Unfortunately, this idea cannot be recommended as a practical method due to the small numerical values of the effects involved (cf. Example P5.2-1), which precludes any quantitative comparison.

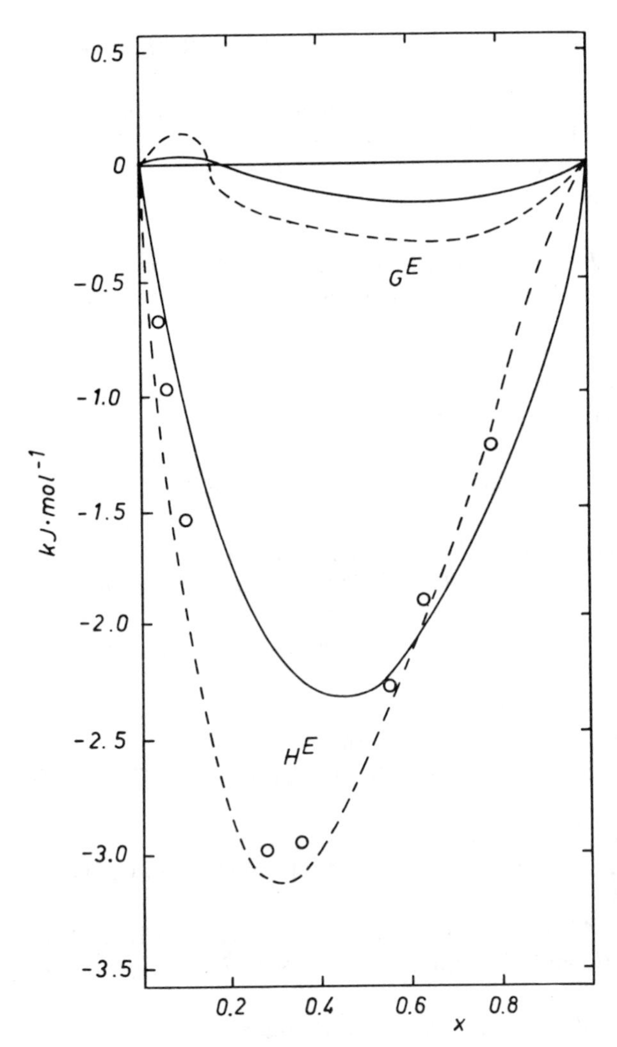

Figure 5.3-4
Excess Gibbs energy (G^E) and enthalpy (H^E) for the system *N*-methylpyrrolidone–water (Melanowski, unpublished data). The solid and broken lines denote the functions calculated from vapor–liquid equilibrium data according to the correlations illustrated in Figures 5.3-2 and 5.3-3, respectively. The circles represent H^E values determined calorimetrically.

Among other thermodynamic quantities, the infinite-dilution activity coefficients (γ^{∞}) are also of value. When properly correlated, reliable vapor–liquid equilibrium data should reproduce infinite-dilution activity coefficients obtained from independent measurements. On the other hand, the simultaneous use of the data for the whole concentration interval and those for finite dilution leads to more reliable results than the separate use of these data.

REFERENCES

Abbott, M. M., Floess, J. K., Walsh, G. E. and Van Ness, H. C., 1975. *AIChE J.* 21: 72.

Anderson, T. F., Abrams, D. S. and Grens, E. A. II, 1978. *AIChE J.* 24: 20.
Barker, J. A., 1953. *Aust. J. Chem.* 6: 207.
Berro, C., Rogalski, M. and Péneloux, A., 1982. *Fluid Phase Equilibria* 8: 55.
Brown, I. and Smith, F., 1954. *Aust. J. Chem.* 7: 296.
Christensen, L. J. and Fredenslund, A., 1975. *AIChE J.* 21: 49.
Christensen, J., Hanks, R. W. and Izatt, R. M., 1982. *Handbook of Heats of Mixing*, Wiley, New York.
Christensen, C., Gmehling, J., Rasmussen, P. and Weidlich, U., 1984–. *Heats of Mixing Data Collection*, Vol. 1–2, *Chemistry Data Series*, Dechema, Frakfurt/Main.
Diaz Peña, M. and Rodriguez Cheda, D., 1970. *An. Quim.* 66: 737.
Dohnal, V. and Fenclová, D., 1985. *Fluid Phase Equilibria* 21: 211.
Fabriès, J. F. and Renon, H., 1975. *AIChE J.* 21: 735.
Ferguson, J. B., Freed, M. and Morris, A. C., 1933. *J. Phys. Chem.* 37: 87.
Figurski, G., Gregorowicz, J. and Malanowski, S., 1990. *Fluid Phase Equilibria* 56: 235.
Gates, J. A., Wood, R. H., Cobos, J. C., Casanova, C., Roux, A. H., Roux-Desgranges, G. and Grolier, J. P. E., 1986., *Fluid Phase Equilibria* 27: 137.
Gierycz, P., Gregorowicz, J. and Malanowski, S., 1988. *J. Chem. Thermodyn.* 20: 385.
Herington, E. F. G., 1947. *Nature* 160: 610.
Ibl, N. V. and Dodge, B. F., 1953. *Chem. Eng. Sci.* 2: 120.
Kémeny, S., Manczinger, J., Skjold-Jørgensen, S., and Toth, K., 1982. *AIChE J.* 28: 20.
Klaus, B. L. and Van Ness, H. C., 1967. *AIChE J.* 13: 1132.
Kollár-Hunek, K., Kémeny, S., Heberger, K., Angyal, P. and Thury, E., 1986. *Fluid Phase Equilibria* 27: 405.
Margules, M., 1895. *Sitzber. Akad. Wiss. Wien, Math. Naturw. (2A)* 104: 1234.
Mixon, F. O., Gumowski, B. and Carpenter, B. H., 1965. *Ind. Eng. Chem. Fundam.*, 4: 455.
Neau, E. and Péneloux, A., 1981. *Fluid Phase Equilibria* 6: 1.
Neau, E. and Péneloux, A., 1982. *Fluid Phase Equilibria* 8: 251.
Nguyen, T. H. and Ratcliff, G. A., 1975. *J. Chem. Eng. Data* 20: 252.
Oracz, P., 1989. Methods of critical evaluation of binary VLE data (preliminary version), a report for the CODATA Task Group on Critical Evaluation and Prediction of Phase Equilibrium Data.
Péneloux, A., 1990. *Fluid Phase Equilibria* 56: 1.
Péneloux, A., Deyrieux, R., Canals, E. and Neau, E., 1976. *J. Chim. Phys.* 73: 706.
Ramalho, R. S. and Delmas, J., 1968. *J. Chem. Eng. Data* 13: 131.
Ratcliff, G. A. and Chao, K. C., 1969. *Can. J. Chem. Eng.* 47: 148.
Redlich, O. and Kister, A. T., 1948. *Ind. Eng. Chem.* 40: 435.
Rogalski, M. and Malanowski, S., 1977. *Fluid Phase Equilibria* 1: 137.
Samuels, M. R., Ulrichson, D. L. and Stevenson, F. D., 1972. *AIChE J.* 18: 1004.

Sayegh, S. G. and Vera, J. H., 1980. *Chem. Eng. Sci.* 35: 2247.
Skjold-Jørgensen, S., 1983. *Fluid Phase Equilibria* 14: 273.
Stevenson, F. D. and Sater, V. E., 1966. *AIChE J.* 12: 586.
Sutton, T. L., MacGregor, J. F., 1977. *Can. J. Chem. Eng.* 55: 602, 609.
Tao, L. C., 1961. *Ind. Eng. Chem.* 55: 307.
Tao, L. C., 1969. *AIChE J.* 15: 362, 15: 460.
Tao, L. C., 1980. In, Malanowski, S. and Szafrański, A. (Eds.), *Vapor–Liquid Equilibria in Multicomponent Systems*, Polish Scientific Publishers, Warszawa.
Ulrichson, D. L. and Stevenson, F. D., 1972. *Ind. Eng. Chem. Fundam.* 11: 287.
Van Ness, H. C. and Abbott, M. M., 1982. *Classical Thermodynamics of Nonelectrolyte Solutions with Applications to Phase Equilibria*, McGraw-Hill, New York.
Van Ness, H. C., Byer, S. M. and Gibbs, R. E., 1973. *AIChE J.* 19: 238.
Van Ness, H. C., Pedersen, F. and Rasmussen, P., 1978. *AIChE J.* 24: 1055.
Van Ness, H. C., Soczek, C. A. and Kochar, N. K., 1967. *J. Chem. Eng. Data* 12: 346.
Wohl, K., 1953. *Chem. Eng. Prog.* 49: 218.

Appendix A

Definitions of Some Important Thermodynamic Concepts

A *system* is that part of the physical world that is the subject of investigation. It is separated from the surroundings by some boundaries. These *boundaries* of the system are also called *walls*, *jackets* or *barriers*.

The *state of a system* is determined by the smallest set of measurable quantities that is necessary and sufficient for the representation of this system. Results of any further measurements performed on this system can be computed by means of functional relations supplied by thermodynamics. The measurable quantities are defined partly by mechanics, partly by the laws of thermodynamics and some of them by electrodynamics. The state of the system defined in this way is also called *a thermodynamic state*.

The *state of equilibrium* is the state toward which a system tends spontaneously or, equivalently, a state in which the thermodynamic quantities characterizing the system are not time dependent and no flows occur.

The *steady state* is a state with nonzero flows and time-independent thermodynamic variables.

The *variables of state* is the necessary and sufficient set of variables that determine the thermodynamic state of the system in equilibrium. A system completely described by the variables of state is also called a *thermodynamic system*. Experience shows that the number of variables that are necessary for a complete representation of an equilibrium state is smaller than that required for any nonequilibrium state of this system.

The *coordinate* (or *phase*) *space* is an abstract vector space whose basis is formed by the variables of state.

A *function of state* is a function that is a potential in the coordinate (phase) space. Sometimes, a definition omitting the principles of vector calculus is used: a function whose differential is a complete differential of variables of state.

The *measurable quantities* (or *properties* or *parameters*) are those macroscopic quantities that can be experimentally measured. Two types of measurable quantities can be distinguished:

1. *Intensive properties*: those that can be measured locally [temperature (T), pressure (P), etc.].
2. *Extensive properties*: those whose measurements involve the amount of the system [volume (V), internal energy (U), etc.].

The value of an extensive property depends on the amount of substance while the value of an intensive property does not.

A *phase* is a part of the system having some distinct boundaries. The phase boundaries are characterized by discontinuities of macroscopic physical and chemical properties.

A substance is called *homogeneous* when its intensive properties (temperature, pressure, concentration etc.) are independent of the spatial position.

If the macroscopic properties change discontinuously within a phase, the phase is *heterogeneous*. If the macroscopic properties of a phase are constant and independent of position, the phase is *homogeneous*.

A system is called a *one-component system* if it contains only one chemical substance. According to the number of chemical substances we deal with *two-component* (*binary*), *three-component* (*ternary*), *four-component* (*quaternary*), *five-component* (*quinary*) etc. systems.

An *open system* can exchange energy as well as mass with the surroundings. A *closed system* can exchange only energy, while an *isolated system* does not exchange energy or mass with its surroundings.

A *homogeneous system* contains only one phase. A *heterogeneous system* contains at least two phases.

A wall is called *adiabatic* if the state of a system can be changed only by a nonthermal process (e.g., mechanical or electrical etc.) Each nonadiabatic wall is called *diathermic* or *heat conducting*.

A *reversible process* is a process after which the system and its surroundings can return to their initial state along any path. A process is *quasi-static* if it proceeds through a sequence of equilibrium states. Each quasi-static process is a reversible one, but the converse statement is not necessarily true. All processes not being quasi-static are called *nonstatic*.

Appendix B

Basic Mathematical Definitions and Theorems

The purpose of this appendix is to give a uniform summary of mathematical definitions and theorems that are necessary to comprehend the course of phase equilibrium thermodynamics. Assuming that readers have the necessary mathematical background, no proofs of theorems are given here that are available in standard textbooks of mathematical analysis. The definitions and theorems are quoted mainly to make uniform the terminology used throughout this book. Those who have problems with understanding the material given below should refer to a textbook of mathematical analysis and vector calculus for physicists (e.g., Margenau and Murphy, 1956).

B.1 SOME FEATURES OF VECTOR SPACES

The concept of phase space is essential for thermodynamics. The phase space is a special case of a vector space.

LINEAR INDEPENDENCE OF VECTORS

Vectors $\mathbf{a}_1, \ldots, \mathbf{a}_k$ are linearly independent when their linear combination vanishes

$$\mathbf{a}_1\lambda_1 + \cdots + \mathbf{a}_k\lambda_k = 0 \tag{B.1-1}$$

if and only if all coefficients λ of the linear combination vanish

$$\lambda_1 = \cdots = \lambda_k = 0 \tag{B.1-2}$$

DIMENSION OF A VECTOR SPACE

The maximum number of linearly independent vectors in a vector space is called the dimension of the vector space. If the dimension of the vector space is n, then exactly n linearly independent vectors exist in this space.

BASIS OF A VECTOR SPACE

A set of vectors containing a maximum number of linearly independent vectors is called the basis of the vector space. Each vector of this vector space can be expressed as a linear combination of the basis vectors.

Let the vectors $\mathbf{a}_1$ from the basis of the space A_n. A vector $\mathbf{x}$ from this space can be expressed as

$$\mathbf{x} = \lambda_1 a_1 + \cdots + \lambda_n a_n \tag{B.1-3}$$

The coefficients λ are unequivocally determined by the basis vectors. They are called the vector coordinates in the space A_n and can be written as

$$\mathbf{x} = (x_1, \ldots, x_n) \tag{B.1-4}$$

SCALAR PRODUCT

The phase space is an example of a Euclidean space. In this space the scalar product can be defined. The scalar product assigns a number to a pair of vectors. It will be denoted by a center dot. The following axioms are satisfied:

1. The scalar product is commutative:

$$\mathbf{a}_1 \cdot \mathbf{a}_2 = \mathbf{a}_2 \cdot \mathbf{a}_1 \tag{B.1-5}$$

2. If λ is a number, then

$$\mathbf{a}_1 \cdot \lambda \mathbf{a}_2 = \lambda(\mathbf{a}_1 \cdot \mathbf{a}_2) \tag{B.1-6}$$

3. A scalar product of a vector and a sum of two vectors is equal to a sum of scalar products:

$$\mathbf{a}_1 \cdot (\mathbf{a}_2 + \mathbf{a}_3) = \mathbf{a}_1 \cdot \mathbf{a}_2 + \mathbf{a}_1 \cdot \mathbf{a}_3 \tag{B.1-7}$$

4. A scalar product of a vector multiplied by itself is always positive or equal to zero when the vector under consideration is a null vector:

$$\mathbf{a} \cdot \mathbf{a} > 0 \quad \text{when } \mathbf{a} \neq 0 \tag{B.1-8}$$

The last axiom makes it possible to define a Euclidean norm that is a measure of the length of the vector

$$\|\mathbf{a}\| = \sqrt{\mathbf{a} \cdot \mathbf{a}} \tag{B.1-9}$$

After defining the norm of a vector, the versor, i.e., a vector having the norm equal to one, can be introduced. The versor, or unit vector, **e** is defined as

$$\mathbf{e} = \frac{\mathbf{x}}{\|\mathbf{x}\|} \tag{B.1-10}$$

ORTHOGONAL AND ORTHONORMAL VECTORS

Vectors are called orthogonal when their scalar product is equal to zero.

It is convenient to use a basis containing exclusively unit vectors. Such a basis is *normalized*. A normalized basis of a vector space for which a scalar product is defined is called an *orthonormal* basis.

It follows from the above definitions that the scalar product of two vectors of an orthonormal basis satisfies the relation

$$\mathbf{a}_i \cdot \mathbf{a}_k = \delta_{ik} \tag{B.1-11}$$

where δ_{ik} is the Kronecker symbol and is equal to zero for $i \neq k$ and one for $i = k$.

The scalar product of two vectors in an orthonormal basis can be represented as a sum of products of corresponding coordinates of these vectors. For example, the scalar product of two vectors $\mathbf{x} = (x_1, \ldots, x_n)$ and $\mathbf{y} = (y_1, \ldots, y_n)$ is

$$\mathbf{x} \cdot \mathbf{y} = \sum_{i=1}^{n} x_i y_i \tag{B.1-12}$$

B.2 ELEMENTS OF DIFFERENTIAL AND INTEGRAL CALCULUS

SCALAR AND VECTOR FIELDS

A scalar field, or a function, assigns a number to each point (vector) of a vector space (or a part of it). Thus, a geometric representation of a function of k variables is a k-dimensional surface in a $(k+1)$-dimensional space.

A vector field assigns a vector to each point (vector) of a vector space (or a part of it). A vector field is a function having vector values.

HYPERSURFACES

Each hypersurface (e.g., a line in a two-dimensional space) can be characterized by a vector **N** that is orthogonal to it. All vectors of this hypersurface are orthogonal (normal) to **N**.

$$\mathbf{N} \cdot (\mathbf{x} - \mathbf{x}^0) = 0 \tag{B.2-1}$$

Equation (B.2-1) represents a hypersurface in a normal form. A linear hypersurface is called a hyperplane.

LINEAR FUNCTIONS

A linear function $h(\mathbf{y})$ where $\mathbf{y}$ is a vector in a k-dimensional space can be expressed as

$$h(\mathbf{y}) = \mathbf{a}_1 \mathbf{y}_1 + \cdots + \mathbf{a}_k y_k \tag{B.2-2}$$

or

$$h(\mathbf{y}) = \mathbf{a} \cdot \mathbf{y} \tag{B.2-3}$$

Thus, a linear function in a Euclidean space can be expressed as a scalar product of two vectors.

DIFFERENTIAL AND GRADIENT OF A FUNCTION

A differential of a function is a linear function of independent variables that approximates the increment of the function. Let us write an equation of a hyperplane approximating the increment of the function $f(\mathbf{x})$ in the point $\mathbf{x}_0$:

$$\Delta f(\mathbf{x}) = f(\mathbf{x}) - f(\mathbf{x}_0) = \mathbf{A} \cdot (\mathbf{x} - \mathbf{x}_0) + \mathrm{o}(\|\mathbf{x} - \mathbf{x}_0\|) \tag{B.2-4}$$

The differential $df(\mathbf{x}, \mathbf{x} - \mathbf{x}_0)$ of the function $f(\mathbf{x})$ is defined as the linear part of the increment $\Delta f(\mathbf{x})$ of the function for the increment $(\mathbf{x} - \mathbf{x}_0)$ of the independent variable.

$$\mathbf{A} \cdot (\mathbf{x} - \mathbf{x}_0) = df(\mathbf{x}, \mathbf{x} - \mathbf{x}_0) \tag{B.2-5}$$

The residuum $\mathrm{o}(\|\mathbf{x} - \mathbf{x}_0\|)$ is of a higher-than-linear order. It is small in that, as $\mathbf{x}$ tends to $\mathbf{x}_0$, it approaches zero more rapidly than the increment of the independent variable $(\mathbf{x} - \mathbf{x}_0)$:

$$\lim_{\mathbf{x} \to \mathbf{x}_0} \frac{\mathrm{o}(\|\mathbf{x} - \mathbf{x}_0\|)}{\|\mathbf{x} - \mathbf{x}_0\|} = 0 \tag{B.2-6}$$

The vector $\mathbf{A}$ is called the gradient of function $f(\mathbf{x})$ in the point $\mathbf{x}_0$. It is frequently denoted by grad $f(\mathbf{x}_0)$.

For example, the differential of a function in a two-dimensional space (see Fig. B.2-1) is

$$\Delta f = df(x_0, x - x_0) + \mathrm{o}(x - x_0) \tag{B.2-7}$$

$$df(x_0, x - x_0) = \left.\frac{df(x)}{dx}\right|_{x_0} \cdot (x - x_0) \tag{B.2-8}$$

where $(df(x)/dx)_{x_0}$ is the derivative of $f(x)$ in x_0

In general, a derivative in a direction is equal to the gradient multiplied by the direction versor. A derivative in the direction of a coordinate axis is

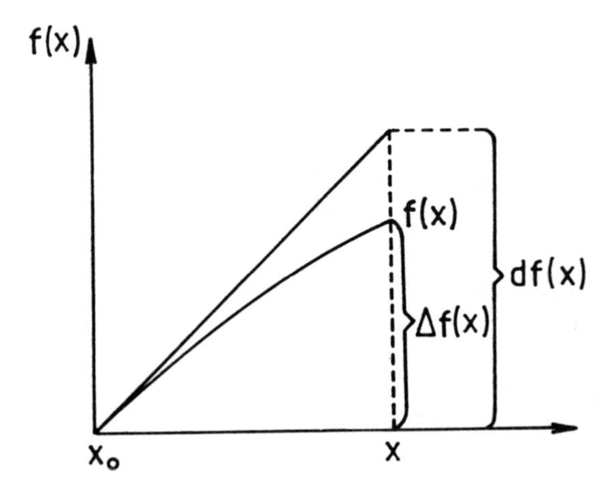

Figure B.2-1
Differential of a function in a two-dimensional space.

called a partial derivative. It can be easily proved that the ith coordinate of a vector $\mathbf{y}$ in the space E_k with a basis $\mathbf{e}_1, \ldots, \mathbf{e}_k$ can be calculated as a scalar product:

$$y_i = \mathbf{y} \cdot \mathbf{e}_i \tag{B.2-9}$$

Therefore, a gradient of function f in an orthonormal basis is a vector whose coordinates are the partial derivatives $\partial f / \partial x_i$:

$$\text{grad } f(x) = \left(\frac{\partial f}{\partial x_1}, \ldots, \frac{\partial f}{\partial x_k} \right) = \mathcal{F} \tag{B.2-10}$$

According to the definition of the vector field, the gradient is also a vector field.

POTENTIAL OF A VECTOR FIELD

A function is called a potential of a vector field if the vector field is the gradient of this function.

WORK IN A VECTOR FIELD

Let us consider a trajectory M in a vector field $\mathcal{F}$ (Fig. B.2-2). A motion along the trajectory M can be traced by a vector $\mathbf{x}(t_i)$ at a moment t_i. The difference of vectors $\mathbf{x}(t_i)$ and $\mathbf{x}(t_{i+\Delta i})$ corresponding to two points t_i and $t_{i+\Delta i}$ on the trajectory defines a vector $\Delta_i \mathbf{x}$

$$\Delta_i \mathbf{x}(t) = \mathbf{x}(t_{i+\Delta i}) - \mathbf{x}(t_i) \tag{B.2-11}$$

The vector $\Delta_i \mathbf{x}$ approximates the segment of the trajectory between t_i and $t_{i+\Delta i}$. Therefore, Eq. (B.2-11) is a parametric representation of the trajectory segment.

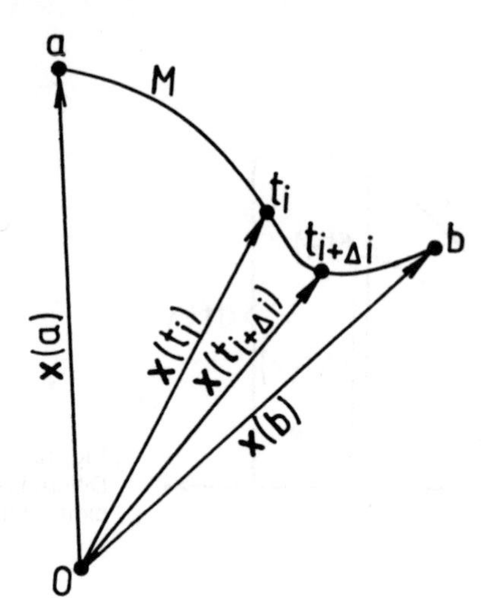

Figure B.2-2
Trajectory in a vector field.

The elementary work A_i in the vector field $\mathscr{F}$ along the trajectory segment is defined as

$$A_i = \mathscr{F}[\mathbf{x}(t)] \cdot \Delta_i \mathbf{x}(t) \tag{B.2-12}$$

This definition requires that the trajectory segment be elementary, i.e., of infinitesimal length $\Delta_i \mathbf{x}(t)$.

The work along the whole curve M can be computed by summing all the elementary works A_i for infinitesimal trajectory segments $\Delta_i \mathbf{x}$:

$$\lim_{\|\Delta_i \mathbf{x}\| \to 0} \sum_{i=1}^{l} \mathscr{F}[\mathbf{x}(t_i)] \cdot \Delta x(t) = \int_{(M)} \mathscr{F}[\mathbf{x}(t)] \cdot d\mathbf{x} \tag{B.2-13}$$

The obtained expression is a curvilinear integral along the curve M.

In an n-dimensional space the vector field $\mathscr{F}$ can be expressed in an explicit form:

$$\mathscr{F}(\mathbf{x}) = \begin{Bmatrix} \mathscr{F}_1(\mathbf{x}) \\ \vdots \\ \mathscr{F}_n(\mathbf{x}) \end{Bmatrix} = \begin{Bmatrix} \mathscr{F}_1[f(t)] \\ \vdots \\ \mathscr{F}_n[f(t)] \end{Bmatrix} \tag{B.2-14}$$

In Eq. (B.2-14) the coordinates of vector $\mathbf{x}$ are functions of the parameter t. The differential $d\mathbf{x}$ takes the form

$$d\mathbf{x} = \begin{Bmatrix} df_1(t) \\ \vdots \\ df_n(t) \end{Bmatrix} = \begin{Bmatrix} \dfrac{df_1}{dt}\, dt \\ \vdots \\ \dfrac{df_n}{dt}\, dt \end{Bmatrix} \tag{B.2-15}$$

Thus, the curvilinear integral along the curve M between $t = a$ and $t = b$ is given by

$$\int_M \mathscr{F}[\mathbf{x}(t)] \cdot d\mathbf{x} = \int_a^b \sum_{l=1}^{n} \mathscr{F}_l(f_1(t), \ldots, f_n(t)) \frac{df}{dt}\, dt \tag{B.2-16}$$

A vector field is conservative if the curvilinear integral is independent of the path of integration, depending only on the initial and final states. It can be easily shown that a vector field is conservative if it is a gradient of a function, i.e.

$$\mathscr{F}(\mathbf{x}) = \text{grad } f(x) \tag{B.2-17}$$

The scalar product of $\mathscr{F}(\mathbf{x})$ and $d\mathbf{x}$ is then equivalent to the total differential of the function f

$$\mathscr{F}(\mathbf{x}) \cdot d\mathbf{x} = \text{grad } f(\mathbf{x}) \cdot d\mathbf{x} = df \tag{B.2-18}$$

Therefore, we obtain

$$\int_{\mathbf{x}_0}^{\mathbf{x}} \mathscr{F}(\mathbf{x}) \cdot d\mathbf{x} = f(\mathbf{x}) \tag{B.2-19}$$

This relationship can be obtained by substituting the definition of the gradient into Eq. (B.2-16).

The following two theorems are valid for a conservative vector field:

1. If a curvilinear integral in a vector field is independent of the path of integration, then the vector field is a gradient of a function.
2. A vector field $\mathscr{F}(\mathbf{x}) = [\mathscr{F}_1(\mathbf{x}), \ldots, \mathscr{F}_n(\mathbf{x})]$ is conservative if the following relationship is satisfied:

$$\frac{\partial \mathscr{F}_k}{\partial x_i} = \frac{\partial \mathscr{F}_i}{\partial x_k} \quad \text{for each } 1 \le i, k \le n \tag{B.2-20}$$

The criterion (B.2-20) can be also written as

$$(\text{rot } \mathscr{F})_{i,k} = \frac{\partial \mathscr{F}_k}{\partial \mathbf{x}_i} - \frac{\partial \mathscr{F}_i}{\partial x_k} = 0 \tag{B.2-21}$$

where rot $\mathscr{F}$ is a vector called the rotation of the vector field $\mathscr{F}$.

EXTREMA OF A FUNCTION

If $\mathbf{x}^0$ is an extremum of a function $\mathscr{F}(\mathbf{x})$, the increment of the function $[\mathscr{F}(\mathbf{x}) - \mathscr{F}(\mathbf{x}^0)]$ in the vicinity of $\mathbf{x}^0$ is given by

$$\mathscr{F}(\mathbf{x}) - \mathscr{F}(\mathbf{x}^0) = \text{o}(\|\mathbf{x} - \mathbf{x}^0\|) \tag{B.2-22}$$

In the limit of $\mathbf{x} \to \mathbf{x}^0$ we obtain

$$\lim_{\|\mathbf{x}-\mathbf{x}^0\| \to 0} \mathscr{F}(\mathbf{x}) - \mathscr{F}(\mathbf{x}^0) = d\mathscr{F}(\mathbf{x}^0) = 0 \tag{B.2-23}$$

The differential $d\mathscr{F}(\mathbf{x}^0)$ in the point $\mathbf{x}^0$ defines a plane tangent to the surface $\mathscr{F}(\mathbf{x})$. Therefore, the condition of the existence of an extremum point $\mathbf{x}^0$ is

$$d\mathscr{F}(\mathbf{x}^0) = 0 \tag{B.2-24}$$

In practice, we frequently deal with constrained extrema when we seek them in a limited part of the surface $\mathscr{F}(\mathbf{x})$ rather than on the whole surface. The constraints are given by

$$\varphi(\mathbf{x}) = 0 \tag{B.2-25}$$

A relative extremum exists when the surface $\mathscr{F}(\mathbf{x})$ is tangent to the surface $\varphi(\mathbf{x})$. To satisfy this condition, the gradient grad $\mathscr{F}(\mathbf{x})$ must be parallel, i.e., linearly dependent on the gradients of the constraints. If the number of constraint equations is k and the dimension of the space is n, we obtain a general equation

$$\text{grad}\ \mathscr{F}(\mathbf{x}^0) = \lambda_1\ \text{grad}\ \varphi_1(\mathbf{x}^0) + \cdots + \lambda_k\ \text{grad}\ \varphi_k(\mathbf{x}^0) \tag{B.2-26}$$

or

$$\text{grad}[\mathscr{F}(\mathbf{x}^0) + \lambda_1\varphi_1(\mathbf{x}^0) + \cdots + \lambda_k\varphi_k(\mathbf{x}^0)] = 0 \tag{B.2-27}$$

Equation (B.2-27) can be rewritten as n scalar equations:

$$\begin{aligned} &\frac{\partial \mathscr{F}}{\partial x_1} + \lambda_1 \frac{\partial \varphi_1}{\partial x_1} + \cdots + \lambda_k \frac{\partial \varphi_k}{\partial x_1} = 0 \\ &\qquad \vdots \\ &\frac{\partial \mathscr{F}}{\partial x_n} + \lambda_1 \frac{\partial \varphi_1}{\partial x_n} + \cdots + \lambda_k \frac{\partial \varphi_k}{\partial x_n} = 0 \end{aligned} \tag{B.2-28}$$

Besides Eqs. (B.2-28) we have k constraint equations:

$$\begin{aligned} &\varphi_1(\mathbf{x}) = 0 \\ &\quad \vdots \\ &\varphi_k(\mathbf{x}) = 0 \end{aligned} \tag{B.2-29}$$

Altogether, we have $n + k$ equations in $n + k$ variables, i.e., n coordinates of the relative extremum $\mathbf{x}^0$ and k indeterminate coefficients λ_i of the linear combination. The above equations make it possible to find the relative extremum $\mathbf{x}^0$. The method described by Eqs. (B.2-28) and (B.2-29) is called the Lagrange method of indeterminate coefficients.

B.3 TRANSFORMATIONS OF COORDINATE SYSTEMS

To transform a system with coordinates $x_1, \ldots, x_n$ into another system with coordinates $p_1, \ldots, p_n$, a total of n functions interrelating the coordinates is required. For the ith coordinate such an equation is

$$x_i = \varphi_i(p_1, \ldots, p_n) \tag{B.3-1}$$

For example, the three-dimensional Cartesian coordinate system can be transformed into the polar coordinate system according to the relations:

$$\begin{aligned} x_1 &= r \sin \vartheta \cos \varphi \\ x_2 &= r \sin \vartheta \sin \varphi \\ x_3 &= r \cos \vartheta \end{aligned} \tag{B.3-2}$$

In the polar coordinate system we have spherical surfaces for $r = \text{constant}$, planes for $\varphi = \text{const}$ and cones for $\vartheta = \text{const}$.

A volume element in an n-dimensional space can be represented in the Cartesian coordinate system as

$$dw_n = dx_1 dx_2 \cdots dx_n \tag{B.3-3}$$

If curvilinear coordinates are used, a volume element should be defined in a different way. For example, let us consider a two-dimensional space represented by curvilinear coordinates p_1 and p_2. Two vectors $\mathbf{x}_1$ and $\mathbf{x}_2$ are localized in the point (p_1^0, p_2^0) of the space. The vectors are given by the following equations:

$$\begin{aligned} \mathbf{x} = \varphi(p_1^0, p_2) &\quad \begin{cases} x_1 = \varphi_1(p_1^0, p_2) \\ x_2 = \varphi_2(p_1^0, p_2) \end{cases} \\ = \varphi(p_1, p_2^0) &\quad \begin{cases} x_1 = \varphi_1(p_1, p_2^0) \\ x_2 = \varphi_2(p_1, p_2^0) \end{cases} \end{aligned} \tag{B.3-4}$$

The vectors defined as

$$\begin{aligned} \frac{\partial \varphi}{\partial p_1} &= \left(\frac{\partial \varphi_1}{\partial p_1}, \frac{\partial \varphi_2}{\partial p_1} \right) \\ \frac{\partial \varphi}{\partial p_2} &= \left(\frac{\partial \varphi_1}{\partial p_2}, \frac{\partial \varphi_2}{\partial p_2} \right) \end{aligned} \tag{B.3-5}$$

are tangent to the curvilinear coordinate axes in the point (p_1^0, p_2^0). The elementary projections of these vectors on the axes p_1 and p_2 are

$$\text{axis } p_1\text{:} \quad \frac{\partial \varphi(p_1^0, p_2^0)}{\partial p_1} \, dp_1$$

$$\text{axis } p_2\text{:} \quad \frac{\partial \varphi(p_1, p_2)}{\partial p_2} \, dp_2 \tag{B.3-6}$$

Therefore, a two-dimensional volume element in the coordinates p_1 and p_2 is

$$dw_2 = \left| \frac{\partial \varphi}{\partial p_1} \, dp_1, \frac{\partial \varphi}{\partial p_2} \, dp_2 \right| = \begin{vmatrix} \dfrac{\partial \varphi_1}{\partial p_1}, & \dfrac{\partial \varphi_2}{\partial p_1} \\ \dfrac{\partial \varphi_1}{\partial p_2}, & \dfrac{\partial \varphi_2}{\partial p_2} \end{vmatrix} dp_1 \, dp_2 \tag{B.3-7}$$

In general, a volume element in an n-dimensional space can be expressed in the curvilinear coordinates $p_1, \ldots, p_n$ as

$$dw_n = \begin{vmatrix} \dfrac{\partial x_1}{\partial p_1}, & \dfrac{\partial x_1}{\partial p_2}, \ldots, & \dfrac{\partial x_1}{\partial p_n} \\ \dfrac{\partial x_2}{\partial p_1}, & \dfrac{\partial x_2}{\partial p_2}, \ldots, & \dfrac{\partial x_2}{\partial p_n} \\ & \vdots & \\ \dfrac{\partial x_n}{\partial p_1}, & \dfrac{\partial x_n}{\partial p_2}, \ldots, & \dfrac{\partial \varphi_n}{\partial p_n} \end{vmatrix} dp_1 \quad dp_2, \ldots, dp_n \tag{B.3-8}$$

The determinant appearing in Eq. (B.3-8) is called the Jacobi determinant, or the Jacobian and can be written in abbreviated notation as $\partial(x_1, \ldots, x_n)/\partial(p_1, \ldots, p_n)$. It can be treated as a coefficient of transformation as it expresses the ratio of the volume elements before and after transformation.

PROPERTIES OF JACOBI'S DETERMINANTS

The following properties of Jacobi's determinants are used in thermodynamic calculations:

(1) $$\frac{\partial(y_1, \ldots, y_n)}{\partial(u_1, \ldots, u_n)} = \frac{\partial(y_1, \ldots, y_n)}{\partial(x_1, \ldots, x_n)} \frac{\partial(x_1, \ldots, x_n)}{\partial(u_1, \ldots, u_n)} \tag{B.3-9}$$

(2) $$\frac{\partial(y_1, \ldots, y_n)}{\partial(x_1, \ldots, x_n)} = \left(\frac{\partial(x_1, \ldots, x_n)}{\partial(y_1, \ldots, y_n)} \right)^{-1} \tag{B.3-10}$$

(3) $$\frac{\partial(x_1, \ldots, x_n)}{\partial(x_1, \ldots, x_n)} = 1 \tag{B.3-11}$$

To illustrate the applications of Jacobians we shall show some two-dimensional special cases that are important in thermodynamics. The first special case is

$$\frac{\partial(x, y)}{\partial(u, y)} = \left(\frac{\partial x}{\partial u} \right)_y \tag{B.3-12}$$

Equation (B.3-12) shows that any derivative can be expressed by means of a Jacobian. A differential can also be written in terms of Jacobians:

$$dz = \frac{\partial(z, y)}{\partial(x, y)}\, dx + \frac{\partial(z, x)}{\partial(y, x)}\, dy \tag{B.3-13}$$

If we introduce another set of independent variables, e.g., r and s, we can obtain a Jacobian describing a transformation from coordinates (z, y) into coordinates (r, s). The differential (B.3-13) then takes the form

$$dz = \frac{\partial(z, y)/\partial(r, s)}{\partial(x, y)/\partial(u, v)}\, dx + \frac{\partial(z, x)/\partial(y, x)}{\partial(y, x)/\partial(u, v)}\, dy \tag{B.3-14}$$

Multiplying both sides by $\partial(y, x)/\partial(u, v)$ and rearranging we get

$$\frac{\partial(z, y)}{\partial(x, y)}\, dx + \frac{\partial(x, z)}{\partial(y, z)}\, dy + \frac{\partial(y, x)}{\partial(z, x)}\, dz = 0 \tag{B.3-15}$$

This is a basic equation for the application of Jacobians to manipulating partial derivatives of thermodynamic quantities.

THE LEGENDRE TRANSFORMATION

A curve or surface is usually teated as a set of points in the coordinate space. Alternatively, it can be represented by a set of planes tangent to the surface. A transformation from the "point" representation to the "tangent plane" representation is called the Legendre transformation. This is a one-to-one transformation, i.e., it preserves all information contained in the original function.

An n-dimensional plane can be represented in an $(n + 1)$-dimensional space E_{n+1} as a function of the variables $(x_1, \ldots, x_n, L)$:

$$L + H - \sum_{i=1}^{n} x_i p_i = 0 \tag{B.3-16}$$

where $p_1, \ldots, p_n, H$ are the coordinates of the plane. An n-dimensional plane L in the $(n + 1)$-dimensional space E_{n+1} is given by

$$L = L(x_1, \ldots, x_n) \tag{B.3-17}$$

According to the definition of the gradient, a plane tangent to (B.3-17) in the point $(x_1^0, \ldots, x_n^0, L)$ is expressed as

$$L - L^0 = \sum_{i=1}^{n} \frac{\partial L(\mathbf{x}^0)}{\partial x_i} (x_i - x_i^0) \tag{B.3-18}$$

or, equivalently,

$$L + \sum_{i=1}^{n} x_i^0 \frac{\partial L(\mathbf{x}^0)}{\partial x_i} - L^0 - \sum_{i=1}^{n} x_i \frac{\partial L(\mathbf{x}^0)}{\partial x_i} = 0 \tag{B.3-19}$$

The coordinates of the plane tangent in point $\mathbf{x}^0$ are

$$p_i^0 = \frac{\partial L(\mathbf{x}^0)}{\partial x_i} \tag{B.3-20}$$

and

$$H^0 = \sum_{i=1}^{n} x_i \frac{\partial L(\mathbf{x}^0)}{\partial x_i} - L^0 \tag{B.3-21}$$

Therefore, for any point of the plane L we obtain

$$H = \sum_{i=1}^{n} x_i \frac{\partial L(\mathbf{x})}{\partial x_i} - L \tag{B.3-22}$$

Derivatives of the function $L(\mathbf{x})$ with the independent variables x_i determine the plane coordinates p_i

$$p_i = \frac{\partial L(\mathbf{x})}{\partial x_i} \tag{B.3-23}$$

An inverse transformation from the **p** coordinates into the **x** coordinates can be performed according to

$$x_i = \frac{\partial H(p)}{\partial p_i} \tag{B.3-24}$$

In thermodynamic applications a partial transformation of coordinates is important. If a function $y = y(x_1, \ldots, x_k, x_{k+1}, \ldots, x_n)$ is to be transformed into the function $f = f(p_1, \ldots, p_k, x_{k+1}, \ldots, x_n)$, then the parameters of the functions are related by

$$p_i = \frac{\partial y}{\partial x_i} \tag{B.3-25}$$

for $i \leq k$

$$-x_i = \frac{\partial f}{\partial p_i} \tag{B.3-26}$$

for $i > k$

$$p_i = \frac{\partial f}{\partial x_i} \tag{B.3-27}$$

A differential form of the (not transformed) function y is

$$dy = \sum_{i=1}^{n} p_i \, dx_i \tag{B.3-28}$$

After transformation, the differential form of the function becomes

$$df = -\sum_{i=1}^{k} x_i \, dp_i + \sum_{j=k+1}^{n} p_j \, dx_j \tag{B.3-29}$$

The corresponding integral forms are

$$f = y - \sum_{i=1}^{k} p_i x_i \tag{B.3-30}$$

and

$$y = f + \sum_{i=1}^{k} x_i p_i \tag{B.3-31}$$

Elimination of the variables y and $x_1, \ldots, x_k$ yields $f(p_1, \ldots, p_k, x_{k+1}, \ldots, x_n)$. Analogously, elimination of f and $p_1, \ldots, p_k$ gives $y(x_1, \ldots, x_k, x_{k+1}, \ldots, x_n)$.

A sufficient condition of the existence of such transformation is

$$\frac{\partial(p_1, \ldots, p_k)}{\partial(x_1, \ldots, x_k)} \neq 0 \tag{B.3-32}$$

In other words, the coefficient expressing the ratio of volume elements before and after transformation has to be different from zero.

In the case of a total transformation (i.e., when $k = n$), Eqs. (B.3-21)–(B.3-23) and (B.3-26) and (B.3-27) reduce to Eqs. (B.3-18)–(B.3-20).

B.4 SOME PROPERTIES OF FUNCTIONS

Gibbs assumed that the fundamental equation was a homogeneous function of all independent variables. Therefore, it is necessary to define a homogeneous function and show some of its properties.

A function $\mathcal{F}$ satisfying the relation

$$\mathcal{F}(t \cdot \mathbf{x}) = t^l \cdot \mathcal{F}(\mathbf{x}) \tag{B.4-1}$$

where t is a parameter, is called a homogeneous function of order l.

EULER'S THEOREM

For a homogeneous function $\mathcal{F}$ of order l defined in an n-dimensional space, the following relation is satisfied:

$$\text{grad}\, \mathcal{F} \cdot \mathbf{x} = \sum_{i=1}^{n} \frac{\partial \mathcal{F}}{\partial x_i} x_i = l\mathcal{F} \tag{B.4-2}$$

Equation (B.4-2) can be obtained by differentiating the definition (B.4-1) for $t=1$. Another differentiation leads to the following lemma:

A derivative of a homogeneous function of order l is a homogeneous function of order $l-1$.

TAYLOR'S SERIES

As the analytical forms of thermodynamic functions are not determined a priori by phenomenological thermodynamics, a purely mathematical representation of these functions as series is frequently convenient. The Taylor theorem is useful for this purpose:

If a function f is k times continuously differentiable over a closed interval $a \leq x \leq b$, its value in the point b can be computed from the series

$$f(b)=f(a)+\frac{b-a}{1!}f^{(1)}(a)+\cdots+\frac{(b-a)^{(k-1)}}{(k-1)!}f^{(k-1)}(a)+R_k \qquad \text{(B.4-3)}$$

where $f^{(m)}(a)$ denotes the mth derivative of the function f in the point a. Assuming that $h=b-a$ and λ and λ' are two numbers from the interval $0 \leq \lambda, \lambda' \leq 1$, the remainder term R_k takes the form

$$R_k=\frac{h^k}{n!}f^{(k)}(a+\lambda h)=\frac{h^k(1-\lambda')^{k-1}}{(k-1)!}f^{(k)}(a+\lambda' h) \qquad \text{(B.4-4)}$$

For $b=x$ and $a=0$ we get a special case of Eq. (B.4-3) called the McLaurin formula:

$$f(x)=f(0)+\frac{x}{1!}f^{(1)}(0)+\cdots+\frac{x^{(k-1)}}{(k-1)!}f^{(k-1)}(0)+R_k \qquad \text{(B.4-5)}$$

If a function is to be expressed as a power series, it is important that R_k vanish in the limit of $k \to \infty$:

$$\lim_{k\to\infty} R_k=0 \qquad \text{(B.4-6)}$$

In this case Eq. (B.4-5) takes the form

$$f(x)=f(0)+\sum_{k=0}^{\infty}\frac{x^k}{k!}f^{(k)}(0) \qquad \text{(B.4-7)}$$

If a function can be expanded in a power series, then only one such expansion exists and is given by the McLaurin formula.

The above formulas are valid for functions of single variables. Analogous expressions can be written for the multidimensional case. If a function $\mathscr{F}$ is differentiable k times over a closed interval $(\mathbf{x}, \mathbf{x}+\mathbf{h})$ of a space U, the following relation holds:

$$\mathscr{F}(\mathbf{x}+\mathbf{h})=\mathscr{F}(\mathbf{x})+\frac{d\mathscr{F}(\mathbf{x})}{1!}+\frac{d^2\mathscr{F}(\mathbf{x})}{2!}+\cdots+\frac{d^{k-1}\mathscr{F}(\mathbf{x})}{(k-1)!}+R_k \qquad \text{(B.4-8)}$$

where $d^k\mathscr{F}(\mathbf{x})$ denotes a differential of the kth order and R_k is defined analogously as in Eq. (B.4-4). An expansion exists if R_k vanishes as $k \rightarrow \infty$.

In practical applications the accuracy of the analytical representation is limited by the accuracy of measurements. Therefore, truncated Taylor series can be applied to represent of measurable properties.

REFERENCE

Margenau, H. and Murphy, G. M., 1956. *The Mathematics of Physics and Chemistry*, 2nd ed., Van Nostrand, Princeton, NJ.

List of Symbols

Latin Letters

$a, b, c, \ldots, A, B, C, \ldots$	parameters of equations
a	molar Helmholtz energy
A	Helmholtz energy
B	second virial coefficient
c	cohesive energy
c	one-third external degrees of freedom
C	third virial coefficient
C_p	heat capacity at constant pressure
C_v	heat capacity at constant volume
dW	a volume element
E	interaction energy
E_k	a thermodynamic potential
f	fugacity
f	number of thermodynamic degrees of freedom
f	scale factor
$\mathcal{F}$	a thermodynamic function
g	molar Gibbs energy
G	Gibbs energy
grad	gradient
h	scale factor
h	Planck constant
h	molar enthalpy
H	enthalpy
H	Henry's constant
k, l, m	binary interaction parameters
K	K factor
K	association constant
N	number of moles
N	number of particles
$\mathbf{N}$	or orthogonal vector

P	pressure
P	parameter
q	normalized surface area
q	contribution to the partition function
Q	heat
Q	configurational integral
Q	number of contacts
r	number of segments or volume parameter
R	gas constant
R	interaction distance
R_H	mean radius of gyration
rot	rotation
t	empirical temperature
s	molar entropy
S	entropy
T	absolute temperature
u, w	cubic EOS parameters
u	molar internal energy
U	internal energy
v	molar volume
V	total volume
w	speed of sound
W	work
x	mole fraction
y	gas phase mole fraction
y	work coordinate
Y	generalized force
Y	polarity factor
z	variable of state
z	compressibility factor
Z	coordination number
Z	partition function

Greek Letters

α	coefficient of thermal expansion
α	factor determining the temperature dependence of a
α	size/shape factor
α	nonrandomness parameter
β	polar factor
γ	activity coefficient
γ	variable exponent in mixing rules
Γ	group activity coefficient
δ	solubility parameter
δ, ϵ, η	cubic EOS parameters

δ	anisotropic parameter
δ	residual of a calculated and an experimental value
δ	variation
δ_{ij}	Kronecker symbol
$\Delta H^0, \Delta S^0, \Delta Cp^0$	standard enthalpy, entropy and heat capacity of association
δQ	average relative deviation of Q
ΔQ	average absolute deviation of Q
ϵ	interaction energy
η	efficiency
ζ	calculated critical compressibility factor
$\zeta_{\alpha\beta}$	binary interaction parameter
$\eta_{\alpha\beta}$	binary interaction parameter
θ	shape factor
θ	$1 - T/T_c$
θ	surface area fraction
κ	isothermal compressibility
κ_s	isentropic compressibility
λ	structural parameter
λ	coefficient of linear combination
λ	indeterminate coefficient
Λ	de Broglie thermal wavelength
Λ	parameters in Wilson equation
μ	chemical potential
μ	Joule–Thomson coefficient
μ	contact of type μ
μ	dipole moment
ν	contact of type ν
ρ	density
σ	intermolecular interaction size parameter
σ	standard error
Σ	z-chart sum (i.e., $z_c - B_c$)
τ	inverse integrating factor
τ	energetic parameters of the NRTL and UNIQUAC equations
ξ	reduced density
ξ	surface fraction
ϕ	shape factor
ϕ	number of phases
ϕ	fugacity coefficient
Φ	volume fraction or segment fraction
Φ	correction for vapor phase nonideality and difference between equilibrium pressure and pure component vapor pressure
Φ_1	Massieu function

Φ_2	Planck function
χ	Halm and Stiel polar factor
χ	mean association number
Ψ	measure of concentration
ω	acentric factor
Ω	grand potential

Sub- and Superscripts: Latin

a, b	substances a and b, respectively
A	pertaining to an associating substance
A_1	pertaining to the monomer of the associating substance
ani	anisotropic contribution
att	attractive contribution
b	boiling point
B	pertaining to an inert substance
c	critical property
c	coordination number
cal	calculated
(ch)	chemical contribution
COM	combinatorial
conf	configurational
dis	dispersive contribution
E	extensive property
E	excess property
exp	experimental
ext	external contribution
f	fusion
G	gas phase
H	homomorph
I	intensive property
i, j, k	components i, j, k
i	number of monomeric units within an associate
id	ideal
int	internal contribution
int	interactional contribution
iso	isotropic contribution
L	liquid phase
m	mixture property
m	melting
M	mixing function
nc	noncentral forces
np	nonpolar
p	polar
(ph)	physical contribution

quac	quasichemical contribution
r	reduced quantity
R	residual quantity
rep	repulsive contribution
res	residual
RES	residual
r_1, r_2	reference fluids 1 and 2, respectively
r	rotational contribution
S	solid phase
sat	saturation
t	triple point
t	translational
v	vibrational contribution
V	vapor phase

Sub- and Superscripts: Greek

α	phase α
α	fluid of interest in corresponding-states treatment
β	phase β
χ	pertaining to a component with mean association number χ

Sub- and Superscripts: Symbols and Numbers

0	reference fluid
0	standard state
*	unsymmetrical normalization
*	apparent (nominal) property (to distinguish from a real one)
∞	infinite dilution

INDEX